D1158367

PLANT REPRODUCTIVE ECOLOGY

PLANT REPRODUCTIVE ECOLOGY

MARY F. WILLSON

Department of Ecology,
Ethology, and Evolution
University of Illinois

A WILEY-INTERSCIENCE
PUBLICATION

JOHN WILEY & SONS

New York • Chichester • Brisbane
Toronto • Singapore

Library of Congress Cataloging in Publication Data:

Willson, Mary F.
Plant reproductive ecology.

"A Wiley-Interscience publication."
Includes bibliographies and index.
1. Plants—Reproduction. 2. Botany—Ecology.
I. Title.
QK825.W54 1983 582'.016 82-24826
ISBN 0-471-08362-3

Printed in the United States of America

10 9 8 7 6 5 4 3 2

PREFACE

The purpose of this book is to present under a single cover some elementary ideas that are relevant to the reproductive ecology of plants. This is a rapidly developing field of study. Many ideas are still controversial, some will soon become obsolete, and still others have not yet entered the arena. Nevertheless—or, perhaps, partly because of this—I think it useful to bring together a variety of approaches, hypotheses, and putative facts in order to provide a springboard for the student who wants a place to begin, an entry into a field of exploration.

I am assuming that readers will have some background in ecology, evolution, and basic botany, so that the fundamentals of these do not need to be introduced. Of course, certain of the things I discuss will be considered too elementary by some readers or too advanced by others; these sections may be skipped or skimmed, accordingly. I have tried to judge the appropriate average level by trying out various kinds of explications on the students in my classes at the University of Illinois; these students are typically advanced undergraduates and beginning graduate students.

My goal is certainly not to present a complete compendium of observations nor a theoretical treatise. I am particularly concerned with presenting a way of looking at plant reproductive ecology that goes beyond case histories or pure modeling to include some suggestions and speculations. These may lead ultimately to some general principles by which we can understand why plants do things the way they do. I hope that students who read this book as a jumping-off place may have the pleasure of contributing their work and imagination to the further progress of the field.

Most terms are defined as the discussion proceeds. Concerning nomenclature: generally I use the specific epithet used by the author in question, even though it may have been altered by subsequent taxonomic revisions.

Just as this manuscript was being typeset, two books on the evolution of sexual systems appeared on my desk: Graham Bell's *The Masterpiece of Nature* (1982, University of California Press) and Eric Charnov's *The Theory of Sex Allocation* (1982, Princeton University Press). These fascinating books appeared too late for inclusion here. They offer rich sources for further explanation of the evolution of sex itself and the diversity of sexual systems.

I wish to thank the following persons for references and constructive commentary on the various chapters: M. Berenbaum, N. Burley, K. Garbutt, D. A. Goldman, K. Hogan, S. J. Kenney, M. Lynch, E. Reekie, S. Rice, and D. W. Schemske. Patti Katusic assisted with reference retrieval and moral support. M. N. Melampy and J. N. Thompson read the entire manuscript and provided useful criticism. The Department of Ecology, Ethology, and Evolution, represented by B. Wright, provided typing services; the Department and the School of Life Sciences covered the cost of the art work. I am grateful to these persons for the use of their photographs credited *in situ:* R. J. Little, K. D. McCrea, and D. W. Schemske. Several authors graciously allowed prepublication access to their work. The biology librarians provided their usual patient and persistent help.

I would like to dedicate this book, first, to those students and colleagues who have challenged traditional interpretations and encouraged new ones. I do not suppose I have answered the challenge; if I can encourage further constructive questioning and the framing of testable, alternative hypotheses, it is enough. The fun lies as much in the challenge as in the answer; without such fun, science is a dry business. In addition, an appreciation is due to Mrs. Mahler, who long ago taught me the pleasures of searching for the first flowers amid the spring snowbanks of the Wisconsin hills.

MARY F. WILLSON

Champaign, Illinois
March 1983

CONTENTS

Chapter 1

LIFE HISTORIES

The life history of an organism encompasses all the stages through which it passes between its birth and its death. We concentrate on the average or expected life history of a (hypothetical) average organism in the population and on deviations from, or variance around, these expectations because, in this way, it becomes possible to compare the patterns found in one population with those of another and to search for the conditions that may have made these patterns adaptive. What is the life expectancy of a typical seed or seedling, and how variable is it, both among individuals in any period of time and through time as well? How many seeds are produced by a typical parent, when in the lifetime is reproductive maturity achieved, and how often in a lifetime are seeds produced? Does each individual that reaches maturity function as a female, or do some function as males or as both sexes? If the individuals are both male and female, are they simultaneously so, or does expression of one sex precede the other, and if so, which comes first?

Although we seek life history patterns at the population level in order to make comparisons with other populations, we are often interested in the degree to which individuals conform to the pattern and the consequences of deviating from the typical pattern, for it is this kind of information that yields insight into why the population pattern exists. The patterns exist at the level of the population, but the fate of individuals is important, not only because collectively they create the population pattern, but because they can tell us what kinds of ecological and evolutionary processes may shape the pattern. And it is on individuals that natural selection acts.

Suppose, for instance, that the number of seeds in a pod of *Asclepias verticillata* ranges from 5 to 60 and that the average number is 40. Perhaps it is of no consequence to the individual just how many seeds are in each pod. On the other hand, it may matter a lot. Predators might differentially consume many-seeded pods (if they could tell which ones they were). Or because each pod is composed of various materials that are probably diverted from other parts of the plant, the cost of constructing a pod to house just a few seeds may be uneconomical; diversion of materials to produce many few-seeded pods might entail slower production of underground stems, storage organs, or whatever. There may be other considerations as well, but these suffice to make the point. In this case, we could investigate the relative predation rates on pods with different numbers of seeds, who the predators are, their foraging behavior, their season of activity, their population and activity levels in different habitats, and so on. Furthermore, if the tendency to pack different numbers of seeds into pods differs among individuals and the difference is genetically controlled, we could discuss the fitness of the individuals with respect to seed-packing traits and the action of natural selection on these traits. If the variation in seed packing is purely a response to microhabitat conditions or age of the plant, but the ability to respond in these ways has a genetic basis, we can still interpret the variation in terms of natural selection. Similarly, we could assess the physiological costs of pod building, add up the costs for individuals with differing seed-packaging tendencies, and place it all in a framework of natural selection. In both instances, the characteristics of individuals are related to both ecological and evolutionary processes and we gain some understanding of why this plant packages seeds the way it does.

Adaptation can be studied by investigating intraspecific variation and sometimes by comparing higher taxa. In some cases, selection (or other processes such as genetic drift) will have exhausted all of the variation within a population, and a trait becomes fixed. Then all individuals in the population exhibit the trait and comparisons are only possible between populations. However, selection has not run its course in all cases, and genetic variation within a population has not been eliminated or reduced to trivial proportions. Then relevant comparisons can be made among conspecific individuals.

Many of the features to be discussed in this essay are indeed variable among individuals, although some are fixed across a whole population or a higher taxon. For most of the intrapopulation variation relevant to our concerns, we do not know the genetic basis. This is a major problem afflicting the study of many ecological and behavioral traits and one that must ultimately be solved. Some biologists would insist that without knowledge of the genetic foundation, all other study is fruitless. I disagree, because I feel we can discern certain processes—or at least the potential for them—even in the absence of genetic understanding. In fact, some comprehension of processes that shape the observed ecological patterns may well indicate which processes and patterns are the most interesting and, therefore, which should be first to be subjected to genetic analysis. Concerning between-taxon differences, we can be reasonably sure that there is a genetic basis, but it becomes more difficult to explore the possibly adaptive nature of the differences because several traits are likely to differ concurrently.

In general, the study of adaptation is fraught with difficulties. There is a parlous tendency to treat each trait separately, although each one occurs as part of an integrated complex—the whole organism—and, furthermore, many traits are genetically and developmentally inextricable. So we read about the adaptive value of large seeds, for example, or of deciduous leaves, of spiny protection of fruits or vegetative parts, of red flowers, or something else, all in isolation from the whole organism and the rest of the genome. However regrettable this tendency is, I think it may be unavoidable, at least at the beginning, simply because too little may be known about the proper context for interpretation. But it is not too much to ask that eventually the dissected organism be reassembled, whereupon the interpretation of its separate bits may need some readjustment. In some cases the reassembly process is well under way, as can be seen in discussions of adaptive compromises or tradeoffs and of cost/benefit ratios. However, a further trap awaits the unwary, and this is the inclination of ecologists to consider only (or mainly) ecological aspects of a trait, and physiologists or morphologists only the corresponding aspects. It is possible for an ecologist to be stumped concerning an explanation for, say, the evolution of opposite versus alternate branching patterns, if the fundamental explanation lies in developmental processes that are related to growth rates and strengths of materials. Conversely, a physiologist might suppose that the colors of immature fruits are a biochemical necessity of the pathway by which the mature fruit color is developed, without noting that colors could have very important consequences related to visually searching predators or dispersal agents.

Still another difficulty arises by way of the assumption made by many investigators that every single trait must be adaptive. This assumption has so many problems it is hard to know where to begin with them. One problem is related to that of isolation of traits, as already mentioned; suppose we are interested in explaining the size of a certain part of a flower, say, an anther. The size of that part could be adaptive itself, but it is also possible that the size of that part is determined because it is developmentally linked to some other part, perhaps the pistil; their sizes are correlated, and selection is actually operating on the size of the pistil. If the advantage of a large pistil outweighs any disadvantages of large anthers, both features will be maintained. That the two features are correlated at all could be adaptive in terms of developmental efficiency, for example, or nonadaptive and merely the result of some event in the past. If there are great disadvantages of having large anthers, but large pistils are still advantageous, any means of breaking the correlation will be favored but may not yet be available because the right mutation has not happened or has been eliminated by chance events. Similarly, a single gene may have pleiotropic effects on different aspects of an organism and may find expression at different times in the life span. When this is true, it is very difficult to eliminate only some of the pleiotropic genes' effects (although modifier genes may diminish their impact in various ways). Because traits are often correlated with others, it is simply unrealistic to take any one of them and suppose that it must be adaptive. But it also seems foolish to suppose that it *isn't,* because that cuts off a whole line of interesting investigation. It is reasonable to presume that a trait *may* be adaptive and then explore various ways in which it might be so. Conclusive proof or disproof, however, is generally out of reach, because one could always argue that just the proper circumstances have not yet been tested. Some biologists argue that therefore the study of adaptation is a waste of time, but it seems to me that if we waited for completely conclusive proof of everything, progress would be exceedingly slow. Much of science proceeds—at least temporarily—along its pathways of investigation without such proofs, operating instead on the basis of workability, plausibility, and lack of contradictions. Sometimes this manner of proceeding leads to dead ends. But surely some rational study of even such a difficult thing as adaptation is better than none.

This essay is focused on possibly adaptive aspects of plant life histories. Of necessity, I often deal with life history features in isolation from each other and from other aspects of the organism. I am aware of the pitfalls, but I see no other way to begin to bring together a body of intriguing material and to encourage the next steps.

THE PRINCIPLE OF ALLOCATION

Every organism allocates its resources to various essential activities, which can be categorized as maintenance, growth, and reproduction. Maintenance includes such survival activities as regulation of water movement, avoidance of predation and disease, baseline (nongrowth) metabolism, and resistance to competitors. Growth obviously refers to an increase in biomass; some organisms grow only or mainly as juveniles, whereas others achieve some growth during virtually every period of their lives. Reproduction includes the acquisition of mates, production of gametes, and parental care.

The proportion of the total resource budget of an organism that is devoted to reproduction is called the reproductive effort (RE). The critical resource may be energy, a mineral nutrient, or even water. If the total resource budget is constant, then an increased allocation to reproduction necessitates a reduction in at least one of the other two activities. But if the resource budget varies through time, such reciprocal adjustments may not always be necessary—augmentation of the budget may be sufficient to allow an increase of reproductive effort without a diminution of the other activities (Figure 1.1).

Natural selection should favor the allocational patterns that maximize fitness of the individual. Thus we should expect to find variation in patterns of allocation among different kinds, ages, or sexes of organisms

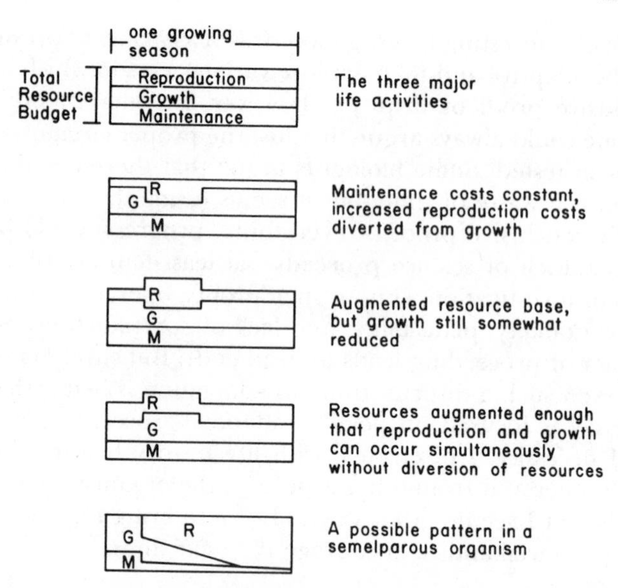

Figure 1.1 Some possible resource budgets, with various patterns of allocation among three primary activities. Semelparous refers to one reproductive episode per lifetime (see later in text).

and in the same kind of organism in different places. For instance, annuals often respond to nutrient stress by reallocating minerals from leaves to inflorescences, but perennials commonly sacrifice both reproduction and growth in favor of maintenance (Chapin 1980). Variation in allocation patterns may be both genetic (e.g., Meagher and Antonovics 1982) and phenotypic (other references below). The resource levels needed for basic maintenance in some circumstances may be extremely high. They may be so high as to preclude reproduction altogether, or high enough to reduce reproductive effort to a small value. In other environments, the costs of maintenance may be relatively low and a large fraction of the resource budget can be allotted to reproduction. A genotype that specified high reproductive effort simultaneously with a need for very high maintenance costs would probably be quickly eliminated from the population by natural selection unless additional resources were available at that time.

In actuality, most organisms probably fall between the extremes of constant budget and precisely reciprocal allocational compensations on the one hand and, on the other, exploitation of temporary surges in resource availability without reciprocal changes in the resource budget. *In part* because of the high cost of reproduction in either energy or materials, organisms typically schedule reproduction to occur at times when resources are abundant. Temperate-zone flowering plants, for instance, commonly make flowers and mature fruits during the long days of summer, when light for carbohydrate synthesis is abundant and warm temperatures permit rapid metabolic rates. Despite the use of times of increased availability of resources, it is very common for trees to have reduced growth rates when resources are shunted to reproduction, and the greater the reproductive effort, the greater the reduction in growth.

COMPONENTS OF REPRODUCTIVE RATE

Three major aspects of reproduction together determine the rate at which an individual contributes genes to future generations; that is, its fitness: "clutch size," age of first reproduction, and frequency of offspring production. Let us take each of these in turn.

Clutch Size

Clutch size in plants, by analogy with the avian world, refers to the number of young produced at one time, or in one physical unit. We can consider a clutch to be all the offspring borne by a plant in one reproductive season or, if they are borne in groups, all the offspring in such a group. Sometimes the group may be the seeds contained in a single mature ovary (of an angiosperm) or perhaps the seeds of a single discrete inflorescence. Such division points are arbitrary, chosen to suit the purposes of the investigation. Sometimes fecundity is used interchangeably with clutch size, as is done here; such usage must be distinguished clearly from lifetime fecundity (all the clutches of a lifetime).

Fecundity is often inversely correlated with the amount of parental care invested in the young (Smith and Fretwell 1974). High fecundity typically necessitates a reduction in parental care per offspring, at the price of offspring survivorship; conversely, large, expensive offspring often necessitate a smaller clutch size. The eventual size of the clutch is thus determined by the effect of parental care on survival of each juvenile and the effect of offspring numbers on the probability that *some* will survive. Sometimes, such tradeoffs can be seen principally in comparisons of different populations or species (e.g., *Plantago,* see below); in other cases, tradeoffs are seen among the individuals of a single population (e.g., *Asclepias exaltata,* Wilbur 1977) or phenotypic adjustments between numbers and sizes of offspring of an individual (e.g., certain *Sesbania* spp., Marshall, 1982).

Within any population of seed plants, average seed size is commonly less variable than seed number, fruit number, or flower number (Harper et al. 1970, Salisbury 1942). Although, in fact, seed size may vary quite a bit, the numerical aspects of seed production usually tend to vary more (but there are exceptions to this generalization). The relative constancy of seed size is thought to be related to the requirements of seedling establishment and survival and subject to strong stabilizing selection within the population (see Chapter 4). Given a certain seed size, clutch size can be adjusted by altering the total number of seeds or their distribution among inflorescences or fruits. However, the number of seeds per ovary in some angiosperms is fixed, which necessitates that all adjustments be made in other reproductive attributes.

Total production of offspring may be constrained, evolutionarily, by the available genetic variants or by counterselection on associated features. There may be insufficient genetic variation in the population for selection to increase fecundity to the "optimum" level or inadequate time for selection to have reached the "optimum." Or perhaps selection for other, developmental or architectural, features renders impossible an increase in fecundity (as, for example, when tiny, spindly branches might preclude large and heavy inflorescences). These constraints have to do with the maximum levels of fecundity achievable and presume that there is genetic control of that maximum. Within genetically determined outer bounds, there are ecological constraints. Fecundity may be limited by resources: light (and energy or photosynthate), some mineral or other essential nutrient, or water. Resource limitation of seed production is very common in plants, as shown by experiments that raise or lower resource levels and by decreases in resource allocation to vegetative parts as a result of reproduction (reviewed in Willson and Burley, in press; Stephenson 1981). For example, application of mineral fertilizer increased the average pod production of *Asclepias verticillata* from 5 to 34 pods per stem (Willson and Price 1980). Also within the genetically set potential, seed production may be constrained by availability of pollen. This limitation seems to be less common than that of nutritional resources, but some instances are documented (see Willson and Burley, in press). For example, hand-pollination of the orchid *Brassavola nodosa* increased fruit production from 12% of flowers to 67% (Schemske 1980). Pollen-limitation can be demonstrated experimentally by providing augmented pollen supplies to receptive surfaces and observing increased seed production. Merely noting that not all female-functioning flowers produce seeds is quite insufficient. Nor is it adequate to provide unlimited pollen and monitor seed set; abortion may intervene during maturation and regulate seed production in accordance with, perhaps, resource levels. Furthermore, seed production on other portions of large individuals and in subsequent seasons must be recorded if one is to eliminate the possibility that enhanced local pollen loads do not merely shift the patterns of resource allocation.

Predation on offspring before they leave the parent may diminish the effective clutch size on an ecological time scale but, on an evolutionary scale, predation probably imposes selection for increases of resource allocation to offspring protection. Therefore, limitations resulting from predation may be related ultimately to resource limitation. Yet another ecological constraint may result from unpredictable seasonal changes—an unusually early frost or dry season—that suddenly make the maturation of offspring impossible or unprofitable. Such constraints also may be linked basically to resources: the detrimental seasonal change could be viewed as making resources unavailable.

Four studies of botanical clutch size are summarized below. They illustrate a variety of approaches to the ecology of clutch size, different functional definitions of clutch size, and varied ecological conclusions.

Ranunculus flammula. This is a perennial, self-incompatible, insect-pollinated buttercup with a wide geographic distribution in North America. It occurs from sea level to 3000 m in the western mountains. The number of carpels and seeds per individual decreases at high latitudes and at high elevations (Johnson and Cook 1968). Average carpel number diminishes from 27 at sea level to 14 per individual at 1700 m. No more than a single seed develops in each carpel, but not all ovules mature into seeds, and seed production declines, over the same elevational gradient, from 24 to 5 per plant. Garden-grown plants from these populations had higher seed set than those in wild populations, but plants from high elevations continued to produce fewer seeds than those from lowland sources, indicating that the differences have a genetic basis. Johnson and Cook report that decreased clutch size was not caused by inefficient pollination reducing the percentage of seed-bearing carpels. Instead, carpel and seed numbers were well correlated with the length of the growing season at different elevations and, presumably, the resources available for seed maturation. The inference is that selection has eliminated (or at least reduced the frequency of) genotypes producing many carpels. A high-elevation plant with many carpels and relatively efficient pollination would produce many seeds; if resources are limiting, perhaps some or all seeds would only partially develop or mature without adequate food reserves. Then whatever resources had been devoted to these inadequate and doomed seeds would be wasted, and these genotypes are at a disadvantage. An alternative would be to abort some seeds and adjust the clutch size to the level of resource availability, a tactic particularly useful if that level is unpredictable (Lloyd et al. 1980, Stephenson 1981). For whatever reason, *R. flammula* emphasized the tactic of reduced carpel number instead, although this change did not account for all of the reduction in seed set.

Asclepias syriaca. This is a rhizomatous perennial of oldfields in eastern North America. The insect-pollinated flowers are distributed in discrete, axillary umbels along an erect stem. Umbels bloom sequentially up the stem and vary greatly in the number of constituent flowers, usually from about 10 to 150 or so. Each flower has two ovaries, so each umbel has the potential to produce twice as many pods as it has flowers. However, no umbel does so. Although the number of flowers pollinated increases directly with umbel size, the number of mature pods often does not. Many young developing pods are aborted, with the result that the number of mature pods per umbel is correlated with umbel size only poorly or not at all (Willson and Price 1977). Pod production in this milkweed is limited not by availability of pollen but by resources (Willson and Price 1980); abortion adjusts the clutch size to the available resources. It is possible that insect predators of seeds could differentially attack large clutches and thus provide selection against large numbers of pods in each group, but (unpublished) evidence so far indicates that this is not likely.

Plantago. Within various species of *Plantago,* Primack (1978) finds that seed weight is often about as variable as the number of seeds per capsule, and there is only seldom a significant inverse relationship between these two features. There are no significant inverse relationships between seeds per capsule and the number of capsules or inflorescences on the plant. In fact, in some cases, positive correlations appear—when a plant increases its fecundity, it does so on all fronts. The lack of tradeoffs among the elements of fecundity within species is interesting, especially in contrast to the situation among species. Here Primack notes a significant inverse correlation of seed weight and seed number per capsule and of inflorescence number per plant and capsules per inflorescence. The calculated regression line is close to a line representing a constant total seed weight per capsule and a constant number of capsules per plant, respectively, and perfect tradeoffs occur between the inversely correlated characters (Figure 1.2). Events associated with speciation have led to genetic changes in the parceling of offspring into "clutches" (within capsules or inflorescences) of different sizes, but then individual phenotypic plasticity adjusts total fecundity without tradeoffs

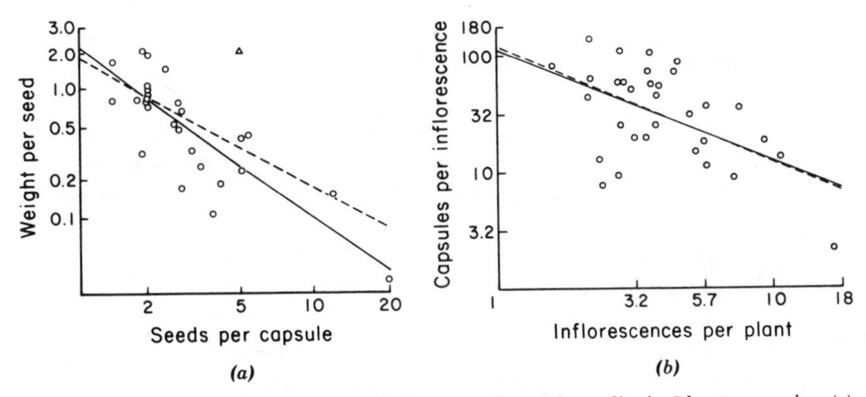

Figure 1.2 Tradeoffs between several elements of total fecundity in *Plantago* species. (*a*) Seed weight and seed number per capsule. (*b*) Capsule and inflorescence number per plant. Redrawn from Primack (1978) by permission of Blackwell Scientific Publications.

between size and number. Even this view is too simplified; such tradeoffs are not always evident among species (Wilbur 1976).

Asteraceae. A survey of clutch size in the Asteraceae (Compositae), tribe Heliantheae, revealed some large-scale trends quite different from those already described (Levin and Turner 1977). Over 1000 species of these sunflower relatives were analyzed; clutch size (ovules per inflorescence or flowering "head") ranged from 1 to 500 (how many normally mature into seeds is not recorded). Overall, clutch sizes of tropical species were significantly smaller than those of temperate or alpine species (an average of 51.2 ovules versus 82.7 and 89.6, respectively); these trends appeared only at the between-genera level, and comparisons within broadly distributed genera exhibited no such differences. The average clutch size of herbaceous perennials (78.1) exceeded that of both annuals (48.6) and woody species (51.9). Annuals had, on the average, lighter seeds (1.7 mg) than either perennial herbs (4.5 mg) or shrubs (11.7 mg). Thus there were no reciprocal adjustments of clutch size and seed size in these three categories of growth form.

Levin and Turner (1977) suggest that lower clutch sizes in tropical composites may be due to the possibly greater risks of attack by herbivorous insect larvae that, once inside a head of seeds, demolish the whole clutch. If these risks are really greater in the tropics, there could be selection favoring genotypes that spread their developing offspring into smaller cohorts, so that fewer are destroyed in each attack: the old principle of "don't put all your eggs in one basket." This hypothesis has not yet been tested, apparently. The low clutch size of annuals compared to perennials in this taxon, and their small seeds, means that the parental expenditure for each clutch is low in the annuals. Whether annuals produce more clutches per season than perennials is not reported. However, this observation should be kept in mind when considering the problem of relative allocation of resources to sexual reproduction; the sexual reproductive effort of annuals often exceeds that of perennials.

Age of First Reproduction

The age of first reproduction can have an enormous effect on the rate at which parental genes enter the gene pool when generations overlap (Cole 1954, Lewontin 1965). A relatively small decrease in this age may produce effects equivalent to a large increase in litter size. For instance, producing 100 offspring at age 9 yields about the same rate of increase of a genotype as 400 young at age 13; a 44% delay of age of first reproduction necessitates a 300% increase in fecundity to achieve the same growth rate for a family lineage (Lewontin 1965). If total fecundity is held constant, then obviously early reproduction enters genes into the gene pool more rapidly: a plant that produces 10^3 offspring at age one and then dies will have, potentially, 10^6 descendants after two reproductive episodes, whereas a similar individual (with perfect survival prospects) that waits until age two will have only 10^3 after the same amount of time. Therefore, a delay in the time of reproductive maturation usually must be accompanied by a compensatingly great increase in a survivorship and age-related fecundity.

When generations do not overlap, however, an early age of first reproduction is not necessarily advantageous (Schaal and Leverich 1981; see also Livdahl 1979). If the probability of a seed surviving decreases steadily through time, an annual plant that flowers early may expose its seeds to greater cumulative mortality risks than one that flowers late, because the seeds of the early flowerer must wait longer before germinating. Under these circumstances, it may be more advantageous to flower later in the season if seedling survivorship is better than that of seeds and if time of germination is similar for all seeds, independent of parental timing.

Both age-related changes in fecundity and the age of first reproduction are important in beginning to understand life histories, and they interact. Many plants continue to grow throughout their lives and, once maturity is reached, fecundity tends to increase with size. Seed production of *Pentaclethra macroloba, Welfia georgii,* and *Astrocaryum mexicanum* increases quite steadily with age (Sarukhan 1980). Average production of mangoes (*Mangifera indica*) increases with age (and size) of tree to about 40 years (Singh 1960). In addition, large and old individuals of many herbaceous plants such as *Asclepias incarnata* have far larger reproductive outputs than small individuals. Larger individuals of three species of *Fucus,* inter- and subtidal brown algae, had significantly more gamete-bearing areas than small individuals (Vernet and Harper 1980). If fecundity is expected to increase with age and size but present reproduction reduces growth, then clearly some "choice" must be made between reproducing early and gaining reproductive success (RS) through short generation times at the price of reducing the age-related gain in fecundity, and delaying reproduction to capitalize on large fecundity at the price of slower input of genes to the gene pool and accumulated risks of mortality through the period of delay. For a delay of reproduction to be advantageous, the age-related gain in fecundity must outweigh the mortality risks and the longer generation times; in order for delay not to be infinite, there must be some point at which this is no longer true and, instead, further increases in fecundity are insufficient to compensate for the costs. Thus there can exist an optimum age of first reproduction at which the benefits exceed the costs by as much as possible. Conditions permitting, variation in the age of first reproduction within the population should tend to converge near this point.

Age is sometimes related to size, but in many cases size has a greater effect on attainment of reproductive maturity than age itself. A wide variety of plants have a minimum size that must be achieved before reproduction is attempted (although attainment of that size does not always guarantee that reproduction occurs (Threadgill et al. 1981). Minimum sizes for reproduction are common in trees and in some herbaceous perennials (e.g., *Uvularia perfoliata,* Whigham 1974; *Viola rostrata* and *V. blanda,* Thompson and Beattie 1981). However, some plants seem to be able to flower at any size (e.g., *Smyrnium olusatrum,* Lovett Doust 1980b). The pawpaw, *Asimina triloba,* has a minimum size for flowering (functioning potentially as a male) and another, higher, minimum for fruiting (Willson and Schemske 1980). The probability of flowering and the number of flowering heads increases with rosette size in the composite *Senecio jacobea* (van der Meijden and van der Walls-kooi 1979) and in the umbellifer *Pastinaca sativa* (Thompson 1978). Moreover, mortality is higher for smaller rosettes. Parallel results obtain for teasel (*Dipsacus sylvestris*; Werner 1975, 1976) and several other species (Gross 1981).

Reproductive Frequency

The number of reproductive attempts in a lifetime varies from one to many. Some organisms reproduce repeatedly and are called *iteroparous* ("polycarpic"). Others reproduce only once and are called *semelparous** ("monocarpic"). Some semelparous plants are annuals and die after one season; others are longer lived, living through two or more years, and reproduce just before dying. Clearly, the latter have the physiological mechanism required for survival through severe seasons; their evolutionary reasons for semelparity are not that extended survival is unachievable. Semelparous plants occur in many families and habitats (Veillon 1971), although circumstances favoring their success may be restricted (see below). A remarkable semelparous life history is reported for *Chaerophyllum prescotti* (Apiaceae): a rosette and tap root are developed during the first years; the plant then goes dormant for years, and when the soil is later disturbed, it shoots up a stalk, flowers, and dies (Rabotnov 1969).

Although these three components of reproduction may often be associated with each other in specific combinations, they can also occur in other combinations, depending on the particular environmental circumstances and evolutionary constraints involved. One of the tasks of this chapter is to outline some of the controlling influences on these life history components and on the life history pattern in general.

THE COSTS OF REPRODUCTION

When a plant initiates reproduction, it begins an investment of considerable magnitude. The reproductive structures require an outlay of materials and metabolic energy. Material costs may include carbon, various minerals, often proteins, and sometimes chemical compounds that deter predators or pathogens. The initial investment is in the structures of vegetative reproduction or, if reproduction is sexual, in the organs that produce and support the gametes and promote fertilization. If reproduction is sexual, the initial costs are followed (for females) by the investment in the maturing embryos, an investment that may be very large.

The diversion of resources to reproduction can affect growth and future reproduction. These delayed costs of reproduction are well documented. Flowering and fruiting are accompanied by decreased growth

*Semele was one of the many mortal lovers of the chief Greek god, Zeus. At the behest of Zeus' jealous and vengeful wife, Hera, silly Semele begged Zeus to appear before her in all his celestial splendor. He came in his chariot of glory, enveloped in lightning and thunder, but the sight was too much for a mere mortal and Semele was consumed by flames. Zeus, however, saved the child from her womb, bearing it in his own thigh until its birth. Because Semele did not give birth normally herself, the colorful term semelparous requires a small leap of imagination in associating reproduction with subsequent death.

in many species of plants (reviewed by Willson and Burley, in press; Kozlowski 1971). Because total clutch size commonly increases with size of parent, decreased growth means that size-related increases in fecundity are likely to be diminished. Current photosynthate is used in seed production of many species such as *Hydrophyllum appendiculatum* (Morgan 1971), but stored materials may be used in others (Mooney and Hays 1973, Rabotnov 1969). Uncommonly large seed crops in *Betula alleghenienisis* and *B. papyrifera* resulted in dwarfed foliage, failure to develop terminal buds, die-back of branches, and reduced growth in height and diameter, followed, not surprisingly, by very low levels of flowering in subsequent years (Gross 1972). These costs to future reproduction often affect an entire plant, but they can also occur on the level of individual branches (Kozlowski 1971). Successful sexual reproduction in *Podophyllum peltatum* seems to decrease rhizome growth and, in turn, future sexual reproduction and to reduce survival as well (Sohn and Policansky 1977). Long ago, Salisbury (1942) noted an inverse relationship of fruiting and stolon formation in *Galeobdolon luteum,* and of flowering and bulbil formation in *Allium carinatum.* In these two cases, and some others cited below, costs may be assessed in terms of alternate forms of reproduction.

Poa annua, despite its name, often is perennial. British populations of this meadow grass were studied in disturbed fields, where mortality of both juveniles and adults was high, and in pastures, where juvenile mortality probably exceeded that of adults. Inbreeding is the rule and each population consisted of a variety of genotypes, but some genetic differentiation of the two kinds of populations was apparent (Law et al. 1977). In general, high rates of reproduction in the first year brought an associated reduction of plant size and reproduction in the second year (Figure 1.3) (Law 1979). In the field population especially, adult survival also diminished, at least in some lineages.

That the flowers and fruits of many species are photosynthetic and provide energy toward their own maintenance and growth (Bazzaz and Carlson 1979, Bazzaz et al. 1979) is intriguing. This may reduce the energetic cost of reproduction for the parent, although it is not likely to affect mineral-nutrient costs. Perhaps it would be interesting to compare the postreproductive growth rates of plants whose fruits are largely self-supporting (in terms of carbon) with those whose fruits are not; experimental manipulations for comparisons of conspecifics should not be difficult.

The act of reproducing might directly increase the plant's susceptibility to physical or biotic agents of death. For several species, the probability of death is greatest during periods of active growth: for example,

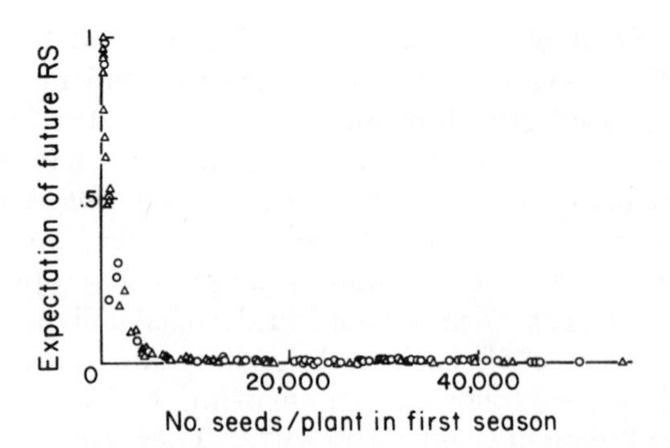

Figure 1.3 The cost of reproduction in *Poa annua*: the estimated expectation of future RS drops sharply with increasing RE. Redrawn from Law (1979) by permission of The University of Chicago Press © 1979 by The University of Chicago.

three species of *Ranunculus* (Sarukhan 1974, 1976; Sarukhan and Harper 1973) and two grasses, *Phleum pratense* and *Festuca pratensis* (Langer 1956, Langer et al. 1964). The buttercups survived well during flowering, but elongation of the flowering tillers in the grasses is a form of growth, and flowering tillers had higher mortality rates than nonflowering tillers (Langer 1956). The scanty evidence can hardly convince us of the importance of direct costs, but little seems to be reported on direct costs of reproduction.

LIFETIME SCHEDULING OF REPRODUCTIVE EFFORT

The scheduling of RE throughout a lifetime is determined to an important degree by the age-specific risks of mortality. Age-specific survivorship is the probability that an individual of age x will be alive (and presumably able to reproduce) at a later time, $x + 1$. The relevant factors include both the age-specific survivorship of adults and the probability that reproduction will be successful, that is, the probability of juvenile survival. Here we are concerned with *extrinsic* mortality factors, including any source of death that is independent of reproduction. (Intrinsic factors, in contrast, are a function of the evolved life history itself; a high risk of death because of a very high RE is one thing we are trying to explain, so it cannot be used as a primary causal factor.) In essence, the lifetime reproductive schedule is constrained through natural selection by the probability of success and the probability of living to try again. Other recent reviews of some of these ideas are provided by Bell (1980), Hirschfield and Tinkle (1975), Horn (1978), Michod (1979), and

Stearns (1980). There has been considerable disagreement concerning the proper way to model age-specific reproductive patterns (e.g., Caswell 1981, Goodman 1982, Schaffer 1981, Yodzis 1981, and references therein); I present a vastly simplified version here in order to avoid both a complex mathematical issue and an unresolved problem best left for further research.

Note that in this discussion mortality schedules mold the reproductive schedules by their effects on contributions of parents to future generations. In no sense can it be said that reproduction is adjusted to balance mortality in some way such that the numbers of individuals remain at some arbitrary level.

If juvenile survivorship is consistently good, a parent can realize a great benefit by investing heavily in reproduction. If juvenile survivorship is regularly low, there is less benefit to be obtained from reproduction and selection may favor lower RE—but only if the lower RE brings with it either an improvement of juvenile survivorship or a compensatory increase in adult longevity and probability of future reproduction. Some organisms may be able to predict the expected levels of juvenile survival from assorted environmental cues and to vary RE appropriately. Lacking predictive powers, other organisms must commit themselves to high or low RE. However, the consequences of variable and unpredictable survival of young are controversial; models have been created to predict low RE and iteroparity, high RE and semelparity, and mixtures of degrees of delayed semelparity. At present, the predicted effects of variable mortality must be left unresolved (e.g., Bell 1980, Hairston et al. 1970, Hastings and Caswell 1979).

When the extrinsic risk of death to adults is high compared to juveniles, then clearly the number of possible reproductive attempts is limited. These circumstances select for relatively high RE. Then any adult that saves up resources for future reproduction at the expense of present reproduction is likely to have fewer present offspring than adults that are less parsimonious. Moreover, because their prospects of living to try again are small, their total output of offspring is also relatively small. A genotype specifying a low RE is likely to have a low fitness when adult survivorship is low. Although short adult life expectancy can select for high RE, the converse (that high longevity favors low RE) is not necessarily true. Other factors may favor high RE despite good prospects of adult survival. When juvenile survival is good and sufficiently large clutches can be produced, individuals producing large clutches may have higher contributions to future generations than those with small clutches, even if fewer clutches are produced. Those with large successful clutches can leave more descendants in a shorter time and

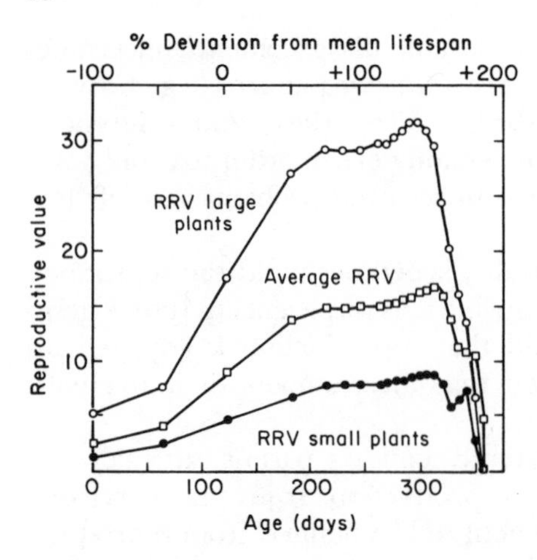

Figure 1.4 Within-season changes in reproductive value of *Phlox drummondii*. The curves are low both initially and terminally because the risks of mortality are considerable. This example also demonstrates the enormous differences between the expectations for large and small individuals: large individuals contributed most of the offspring to the next generations of this population. Redrawn from Leverich and Levin (1979) by permission of The University of Chicago Press © by The University of Chicago.

thus out-reproduce those with small clutches. Whenever this is true, high RE could be favored even if adult life were potentially long. Obviously, the balance between present and future reproduction in any particular case will depend on how well juveniles survive, how big a clutch is possible with available resources, and how much the production of a large clutch now might reduce the ability to do so again later (Williams 1966). Present reproduction may detract from future reproduction by using up resource reserves or increasing exposure to predators, pathogens, or physical damage. At this point, intrinsic costs as well as extrinsic risks are involved.

To explore the interaction between present and future reproduction in a stable population, we use the notion of reproductive value (RV; Pianka 1976, Pianka and Parker 1975, Schaffer 1974, Williams 1966). This is an organism's age-specific expectation of reproducing. Thus at any age t, the fecundity m is represented by m_t. The survivorship l is a function of age, which varies from x to y, including t. RV then depends on the probability of living to age t and the expected fecundity at that age m_t. An example of within-season changes in RV is shown in Figure 1.4.

We can divide RV into present reproduction (M) and residual reproductive value (RRV), which is the expected value of future reproduction:

$$\text{RV} = \text{M} + \text{RRV}.$$

More specifically, if an organism is now age x, its present fecundity is m_x, and its RRV is the sum of expected age-specific future fecundity modified by the probability of surviving to each successive age:

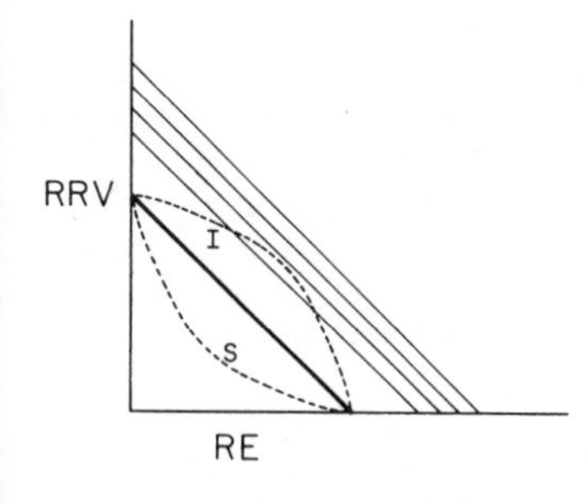

Figure 1.5 Possible relationships between RE and RRV, in which the cost of RE is reflected in a change in RRV. Changes in the cost help determine the life history pattern. Modified from Pianka and Parker (1975) by permission of The University of Chicago Press © 1975 by The University of Chicago.

$$\text{RV} = m_x + \sum_{t=x+1}^{y} \frac{l_t}{l_x} m_t.$$

One can imagine patterns of allocation in which RE (and hence M) is so high that RRV is reduced to zero: no resources are left for continued growth and maintenance. The organism is semelparous. Others may achieve high fitness by having lower RE for present reproduction, maintaining a higher RRV, and reproducing iteroparously. The shape of the relationship between RE and RRV determines whether an organism should be semelparous or iteroparous.

Consider the diagonal lines in Figure 1.5. Each diagonal represents the condition in which RRV is reduced equally for every increment in RE. There is a whole family of possible diagonals, each representing a different total RV. Lifetime RV will be maximal where the highest diagonal is tangent to the curve relating RE and RRV. If the curve is convex upward, the decrement in RRV is slow at first, but for higher values of RE, the decrease is ever more rapid. This means that, at some intermediate value of RE, RRV will have decreased so little that their sum exceeds the sum of all other possible combinations (i.e., RE + RRV is a maximum). This shape of the curve results in iteroparity. On the other hand, when the curve is concave upward and increasing RE has ever-diminishing costs in RRV, selection should favor either very high RE and zero RRV (semelparity), or no present RE and a delay of reproduction.

Several kinds of factors could produce a curve that is concave upward and selection for semelparity. A large RE (and concomitantly large M) might be disproportionately more successful than small RE; for example, large seed crops might satiate seed predators so that some seeds escape whereas small crops are totally demolished (see Chapter 4), or large inflorescences might attract relatively more pollinators and enjoy a higher percentage of pollination (e.g., Schaffer and Schaffer 1977,

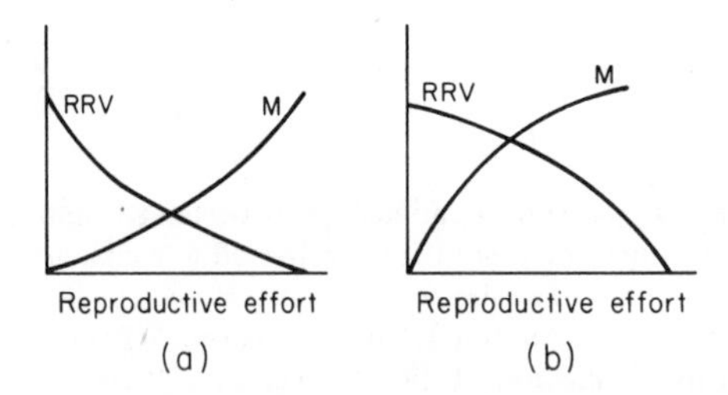

Figure 1.6 Changes in M and RRV as a function of RE. *(a)* The condition favoring semelparity. *(b)* The condition favoring iteroparity. Modified from Schaffer (1974) by permission of Duke University Press. Copyright 1974, The Ecological Society of America.

1979) or a higher percentage of flowers that contribute pollen to others (S. Sutherland pers. comm.). Perhaps even a small seed crop has a high cost in energy or materials and the cost of further increasing the seed crop is reduced.

It is possible to imagine, furthermore, that M is not related linearly to RE. Perhaps, for example, large seed crops entail increasingly costly protective or support devices. Or, possibly, a considerable RE is required to produce even a rather small seed crop if, for instance, the flowers are borne on a stout, costly stalk that must be produced regardless of the size of the seed crop. These situations are depicted in Figure 1.6. If the curves for both M and RRV are concave upward (Figure 1.6*a*), their sum is greatest at the extremes, and the optimal RE is either 100% or 0% (semelparity). If both curves are convex upward (Figure 1.6*b*), the sum of M + RRV is greatest at intermediate values of RE, and the organism should be iteroparous.

We can consider that the life history pattern is the result of an evolutionary "decision": if semelparity is to evolve, it must be possible to increase RE sufficiently to compensate for the loss of iteroparity (e.g., Bell 1980). Such an increase entails expenditures of resources, which must be available in the environment or by internal reallocations. When semelparity is favored, selection may favor internal reallocations of resources that further increase RE at the cost of maintenance functions. The act of reproduction then entails the suicide of the parent. Thus intrinsic life history features can reinforce the effects of extrinsic environmental factors.

Several factors may influence tradeoffs between RE and future reproduction. If reproductive output is limited by availability of male gametes instead of resources, then high RE is likely to have reduced effects on future reproduction. (This might result in selection for higher RE, especially if, for example, producing more flowers increases visitation by animal pollinators; eventually, the system might approach resource limitation.) If the resources that limit reproduction are different from those limiting growth and maintenance, no tradeoffs may occur. Although energy is perhaps the commonly shared currency, there may be cases in which reproduction is limited by a mineral but growth and maintenance are limited by energy. Changes in size and in protective devices may alter the expectation of life and reduce or delay the effects of extrinsic risks. In so doing, selection may adjust RE in response to the potentially conflicting demands of growth and maintenance. If, for instance, plants are subject to high rates of herbivory or intense competition from other plants, greater allocation to mechanisms of resistance may be favored at the expense of RE. Production of biochemical or morphological defenses may be costly. Competition among established individuals can place a premium on stem growth, to prevent overtopping and shading; on leaf production, to capture as much light as possible, or on root growth, to improve nutrient capture. Success in resisting herbivores or competitors may improve life expectancy but reduce RE and clutch size (assuming the resource budget has stayed the same). Clearly, selection should favor such reallocation if lifetime reproductive success is enhanced.

The interaction between RE or clutch size and reproductive schedule can be explored abstractly by means of a bit of algebra. How great an increase of RE is necessary for an annual to leave as many offspring as a perennial? We can compare clutch size for two otherwise equivalent genotypes, one an annual and the other a perennial that reproduces iteroparously (e.g., Schaffer and Gadgil 1975, Young 1981). The yearly rate of increase of each type is called λ. For an annual, the clutch size (M_A) multiplied by the probability of those young surviving (S_Y) determines λ_A. For an iteroparous perennial, λ_P is determined not only by size of seed crop and offspring survival but also the fraction of mature individuals surviving to the next year (S_M). Thus:

$$\lambda_A = S_Y M_A$$

$$\lambda_P = S_Y M_P + S_M.$$

Then if S_Y is the same for annuals and perennials, an annual pattern can outproduce an iteroparous pattern when

$$\lambda_A > \lambda_P$$

or when

$$M_A > M_P + \frac{S_M}{S_Y}.$$

The term S_M/S_Y is a measure of the relative magnitude of juvenile and adult survival. For many perennial plants, S_M is much larger than S_Y, so the second term on the right-hand side of the equation is likely to be large. This means that M_A should be considerably larger than M_P in most cases, if the annual habit is to succeed. However, any conditions that improve S_Y compared to S_M could make annualness easier to accomplish.

Some perennials do not reproduce iteroparously; they are long-lived but semelparous with delayed reproduction. Some reproduce in their second year and are called biennials; others may delay reproduction far longer. What fecundity is required of such a life history pattern to equal that of an iteroparous perennial or of an annual?

As we have seen, $\lambda_A = S_Y M_A$. After n generations, an annual genotype would have $(S_Y M_A)^n$ descendants. To match or better this rate of family growth, a biennial must produce at least $M_B = (S_Y M_A)^2$ offspring every two years; a plant that delays 10 years must make at least $M_D = (S_Y M_A)^{10}$. The semelparous delayers encounter some mortality between the first year and reproductive maturity; this is represented by S_D, the yearly probability of survival during the period of delay. As before, S_Y is taken as equal for both annuals and delayers. Then the number of surviving young produced by a delayer is equal to $S_Y S_D^{n-1} M_D$. We can ask what value of M_D is sufficient to compensate for extended prereproductive mortality and to achieve a rate of family growth greater than that of an otherwise equivalent annual. Then

$$S_Y S_D^{n-1} M_D > (S_Y M_A)^n$$

and

$$M_D > \frac{S_Y}{S_D} M_A^n.$$

The larger is S_D with respect to S_Y, the lower M_D needs to be for equal rates of family increase to be achieved. By similar considerations (Hart 1977), the fecundity of a biennial necessary to exceed the rate of increase for a perennial is

$$M_B > \frac{S_Y}{S_D} M_P^2 + 2M_P + \frac{S_D}{S_Y}.$$

When $n > 2$, the expression is still more complex.

In general, minimum M_B or M_D must be considerably larger than M_A and M_P to be favored by natural selection, essentially because the advantages of both early maturation and of iteroparity are lost. For various values of S_Y and S_D, Figure 1.7 shows the size of seed-crop needed by an annual and a biennial to match the rate of increase of a perennial (whose seed-crop size is held at an arbitrary constant for purposes of comparison). Note that, as expected from discussion above, annuals are likely to

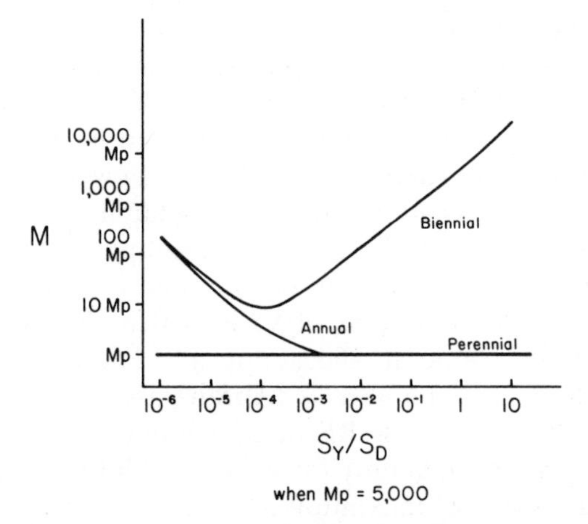

Figure 1.7 Fecundities required by an annual and a biennial to match the rate of increase of a perennial. Redrawn from Hart (1977) by permission of The University of Chicago Press © 1977 by The University of Chicago.

be successful only when S_Y is not too small. Biennials approach the other curves only in a narrow range of values; at either extreme, M_B becomes very high and may not be attainable (Hart 1977).

OBSERVED PATTERNS OF REPRODUCTIVE ALLOCATION

It must be noted at this juncture that the theoretical verbal models just summarized are primarily exercises in logic. The data available to support the predictions from the logical models are a mixed bag, however, and they only sometimes are in accord with expectations. Comparisons within species reduce the number of confounding variables and may provide the best indicators of adaptive shifts in allocational patterns. In some cases differences in allocation are found in different genotypes; in others the changes result from phenotypic plasticity. As noted by Harper (1977), phenotypic and genotypic control may offer very different possibilities; Bateson (1963) and others suggest that genetic control should prevail when patterns of variation are predictable, but phenotypic flexibility should be favored when unpredictability is the rule.

Before presenting some of the data, it is appropriate to discuss some cautions. Most assays of reproductive effort use dry-weight measures of amount of tissue associated with reproduction relative to the amount in the whole plant. Such measures, although practical, do not take into account the metabolic costs of producing these tissues. These costs can be expected to vary considerably with the chemical constitution of the tissues (Sinclair and deWit 1975) and whether the same metabolic cycles can simultaneously produce materials to be used in both reproductive and nonreproductive portions of the plant (Calow 1979). These measures also neglect the allocation of minerals, vitamins, or defensive compounds to different parts of the plant. For example, nitrogen and phosphorus may be critical resources to be allocated in some plants (Lovett Doust 1980a,b); mineral content of various plant parts also can vary among populations. Mineral allocations are known to differ from biomass allocations in some species (Abrahamson and Caswell 1982). Furthermore, some studies use RE values summed over a whole season whereas others present data only for a single point in time. Even if the time of sampling is that of maximum RE, much reproductive allocation may occur beyond this time, so that peak RE can underestimate the actual total by a considerable amount. Differences in methods can make good comparisons difficult. It is not always easy to separate reproductive from vegetative parts of plants, especially if the flowering stem bears leaves or when the whole plant can be seen as part of the reproductive effort, as in the case of annuals (Thompson and Stewart 1981). It is really only proper to assay the entire reproductive effort and not just output of offspring, which is largely the measure of female performance. In addition, most (but not all) studies of comparative RE emphasize the allocation to the production of offspring through flowers; such offspring are usually produced sexually, but sometimes apomictically. Relatively few studies assess the allocation to vegetative propagation in making broad comparisons between habitats or frequency of reproduction. Furthermore, sexual allocation to seeds and fruits may be limited by pollinator activity in some cases; habitat differences in RE may then be related to pollinators rather than mortality patterns or competition (Bierzychudek 1981).

Most existing estimates of RE are based on the biomass of reproductive parts as a proportion of whole-plant (or sometimes total above-ground) biomass. Clearly, however, this is not quite the proper index for perennials, which maintain at least some of their biomass from year to year. For these plants, the relevant index of RE is made on an annual basis: allocation to reproductive parts as a proportion of annual increment in total biomass (if, indeed, carbon is the relevant resource). The estimate of RE for perennials by this means will generally be higher than estimates using the whole plant. (However, RE estimates based solely on above-ground parts will be affected less for herbaceous perennials that die back to ground level each year than for perennials that do not exhibit annual die-back.) With these caveats in mind, let us look at the information available.

One common expectation is that semelparous plants have higher RE than iteroparous ones. Counting only floral and fruiting RE, Abrahamson (1979a) found that the average RE of annuals was significantly greater than that of perennials growing in field habitats (20% > 12%). Similarly, Struik (1965) noted that the average RE of annuals exceeded that of perennials in forest (25% > 9%) and in open habitats (28% > 10%). With a different estimate of RE, perennial *Plantagos* averaged 1.6 mg seeds per 10 cm^2 of leaves, whereas annuals averaged 2.3, significantly more (Primack 1979). Thus the predictions seem to be upheld (see also Kawano 1975).

However, there are numerous cases of high RE by perennials. For example, the perennial palm *Astrocaryum mexicanum* allocates about 37% of the yearly production of biomass to reproduction; over an entire lifetime the average RE was estimated at about 32% (Sarukhan 1980). *Lupinus arboreus* has an RE as much as 26% (Pitelka 1977), *Claytonia virginica* up to 27% (Schemske et al. 1978), *Solidago speciosa* up to 35% (Abrahamson and Gadgil 1973), *Arisaema atrorubrus* 39% and *Hieracium floribundum* 34% (Abrahamson 1979a), and *Ranunculus bulbosus* and *R. acris* up to almost 60% (Sarukhan 1976), to present only some of the higher values for perennials. On the other hand, some annuals seem to have low RE: *Abutilon theophrasti,* 3% (Abrahamson 1979a), *Polygonum minimum,* ≤ 10% (Hickman 1977), *Impatiens capensis,* < 5% (Abrahamson and Hershey 1977), *Erodium cicutarium,* < 11% (Jaksic and Montenegro 1979). We do not know if these are unusual populations of the species or if these plants are true exceptions to the "rule." Most of these annuals with apparently low RE were sampled at only one point in time, and seasonal RE was undoubtedly greater; however, the same can be said for some of the high RE perennials, thus it seems that the contrasts are not explained by sampling differences. For some species, estimates differ greatly; for *Amphicarpa bracteata,* neglecting the subterranean self-fertilizing flowers, Struik (1965) found that RE reached as high as 41% in one location but only 15% in another. Many species show great flexibility; we will see examples of this below. Several biennials have high RE (*Daucus carota* 39%, *Melilotus alba* 22%, Abrahamson 1979a; *Dipsacus sylvestris* 27%, Caswell and Werner 1978; *Smyrnium olusatrum* 25–35%), but at least one has low RE (*Pastinaca sativa* 12%, Lovett Doust 1980a). Salisbury (1942) provides data for several genera in which seed production of perennials far exceeds that of annuals; if we assume that flowering effort parallels female RS for these species, the comparison enhances the conclusion that the disparity of RE in semelparous and iteroparous plants is anything but clearcut.

Many perennials propagate vegetatively through rhizomes, stolons, adventitious plantlets, or bulbils. In some cases, the allocation to this reproductive mode can be very high: up to 65% in *Trientalis borealis* (Anderson and Loucks 1973), between 40 and 80% for some *Allium* (Kawano and Nagai 1975), 49% in *Ranunculus repens* (Sarukhan 1976), about 40% in *Podophyllum peltatum* (Sohn and Policansky 1977), 26% in *Achillea millefolium* (Bostock and Benton 1979), 23% in *Tussilago farfara* (Bostock and Benton 1979, Ogden 1974). If flowering and vegetative RE are summed, in many instances the total RE for perennials falls in the usual range for annuals. Although the biological functions of vegetative propagation and reproduction by some kind of seed are undoubtedly quite different in some respects, the simple expectation of high RE for annuals and low RE for perennials is instantly rendered more complex. Should vegetative propagation be considered as part of present RE or as contributing to future reproduction? Although a vegetatively produced individual is genetically like its parent and thus perhaps an extension of it, because the new entity is capable of an independent existence, in one sense it can be considered as part of present RE and, indeed, as a form of reproduction, not just propagation (Abrahamson 1980).

Even within the same form of life history, shifts in RE might occur in response to expected mortality patterns or environmental constraints on allocational patterns. Closed, densely occupied habitats might favor greater relative allocation to growth and maintenance and hence lower RE than open, sparsely occupied habitats, because competition for three-dimensional space places a premium on capturing light and other resources. Furthermore, open habitats are sometimes those subject to frequent disturbance that decimates the existing populations of adult plants and, therefore, for this reason also, selection might favor relatively high RE in such conditions. The relative importance of mortality schedules and of possible competitive pressure is seldom disentangled, and neither kind of factor is usually measured.

Abrahamson (1979a) found that the average RE of perennials at flowering was greater in field than forest (12% > 8%). Comparing a one-year field, a ten-year field, and a forest, Newell and Tramer (1978) found that the average RE in the very early successional field (24%) exceeded that in both other habitats (5% each). The relatively low RE in the older field provides a cautionary note to facile comparisons of field and forest.

The several studies summarized below demonstrate that the predictions of increased RE with possibly increased adult mortality or with

Table 1.1 Differences in average RE for *Chamaesyce hirta* on gradients of nutrient levels and plant density (Snell and Burch 1975)

Nutrient Level	Density			
x	2 Plants/Flat %	8/Flat %	32/Flat %	64/Flat %
1	14	8	7	7
2	27	18	11	10
4	32	29	25	14
8	26	25	22	16

apparently diminished competitive pressure are not upheld as generally true for comparisons within species (see also Hickman 1975, Soule and Werner 1981). At this point it is difficult to say whether the predictions are fundamentally wrong or merely too simplistic.

Experiments with *Chamaesyce hirta,* an annual, showed that densely grown plants had lower RE than sparsely grown plants; very low nutrient levels also reduced RE at all densities (Table 1.1, Snell and Burch 1975). Dandelion (*Taraxacum officinale*) genotypes growing in disturbed sites allocated more material to flowers and seeds than did genotypes in less disturbed sites (Gadgil and Solbrig 1972; Solbrig and Simpson 1974). An annual grass, *Coix Ma-yuen,* generally responded to increased availability of nitrogen by increasing RE; decreased planting density had a similar effect (Kawano and Hayashi 1977). Interestingly, however, the risk of death, apparently due to competition, also increased at high N levels. The data from *Chamaesyce, Taraxacum,* and *Coix* (in part) tend to support the predictions, but the following examples are divergent.

Polygonum cascadense, an alpine annual, is phenotypically flexible in its RE, allocating an average of 58% of its biomass to reproduction in a dry, open, gravelly field, but only 38% in a moister, richer meadow (Hickman 1975). In contrast, Hickman (1977) also found that *Polygonum kelloggii* growing in relatively moist environments in company with a number of other species, allocated more biomass to reproduction than in dry, less occupied environments. Greenhouse-grown individuals of both species from both extremes showed no such differences. *Polygonum minimum* exhibited no significant shifts in RE along a similar gradient; *P. douglasii* indicated possible genotypic differences that, however, were not manifest in field populations.

Solidago speciosa growing in dry oldfields and *S. rugosa* in wet meadows in Massachusetts both allocated more above-ground biomass to flowers than their conspecific counterparts growing in a forest (Abrahamson and Gadgil 1973, Gadgil and Solbrig 1972), especially in the summer and fall; however, subterranean parts and vegetative propagation were ignored. Four of five species of *Solidago* (including *S. speciosa*) studied by Werner and Platt (1976) in Michigan and Iowa produced fewer but larger seeds in prairie than in oldfield, and the total weight of seeds per stem was far less in prairie. These observations tend to support the trends found in Massachusetts. However, sexual and vegetative RE of *Solidago canadensis* growing in a recently established oldfield and in an older field with other perennial herbs were indistinguishable, showing no shifts in allocational patterns, except the rhizomes were longer in the newer field (Bradbury and Hofstra 1976).

Sexual RE (above-ground parts only) of *Andropogon scoparius* first increased and then decreased with increasing age of the field (Roos and Quinn 1977). Reproductive effort of *Potentilla recta* declined with increasing habitat maturity in one year but not in another (Soule and Werner 1981). The perennial *Chamaenerion angustifolium,* when it flowered, showed no shift of RE with treatments of mineral fertilizer, merely failing to flower at all in some cases (van Andel and Vera 1977). A similar lack of plasticity in RE occurred in *Senecio sylvatica* (van Andel and Vera 1977), *S. vulgaris* (Harper and Ogden 1970), *Polygonum minimum* (Hickman 1977), and some populations of *Tussilago farfara* (see below).

Kawano and Musada (1980) studied a perennial lily, *Heloniopsis orientalis,* along an elevational gradient in Japan. Sexual reproductive effort (measured at time of fruit maturation) increased with elevation, from 10% at 100 m to 26% at 2600 m. Habitats of the plant also shifted along the altitudinal gradient; at 900 m and below, *Heloniopsis* was sampled from deciduous and coniferous forests, but at higher elevations it grew in open alpine moor and grassland. Habitat differences along the elevational gradient are likely to account for some of the allocational shifts; for instance, the two high-elevation, open sites supported much smaller plants that had high RE. To what aspect of the environment the plants are responding is not known; some possibly relevant factors are shown in Table 1.2. Despite the increase of RE, plants at higher elevations produced fewer seeds, and seed number was inversely correlated with RE. Thus the cost per seed increased by a factor of six as altitude increased. *Heloniopsis* produces adventitious plantlets at the leaftips; when the leaf droops, the plantlets touch the ground and can establish themselves. Plants growing under forest canopy produced plantlets far more frequently than did those of open alpine areas (Table 1.2).

Table 1.2 Differences in RE for *Helionopsis orientalis* on an elevational gradient, and some associated ecological conditions (Kawano and Masuda 1980)

Altitude (m)	Habitat	Maximum length of growing season (days)	Maximum light intensity (relative) (%)	Total average plant weight (g)	Sexual RE (%)	Frequency of plantlets (% of all individuals)
100	Deciduous forest	330	15	16	10	82
200	Deciduous and conifer forest	300	20	8	13	73
900	Deciduous and conifer forest	210	40	7	15	50
1900	Alpine moor and grassland	180	100	2	22	19
2600	Alpine meadow and pine thickets	135	100	2	26	20

If both sexual and vegetative RE are considered as part of total RE, the relationship between the two modes of reproduction is relevant here. Williams (1975) predicted an emphasis on sexual reproduction and long-distance dispersal of propagules at high density and greater vegetative reproduction at low densities. Implied by this prediction is an expectation that, to some extent, the proportions of sexual and asexual RE will be inversely correlated. Vegetative propagules have some advantages over seeds in that they have a tremendous head start in development and growth, often can mature earlier, and may have better juvenile survivorship (e.g., Abrahamson 1980, Amor 1974, Lee and Harmer 1980, Sarukhan 1976, Whigham 1974). But they are very similar, often identical, to the parent (unless somatic mutation occurs), and may be better suited for filling up space around the parent than for movement to sites of unpredictable suitability. What is the evidence for inverse correlations of vegetative and sexual RE?

Rubus vestitus plants producing the greatest numbers of inflorescences tended to produce the least vegetative propagation (Kirby 1980), as predicted. However, *Aster acuminatus* showed no change in vegetative RE over a range of densities and light regimes; sexual RE did not change with density but increased with increasing light availability (Pitelka et al. 1980). Contrary to expectations, allocation to vegetative propagation in *Mimulus primuloides* changed little with environmental "harshness" (increasing elevation), but allocation to sexual RE increased at higher altitudes (although the ensuing RS declined; Douglas 1981). In addition, several turf grasses are said to decrease seed output at high density (Law et al. 1979), the reverse of the "expected" trend. Biomass allocation to tubers and to inflorescences of *Cyperus rotundus* were similar at low and medium densities, but at high density, allocation to tubers was considerably greater, again in contrast to the theoretical prediction (Williams et al. 1977).

Ogden's (1974) study of *Tussilago farfara* indicated that this species shifts its RE in sexual reproduction (measured as seed production) very little in response to experimental manipulations of density or nutritional level. Far different was the response of vegetative propagation; rhizome formation was greatest when soil fertility was low and at low densities (Table 1.3). Total RE was greatest in fertile soil and at lower densities. In contrast, Bostock (1980) reported considerable variation in sexual and vegetative RE among *T. farfara* populations but found no response to soil nutrients. Plants from what was considered to be the most severe environment characteristically allocated less biomass to seed production and more to vegetative propagation. The populations differed in phenotypically plastic responses—plants from certain areas exhibited more shift in allocation pattern than others.

Table 1.3 Variation in average RE of *Tussilago farfara* (Ogden 1974)

	% Sexual (Seed) RE		% Vegetative RE		% Total RE	
Treatment	Year 1	Year 2	Year 1	Year 2	Year 1	Year 2
Soil fertility						
Low (sand and clay)	7.1	7.2	10.2	10.2	17.3	17.4
High (compost)	5.0	8.1	22.9	18.0	27.9	26.1
Density in field						
Low	6.6	4.9	11.8	15.2	18.4	20.1
Medium	6.3	4.5	13.4	14.7	19.7	19.2
High	6.6	4.9	6.9	10.0	13.5	14.9
Density in trays						
Low	5.7		11.1		16.6	
High	3.4		4.0		7.3	

Allocation to vegetative propagation can vary independently of flowering RE, as in *Tussilago* and *Aster.* Furthermore, two species of *Rubus* decreased total RE and vegetative propagation at higher densities, although flowering RE was unchanged (e.g., Abrahamson 1979b); a similar pattern was observed for wild strawberries (*Fragaria virginiana*); Holler and Abrahamson 1977). However, in other cases there is a positive association. Wild strawberries responded to the experimental removal of neighbors by a three-fold increase of both vegetative and sexual RE, although the increase in flower production was not evident until the following year because buds are set the year before flowering (Smith 1972). Thomas and Dale (1974) suggest a positive correlation of stolon and flower production in *Hieracium floribundum,* both reproductive features associated, at least to some degree, with rosette size.

Because the available resource budgets, the genetic variation, the phenotypic responses to various limiting factors (Grace and Wetzel 1981), and the selection pressures all differ among populations and species, it is impossible to judge the significance of this variation in the relationship of sexual and asexual RE; detailed experiments in specific ecological contexts are required to unravel things. At least it should be clear that the consequence of failing to record shifts in vegetative reproduction as well as sexual is that patterns of total RE cannot be discerned because sexual and asexual modes of reproduction have no consistent relationship to each other.

The great variability of results emerging from existing studies of RE raises a general interpretational problem. If RE does not change with changing environmental regime, is it because RE has no adaptive value, because it is fixed for historical reasons, because its very fixity has some unknown adaptive value, or because the range of conditions and the factors tested were insufficient to elicit a response? Are some kinds of plants very flexible and others less so? Even if RE does shift in different conditions, the direction of the shift may vary with the position of the plant relative to its optimal environmental conditions (Soule and Werner 1981). Furthermore, we cannot presume that the end-product itself is at some optimum value; a given RE may reflect merely the best the individuals could do under the circumstances.

THE EVOLUTION OF SENESCENCE

Age-specific mortality rates are central to the evolution of senescence as well as reproductive schedules (Edney and Gill 1968, Hamilton 1966, Medawar 1957, Williams 1957). Senescence is the phenotypic manifestation of deleterious effects accumulating in old age. As we have seen, every species is subject to an array of extrinsic causes of death unrelated to age—death from starvation, drowning, desiccation, sometimes predation or pathogens, and so on. As time passes, the cumulative probability of survival becomes correspondingly low. Even if every individual is *potentially* immortal, the extrinsic hazards set a limit to longevity. Because the extrinsic hazards facing every population are different, the time course of senescence and the expected longevity must differ among species.

Elimination of many genotypes from the population, by the accumulated risks, reduces the variation on which selection can act and thereby diminishes the effectiveness of selection. Further, the greatest portion of the contribution to future generations of any individual generally comes early in its reproductive life. Since the genetic contribution of older individuals is thus necessarily less than of those just reaching maturity, the efficacy of natural selection is reduced at greater ages. This permits the accumulation of harmful effects in old age, the effects of deleterious genes or the negative effects of pleiotropic genes whose favorable effects in early life outweigh their later harmful results. Modifier genes may postpone the expression of deleterious effects, but the extrinsic hazards reduce their effectiveness as well. Therefore, deleterious effects accumulate at older ages, resulting in senescence. And because selection is weaker in later stages of a life history, we should expect to see greater between-individual variation in those stages than in earlier ones (Charlesworth and Leon 1976, Leon 1976).

Senescence has been a troublesome evolutionary problem partly because people have tended to confuse the causes with the symptoms. A postreproductive period, if such indeed exists, in an organism's lifetime is a symptom; deleterious effects pile up and cause a postreproductive period, not vice versa. Susceptibility to disease or injury is also symptomatic; neither disease nor injury are causes of senescence. It is possible that cumulative DNA damage in the somatic cells of multicellular organisms contributes to senescent decline (Gensler and Bernstein 1981). Yet at least to some degree the failure of repair mechanisms may be seen as a consequence of weaker selection at older ages, as described above.

Senescence in plants is particularly knotty because, like many invertebrate animals and prokaryotes, there is often a capacity for vegetative propagation. Harper (1977) remarked that plants with clonal growth seldom show signs of senescence. He cited as an example the case of bracken fern in which clones established about 1500 years ago are still spreading and distributing spores (Oinonen 1967). Clones of creosote bush (*Larrea tridentata*) exceed 10,000 years of age (Vasek 1980), and aspen clones may do so too. However, long-lived trees such as the bristlecone pine (*Pinus longaeva*) are known to live several thousand years and some other conifers can achieve ages over 1000 years. Although they all presumably exhibit signs of senescence and some may lack means of vegetative propagation, perhaps then the ages of bracken ferns and creosotebush are less impressive and not necessarily a function of their vegetative spread. Furthermore, *Lemna minor* forms clones, but each individual frond lives only five to six weeks and exhibits signs of senescence (Ashby et al. 1949). However, clones in which the "offspring" become separate from the "mother" need not behave like those in which some connection is retained and in which the entire clone functions as an individual, because death affects separate entities independently (Charlesworth 1980). Despite the fact that most plants have indeterminate growth and grow more or less continually throughout their lives with potentially increasing fecundity, and that clonal plants may achieve particularly great increases of fecundity with age, Hamilton (1966) argued that senescence is eventually inevitable. We are presented with an apparent dearth of evidence for what is claimed to be a logical necessity.

One possible and partial escape from this dilemma may be offered by developmental genetics (or either the logic or the evidence could be wrong!). Genes are turned on and off at specific points in the development of an individual. It is likely that this process is geared to a sequence of events, not just to elapsed time. Therefore, we should expect that detrimental genetic effects will turn on at times triggered by the developmental sequence. Every time a plant produces a vegetative propagule that is potentially capable of independent existence, the sequence begins almost anew. This process perhaps could delay senescence; whether it could prevent it is moot. It then becomes interesting to wonder why individuals cannot reset the developmental clock within themselves, as new cells are formed.

Another curious observation is that maternal age can influence the life history of offspring and that this effect can be transmissible from one generation to the next. The effect is often cumulative over several generations, but it is also sometimes reversible. These phenomena are usually called "Lansing effects" (reviewed by Lints 1978). A botanical example of this is provided by studies of *Lemna minor,* a tiny, clonal, aquatic flowering plant (Ashby and Wangermann 1951, 1954, Ashby et al. 1949, Wangermann and Ashby 1951). Each frond produces several daughter fronds, the average number of daughters varying among clones. Successive daughters of a single mother frond are progressively smaller and are composed of fewer cells. Smaller daughter fronds have shorter lifespans and fewer daughters themselves. If the lifespan of the mother frond is altered by environmental conditions, this alteration produces a corresponding change in the daughter-series. Thus it seems clear that some factor is transmitted from mother to daughter and that clonal propagation does not always reset the developmental clock. Nevertheless, *Lemna minor* individuals do not become infinitely small; small daughters can produce first-daughters larger than themselves, and these daughters do likewise until the size of the original clone-mother is restored. Thus some (apparently still unknown) event counters the effect of maternal age, and the clock is eventually turned back (but probably not to time-zero, since clones of most organisms seem to die eventually). Lansing effects are studied with respect to mechanisms of senescence, but in the case of *L. minor,* it should be noted that there seems to be an implicit assumption that small daughters are somehow less valuable than large ones and merely the unfortunate result of increasing age. Perhaps an examination of possible adaptive aspects of producing small offspring at certain stages of the life history would be appropriate.

The perhaps high frequency of somatic chimeras among many species of plants (Whitham and Slobodchikoff 1981) may offer some new insights into plant senescence. These authors note that plants consist of repetitive units, many of which are capable of reproduction (sexual or asexual). Somatic mutations occur rather frequently in plants and can be passed into the reproductive propagules by several means. Thus even vegetatively propagated individuals or stems may differ from the parent stock and could, conceivably, possess genes that delay senescence. If these ramets reproduce more successfully than others, and there is a continual process of somatic mutation with subsequent selection among ramets, it seems clear that the chimeric nature of many plants might contribute to longevity and the postponement of senescence. The magnitude of the importance of somatic mutation would depend on the frequency of favorable mutations, which is likely to be low, and the (unknown) frequency of somatic mutation relative to gametic mutation. The evolution of senescence remains a very poorly understood matter.

COMPLEX LIFE CYCLES IN PLANTS

The life cycles of many plants consist of two or even more distinct parts, each with a separate life history. Typically, one part is the haploid gametophyte and the other the diploid sporophyte. As the names imply, sporophytes produce spores, which grow up into gametophytes, and gametophytes produce gametes, two of which fuse to form a zygote, which grows up into a sporophyte. This basic scheme is modified in numerous ways; either the gametophyte or sporophyte generation may be emphasized, one or the other may be suppressed, dormant, or parasitic on the other, or sometimes a second diploid phase exists. Some of these variants are diagrammed in Figure 1.8. In many green algae, such as *Chlamydomonas,* the active alga is haploid. It produces more haploid cells by mitosis. Several asexual multiplications may occur, followed by the production of gametes and zygote formation. The zygote is dormant and resistant to a variety of environmental conditions. Eventually it undergoes meiosis, producing more haploid, active cells. At the other extreme are such things as the brown alga, *Fucus*; this exists as a conspicuous diploid that produces gametes that fuse, and the zygote grows up into a new diploid. Seed plants are similar to *Fucus* in some respects. The highly conspicuous sporophyte bears tiny parasitic gametophytes that have no independent life of their own.

Thus at one extreme we find the haploid generation dominant, and at the other the diploid is dominant. In between is a variety of other conditions. *Bryopsis plumosa,* a green alga, is interesting because it exhibits geographic variation in the life cycle. Off the coast of the Netherlands, zygotes develop directly into the plumose, diploid adult; near France, the zygotes form the branched entity that produces spores, which then give rise to the plumose forms (see Bold et al. 1980). *Polysiphonia* (a red alga) has two diploid phases. The first one is borne on the haploid female and gives rise to diploid spores that form a free-living sporophyte, which produces a second set of spores, haploid this time, that grow into the haploid gametophytic individuals. In mosses and

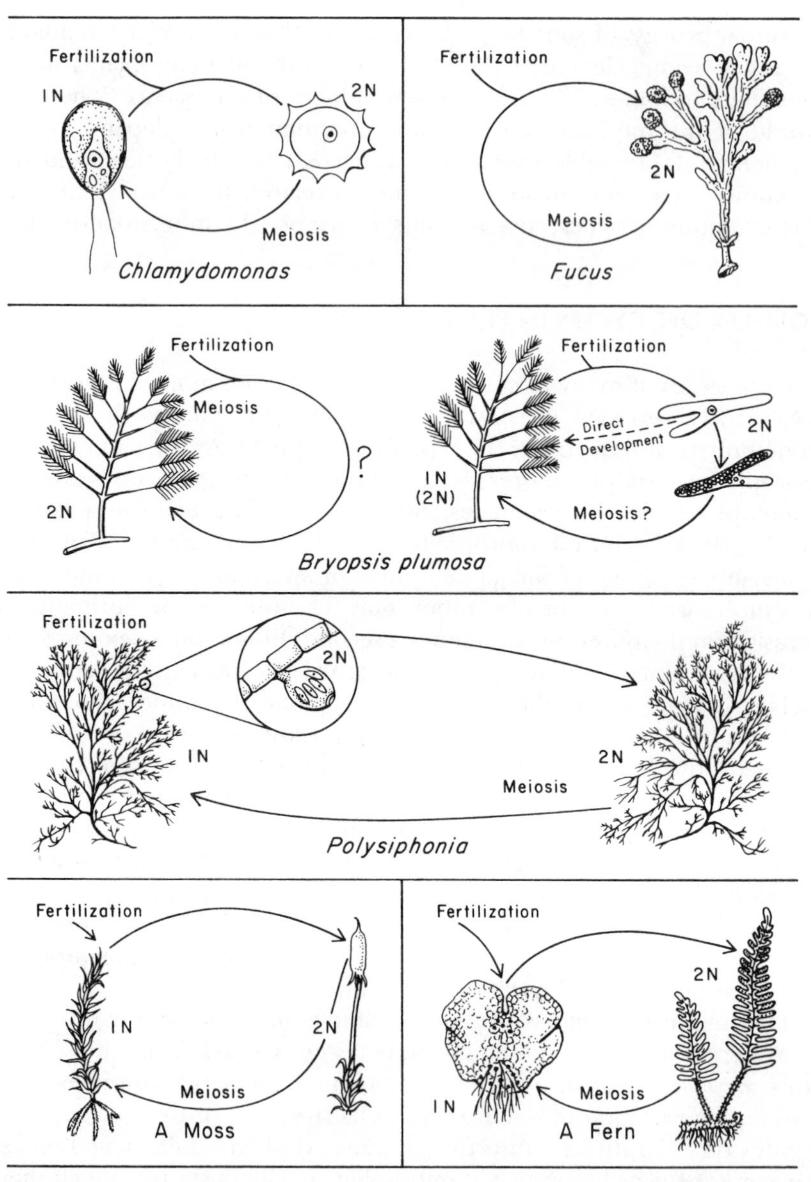

Figure 1.8 Some complex life cycles in plants. The top two panels illustrate the extremes of haploid "dominance" (*Chlamydomonas*) and diploid "dominance" (*Fucus*). The remaining panels present examples in which both generations are more or less active.

ferns the sporophyte is borne on the gametophyte, but in ferns the gametophyte eventually disappears. The questions to be addressed here are why have two (or more) generations, why are they of differing levels of ploidy, and what might determine their relative dominance or conspicuousness. The following discussion derives chiefly from Willson (1981).

The existence of two generations usually suggests that each lives in a different environment, spatially or seasonally. The two generations may be alike in form (isomorphic) or unlike (heteromorphic). A few algae such as *Ulva* are intriguing in that isomorphic generations sometimes are found side by side on the same rocks. Although this observation hints that the two morphs are not using different environments, possible seasonal differentiation in times of peak activity or reproduction have apparently not been sought.

The advantages of occupying two different environments include the possibility of rapid growth and asexual multiplication in a temporarily rich environment. A genotype that can capitalize on environmental heterogeneity in this way may (1) outcompete other genotypes in capturing resources and be able to achieve high rates of multiplication, (2) therefore, have more propagules available for dispersal, and (3) if it is the haploid phase that multiplies asexually, have more gametes available to participate in fertilization. The alternating generation may occupy an environment that facilitates dispersal and colonization of new sites.

But why use another, separate, entity instead of a transforming larva, as do most animals with complex life cycles? A larval stage typically gives the ensuing adult a size advantage, because the new adult is larger than it would have been if it came directly from an egg. Nevertheless, larval plants are almost unknown. Two things require explanation: why are larval plants so rare and, then, why use a differing ploidy to exploit a second environment. The means by which plants exploit their environment may be inherently less varied than those available to animals. For instance, differences in feeding and foraging ecology of animals are associated with tremendous differences in morphology and behavior that often demarcate large taxa as well as stages of the life cycle. Such differences seem less extreme in plants—their methods of obtaining nutrition may limit their ability to diverge in form. Thus perhaps a larval plant seldom could be sufficiently different from the adult to be termed a proper larva. In the absence of a larva, one available way to capitalize on a second environment might be to develop independent entities that differ in ploidy. In general, haploids can be expected to multiply (asexually) faster than diploids because they have smaller cell volumes and greater surface/volume ratios. Therefore, haploids may be useful whenever rapid multiplication is particularly advantageous; this might be true especially for plants, which have limited mobility and few behavioral adjustments (unlike animals) that can increase juvenile survival. The argument that a haploid generation allows the elimination of exposed, deleterious genes is somewhat circular because many of these genes are only deleterious in the haploid state.

The relative dominance of each generation may be, in part, a result of selection for increased fecundity. Larger organisms are generally more fecund than smaller ones of the same kind. So we might expect greater body size in the life-cycle generation most subject to selection for high fecundity. Asexual multiplication of the genotype might be an alternative means of achieving high fecundity. Furthermore, greater height can increase dispersal of air-borne propagules (see also Keddy 1981). In some cases, the size of a gametophyte may be constrained by the need to have access to water-borne gametes as in mosses and ferns, but it also reflects the need to support and nourish the sporophyte, especially in mosses. Large body sizes and increased fecundity can also increase the variability of sexually-produced progeny: the more cells are dividing meiotically, the more genetic variation can be expressed in the progeny (see the section on evolution of sex in Chapter 2). Small bodies may elude herbivores more successfully, so that small entities are favored in environments with high risks of herbivory. The ecology of alternating generations has been little investigated in plants, so these suggestions are untested.

LITERATURE CITED

Abrahamson, W. G. 1979a. Patterns of resource allocation in wildflower populations of fields and woods. *Am. J. Bot.* **66:** 71–79.

Abrahamson, W. G. 1979b. A comment on vegetative and seed reproduction in plants. *Evolution* **33:** 517–519.

Abrahamson, W. G. 1980. Demography and vegetative reproduction. *Bot. Monogr.* **15:** 89–106.

Abrahamson, W. G. and H. Caswell. 1982. On the comparative allocation of biomass, energy, and nutrients in plants. *Ecology* **63:** 982–991.

Abrahamson, W. G. and M. Gadgil. 1973. Growth form and reproductive effort in goldenrods (*Solidago,* Compositae). *Am. Nat.* **107:** 651–661.

Abrahamson, W. G. and B. J. Hershey. 1977. Resource allocation and growth in *Impatiens capensis* (Balsaminaceae) in two habitats. *Bull. Torrey Bot. Club* **104:** 160–164.

Amor, R. L. 1974. Ecology and control of blackberry (*Rubus fruticosus* L. *agg.*), II: Reproduction. *Weed Res.* **14:** 231–238.

Anderson, R. C. and O. L. Loucks. 1973. Aspects of the biology of *Trientalis borealis* Raf. *Ecology* **54:** 798–808.

Ashby, E. and E. Wangermann. 1951. Studies in the morphogenesis of leaves. VII, Part II: Correlative effects of fronds in *Lemna minor*. *New Phytol.* **50:** 200–209.

Ashby, E. and E. Wangermann. 1954. The effects of meristem aging in the morphology and behavior of fronds in *Lemna minor*. *Ann. N.Y. Acad. Sci.* **57:** 476–483.

Ashby, E., E. Wangermann, E. J. Winter. 1949. Studies in the morphogenesis of leaves, III: Preliminary observations on vegetative growth in *Lemna minor*. *New Phytol.* **48:** 374–381.

Bateson, G. 1963. The role of somatic change in evolution. *Evolution* **17:** 529–539.

Bazzaz, F. A. and R. W. Carlson. 1979. Photosynthetic contribution of flowers and seeds to reproductive effort of an annual colonizer. *New Phytol.* **82:** 223–232.

Bazzaz, F. A., R. W. Carlson, and J. L. Harper. 1979. Contribution to reproductive effort by photosynthesis of flowers and fruits. *Nature* **279:** 554–555.

Bell, G. 1980. The costs of reproduction and their consequences. *Am. Nat.* **116:** 45–76.

Bierzychudek, P. 1981. Pollinator limitation of plant reproductive effort. *Am. Nat.* **117:** 838–840.

Bold, H. C., C. J. Alexopoulos, and T. Delevoryas. 1980. *Morphology of Plants and Fungi* (4th ed.). Harper & Row, New York.

Bostock, S. J. 1980. Variation in reproductive allocation in *Tussilago farfara*. *Oikos* **34:** 359–363.

Bostock, S. J. and R. A. Benton. 1979. The reproductive strategies of five perennial compositae. *J. Ecol.* **67:** 91–108.

Bradbury, I. K. and G. Hofstra. 1976. The partitioning of net energy resources in two populations of *Solidago canadensis* during a single developmental cycle in southern Ontario. *Can. J. Bot.* **54:** 2449–2456.

Calow, P. 1979. The cost of reproduction—a physiological approach. *Biol. Rev.* **54:** 23–40.

Caswell, H. 1981. Reply to comments by Yodzis and Schaeffer. *Ecology* **62:** 1685.

Caswell, H. and P. A. Werner. 1978. Transient behavior and life history analysis of teasel (*Dipsacus sylvestris* Huds.). *Ecology* **59:** 53–66.

Chapin, F. S. 1980. The mineral nutrition of wild plants. *ARES* **11:** 233–260.

Charlesworth, B. 1980. The cost of meiosis with alternation of sexual and asexual generations. *J. Theor. Biol.* **87:** 517–528.

Charlesworth, B. and J. A. Leon. 1976. The relation of reproductive effort to age. *Am. Nat.* **110:** 449–459.

Cole, L. C. 1954. The population consequences of life history phenomena. *Q. Rev. Biol.* **22:** 283–314.

Douglas, D. A. 1981. The balance between vegetative and sexual reproduction of *Mimulus primuloides* (Scrophulariaceae) at different altitudes in California. *J. Ecol.* **69:** 295–310.

Edney, E. B. and R. W. Gill. 1968. Evolution of senescence and specific longevity. *Nature* **220:** 281–282.

Gadgil, M. and O. T. Solbrig. 1972. The concept of *r*- and *K*-selection: evidence from wild flowers and some theoretical considerations. *Am. Nat.* **106:** 14–31.

Gensler, H. L. and H. Bernstein. 1981. DNA damage as the primary cause of aging. *Q. Rev. Biol.* **56:** 279–303.

Goodman, D. 1982. Optimal life histories, optimal notation, and the value of reproductive value. *Am. Nat.* **119:** 803–823.

Grace, J. B. and R. G. Wetzel. 1981. Phenotypic and genotypic components of growth and reproduction in *Typha latifolia*: Experimental studies in marshes of differing successional maturity. *Ecology* **62:** 789–801.

Gross, H. L. 1972. Crown deterioration and reduced growth associated with excessive seed production by birch. *Can. J. Bot.* **50:** 2431–2437.

Gross, K. L. 1981. Predictions of fate from rosette size in four "biennial" plant species: *Verbascum thapsus, Oenothera biennis, Daucus carota,* and *Tragopogon dubius. Oecologia* **48:** 209–213.

Hairston, N. G., D. W. Tinkle, and H. M. Wilbur. 1970. Natural selection and the parameters of population growth. *J. Wildl. Manage.* **34:** 681–690.

Hamilton, W. D. 1966. The moulding of senescence by natural selection. *J. Theor. Biol.* **12:** 12–45.

Harper, J. L. 1977. *Population Biology of Plants.* Academic, New York.

Harper, J. L. and J. Ogden. 1970. The reproductive strategy of higher plants, I: The concept of strategy with special reference to *Senecio vulgaris* L. *J. Ecol.* **58:** 681–698.

Harper, J. L., P. H. Lovell, and K. G. Moore. 1970. The shapes and sizes of seeds. *ARES* **1:** 327–356.

Hart, R. 1977. Why are biennials so few? *Am. Nat.* **111:** 792–799.

Hastings, A. and H. Caswell. 1979. Role of environmental variability in the evolution of life history strategies. *PNAS(USA)* **76:** 4700–4703.

Hickman, J. C. 1975. Environmental unpredictability and plastic energy allocation strategies in the annual *Polygonum cascadense* (Polygonaceae). *J. Ecol.* **63:** 689–701.

Hickman, J. C. 1977. Energy allocation and niche differentiation in four co-existing annual species of *Polygonum* in western North America. *J. Ecol.* **65:** 317–326.

Hirshfield, M. F. and D. W. Tinkle. 1975. Natural selection and the evolution of reproductive effort. *PNAS(USA)* **72:** 2227–2231.

Holler, L. C. and W. G. Abrahamson. 1977. Seed and vegetative reproduction in relation to density in *Fragaria virginiana* (Rosaceae). *Am. J. Bot.* **64:** 1003–1007.

Horn, H. S. 1978. Optimal tactics of reproduction and life-history. *In* J. R. Krebs and N. B. Davies (eds.), *Behavioral Ecology.* Sinauer, Sunderland, Mass.

Jaksic, F. M. and G. Montenegro. 1979. Resource allocation of Chilean herbs in response to climatic and microclimatic factors. *Oecologia* **40:** 81–89.

Johnson, M. P. and S. A. Cook. 1968. "Clutch size" in buttercups. *Am. Nat.* **102:** 405–411.

Kawano, S. 1975. The productive and reproductive biology of flowering plants, II: The concept of life history strategy in plants. *J. College Liberal Arts, Toyama Univ.* (Japan) **8:** 51–86.

Kawano, S. and S. Hayashi. 1977. Plasticity in growth and reproductive energy allocation of *Coix Ma-yuen* Roman. cultivated at varying densities and nitrogen levels. *J. College Liberal Arts, Toyama Univ.* (Japan) **10:** 61–92.

Kawano, S. and J. Musada. 1980. The productive and reproductive biology of flowering plants, VII: Resources allocation and reproductive capacity in wild populations of *Heloniopsis orientalis* (Thunb.) C. Tanaka (Liliaceae). *Oecologia* **45:** 307–317.

Kawano, S. and Y. Nagai. 1975. The productive and reproductive biology of flowering plants, I: Life history strategies of three *Allium* species in Japan. *Bot. Mag. Tokyo* **88:** 281–318.

Keddy, P. A. 1981. Why gametophytes and sporophytes are different: Form and function in a terrestrial environment. *Am. Nat.* **118:** 452–454.

Kirby, K. J. 1980. Experiments on vegetative reproduction in bramble (*Rubus vestitus*). *J. Ecol.* **68:** 513–520.

Kozlowski, T. T. 1971. *Growth and Development of Trees,* Vol. 2. Academic, New York.

Langer, R. H. M. 1956. Growth and nutrition of timothy (*Phleum pratense*), I: The life history of individual tillers. *Ann. Appl. Biol.* **44:** 166–187.

Langer, R. H. M., S. M. Ryle, and O. R. Jewiss. 1964. The changing plant and tiller populations of timothy and meadow fescue swards, I: Plant survival and the pattern of tillering. *J. Appl. Ecol.* **1:** 197–208.

Law, R. 1979. The costs of reproduction in an annual meadow grass. *Am. Nat.* **113:** 3–16.

Law, R., A. D. Bradshaw, and P. D. Putwain. 1977. Life-history variation in *Poa annua. Evolution* **31:** 233–246.

Law, R., A. D. Bradshaw, and P. D. Putwain. 1979. Reply to W. G. Abrahamson. *Evolution* **33:** 519–520.

Lee, J. A. and R. Harmer. 1980. Vivipary, a reproductive strategy in response to environmental stress? *Oikos* **35:** 254–265.

Leon, J. A. 1976. Life histories as adaptive strategies. *J. Theor. Biol.* **60:** 301–335.

Leverich, W. J. and D. A. Levin. 1979. Age-specific survivorship and reproduction in *Phlox drummondii. Am. Nat.* **113:** 881–903.

Levin, D. A. and B. L. Turner. 1977. Clutch size in the Compositae. *In* B. Stonehouse and C. Perrins (eds.), *Evolutionary Ecology,* University Park Press, Baltimore, Md.

Lewontin, R. C. 1965. Selection for colonizing ability. *In* H. G. Baker and G. L. Stebbins (eds.), *The Genetics of Colonizing Species.* Academic, New York.

Lints, F. A. 1978. *Genetics and Ageing.* Karger, Basel.

Livdahl, T. P. 1979. Environmental uncertainty and selection for life-cycle delays in opportunistic species. *Am. Nat.* **113:** 835–842.

Lloyd, D. G., C. J. Webb, and R. B. Primack. 1980. Sexual strategies in plants, II: Data on the temporal regulation of maternal investment. *New Phytol.* **86:** 81–92.

Lovett Doust, J. 1980a. A comparative study of life history and resource allocation in selected Umbelliferae. *Biol. J. Linn. Soc.* **13:** 139–154.

Lovett Doust, J. 1980b. Experimental manipulation of patterns of resource allocation in the growth cycle reproduction of *Smyrnium olusatrum* L. *Biol. J. Linn. Soc.* **13:** 155–166.

Marshall, D. L. 1982. The nature and consequences of variation in the components of yield in *Sesbania macrocarpa, S. drummondii,* and *S. vesicaria.* Ph.D. Thesis, University of Texas.

Meagher, T. R. and J. Antonovics. 1982. Life history variation in dioecious plant populations: A case study of *Chamaelirium luteum. In* H. Dingle and J. Hegmann (eds.), *Evolution and Genetics of Life Histories.* Springer-Verlag, New York.

Medawar, P. B. 1957. *The Uniqueness of the Individual.* Methuen, London.

Michod, R. E. 1979. Evolution of life histories in response to age-specific mortality factors. *Am. Nat.* **113:** 531–550.

Mooney, H. A. and R. I. Hays. 1973. Carbohydrate storage cycles in two Californian Mediterranean climate trees. *Flora* **162:** 295–304.

Morgan, M. D. 1971. Life history and energy relationships of *Hydrophyllum appendiculatum. Ecol. Monogr.* **41:** 329–349.

Newell, S. J. and E. J. Tramer. 1978. Reproductive strategies in herbaceous plant communities during succession. *Ecology* **59:** 228–234.

Ogden, J. 1974. The reproductive strategy of higher plants, II: The reproductive strategy of *Tussilago farfara* L. *J. Ecol.* **62:** 291–324.

Oinonen, E. 1967. Sporal regeneration of bracken in Finland in the light of the dimensions and age of its clones. *Acta For. Fenn.* **83**(1): 3–96.

Pianka, E. R. 1976. Natural selection of optimal reproductive tactics. *Am. Zool.* **16:** 775–784.

Pianka, E. R. and W. S. Parker. 1975. Age-specific reproductive tactics. *Am. Nat.* **109:** 453–464.

Pitelka, L. F. 1977. Energy allocation in annual and perennial lupines (*Lupinus:* Leguminosae). *Ecology* **58:** 1055–1065.

Pitelka, L. F., D. S. Stanton, and M. O. Peckenham. 1980. Effects of light and density on resource allocation in a forest herb, *Aster acuminatus* (Compositae). *Am. J. Bot.* **67:** 942–948.

Primack, R. B. 1978. Regulation of seed yield in *Plantago. J. Ecol.* **66:** 835–847.

Primack, R. B. 1979. Reproductive effort in annual and perennial species of *Plantago* (Plantaginaceae). *Am. Nat.* **114:** 51–62.

Rabotnov, T. A. 1969. On coenopopulations of perennial herbaceous plants in natural coenoses. *Vegetatio* **19:** 87–95.

Roos, F. H. and J. A. Quinn. 1977. Phenology and reproductive allocation in *Andropogon scoparius* (Graminae) populations in communities of different seral stages. *Am. J. Bot.* **64:** 535–540.

Salisbury, E. J. 1942. *The Reproductive Capacity of Plants.* Bell, London.

Sarukhan, J. 1974. Studies on plant demography: *Ranunculus repens* L., *R. bulbosus, R. bulbosus* L., and *R. acris* L. *J. Ecol.* **62:** 151–177.

Sarukhan, J. 1976. On selective pressures and energy allocation in populations of *Ranunculus repens* L., *R. bulbosus* L., and *R. acris* L. *Ann. Missouri Bot. Gard.* **63:** 250–308.

Sarukhan, J. 1980. Demographic problems in tropical systems. *Bot. Monogr.* **15:** 161–188.

Sarukhan, J. and J. L. Harper. 1973. Studies on plant demography: *Ranunculus repens* L., *R. bulbosus* L. and *R. acris* L. *J. Ecol.* **61:** 675–716.

Schaal, B. A. and W. J. Leverich. 1981. The demographic consequences of two-stage life cycles: Survivorship and the time of reproduction. *Am. Nat.* **118:** 135–138.

Schaffer, W. M. 1974. Selection for optimal life histories: The effects of age structure. *Ecology* **55:** 291–303.

Schaffer. W. M. 1981. On reproductive value and fitness. *Ecology* **62:** 1683–1686.

Schaffer, W. M. and M. D. Gadgil. 1975. Selection for optimal life histories in plants. *In* M. L. Cody and J. M. Diamond (eds.), *Ecology and Evolution of Communities.* Belknap, Cambridge, Mass.

Schaffer, W. M. and M. V. Schaffer. 1977. The adaptive significance of variations in reproductive habit in the Agavaceae. *In* B. Stonehouse and C. Perrins (eds.), *Evolutionary Ecology.* University Park Press, Baltimore, Md.

Schaffer, W. M. and M. V. Schaffer. 1979. The adaptive significance of variations in reproductive habit in the Agavaceae, II: Pollinator foraging behavior and selection for increased reproductive expenditure. *Ecology* **60:** 1051–1069.

Schemske, D. W. 1980. Evolution of floral display in the orchid *Brassavola nodosa. Evolution* **34:** 489–493.

Schemske, D. W., M. F. Willson, M. N. Melampy, L. J. Miller, L. Verner, K. M. Schemske, and L. B. Best. 1978. Flowering ecology of some spring woodland herbs. *Ecology* **59:** 351–366.

Sinclair, T. R. and C. T. deWit. 1975. Photosynthate and nitrogen requirements for seed production by various crops. *Science* **189:** 565–567.

Singh, L. B. 1960. *The Mango.* Leonard Hill, London.

Smith, C. C. 1972. The distribution of energy into sexual and asexual reproduction in wild strawberries (*Fragaria virginiana*). *Midwest Prairie Conf. Proc.* **3:** 55–60.

Smith, C. C. and S. Fretwell. 1974. The optimal balance between size and number of offspring. *Am. Nat.* **108:** 499–506.

Snell, T. W. and D. G. Burch. 1975. The effects of density on resource partitioning in *Chamaesyce hirta* (Euphorbiaceae). *Ecology* **56:** 742–746.

Sohn, J. J. and D. Policansky. 1977. The costs of reproduction in the mayapple *Podophyllum peltatum* (Berberidaceae). *Ecology* **58:** 1366–1374.

Solbrig, O. T. and B. B. Simpson. 1974. Components of regulation of a population of dandelions in Michigan. *J. Ecol.* **62:** 473–486.

Soule, J. D. and P. A. Werner. 1981. Patterns of resource allocation in plants with special reference to *Potentilla recta* L. *Bull. Torrey Bot. Club* **108:** 311–319.

Stearns, S. C. 1980. A new view of life-history evolution. *Oikos* **35:** 266–281.

Stephenson, A. G. 1981. Flower and fruit abortion: Proximate causes and ultimate functions. *ARES* **12:** 253–279.

Struik, G. J. 1965. Growth patterns of some native annual and perennial herbs in southern Wisconsin. *Ecology* **46:** 401–420.

Thomas, A. G. and H. M. Dale. 1974. Zonation and regulation of old pasture populations of *Hieracium floribundum. Can. J. Bot.* **52:** 1451–1458.

Thompson, D. A. and A. J. Beattie. 1981. Density-mediated seed and stolon production in *Viola* (Violaceae). *Am. J. Bot.* **68:** 383–388.

Thompson, K. and A. J. A. Stewart. 1981. The measurement and meaning of reproductive effort in plants. *Am. Nat.* **117:** 205–211.

Thompson, J. N. 1978. Within-patch structure and dynamics in *Pastinaca sativa* and resource availability to a specialized herbivore. *Ecology* **59:** 443–448.

Threadgill, P. F., J. M. Baskin, and C. C. Baskin. 1981. The ecological life cycle of *Frasera caroliniensis,* a long-lived monocarpic perennial. *Am. Midl. Nat.* **105:** 277–289.

van Andel, J. and F. Vera. 1977. Reproductive allocation in *Senecio sylvaticus* and *Chamaenerion angustifolium* in relation to mineral nutrition. *J. Ecol.* **65:** 747–758.

van der Meijden, E. and R. E. van der Walls-kooi. 1979. The population ecology of *Senecio jacobaea* in a sand dune system, I. Reproductive strategy and the biennial habit. *J. Ecol.* **67:** 131–153.

Vasek, F. C. 1980. Creosote bush: Long-lived clones in the Mohave Desert. *Am. J. Bot.* **67:** 246–255.

Veillon, J. M. 1971. Une Apocynacée monocarpique de Nouvelle-Calédonie *Cerberiopsis candelabrum* Vieill. *Adansonia* **11:** 625–639.

Vernet, P. and J. L. Harper. 1980. The costs of sex in seaweeds. *Biol. J. Linn. Soc.* **13:** 129–138.

Wangermann, E. and E. Ashby. 1951. Studies in the morphogenesis of leaves, VII, Part I: Effects of light intensity and temperature on the cycle of ageing and rejuvenation in the vegetative life history of *Lemna minor. New Phytol.* **50:** 186–199.

Werner, P. A. 1975. Predictions of fate from rosette size in teasel (*Dipsacus fullorum* L.). *Oecologia* **20:** 197–201.

Werner, P. A. 1976. Ecology of plant populations in successional environments. *Syst. Bot.* **1:** 246–268.

Werner, P. A. and W. J. Platt. 1976. Ecological relationship among co-occurring goldenrods (*Solidago:* Compositae). *Am. Nat.* **110:** 959–971.

Whigham, D. 1974. An ecological life history of *Uvularia perfoliata. Am. Midl. Nat.* **91:** 343–359.

Whitman, T. G. and C. N. Slobodchikoff. 1981. Evolution by individuals, plant-herbivore interactions, and mosaics of genetic variability: The adaptive significance of somatic mutations in plants. *Oecologia* **49:** 287–292.

Wilbur, H. M. 1976. Life history evolution in seven milkweeds of the genus *Asclepias. J. Ecol.* **64:** 223–240.

Wilbur, H. M. 1977. Propagule size, number, and dispersion pattern in *Ambystoma* and *Asclepias. Am. Nat.* **111:** 43–68.

Williams, G. C. 1957. Pleiotropy, natural selection, and the evolution of senescence. *Evolution* **11:** 398–411.

Williams, G. C. 1966. Natural selection, the costs of reproduction, and a refinement of Lack's principle. *Am. Nat.* **100:** 687–690.

Williams, G. C. 1975. *Sex and Evolution.* Princeton University Press, Princeton, N.J.

Williams, R. D., P. C. Quimby, and K. E. Frick. 1977. Intraspecific competition of purple nutsedge (*Cyperus rotundus*) under greenhouse conditions. *Weed Sci.* **25:** 477–481.

Willson, M. F. 1981. The evolution of complex life cycles in plants: A review and an ecological perspective. *Ann. Missouri Bot. Gard.* **68:** 275–300.

Willson, M. F. and N. Burley (in press). Mate choice in plants: Tactics, mechanisms, and consequences. Princeton University Press, Princeton, N.J.

Willson, M. F. and P. W. Price. 1977. The evolution of inflorescence size in *Asclepias* (Asclepiadaceae). *Evolution* **31:** 495–511.

Willson, M. F. and P. W. Price. 1980. Resource limitation of fruit and seed production in some *Asclepias* species. *Can. J. Bot.* **58:** 2229–2233.

Willson, M. F. and D. W. Schemske. 1980. Pollinator limitation, fruit production, and floral display in pawpaw (*Asimina triloba*). *Bull. Torrey Bot. Club* **107:** 401–408.

Yodzis, P. 1981. Concerning the sense in which maximizing fitness is equivalent to maximizing reproductive value. *Ecology* **62:** 1681–1682.

Young, T. P. 1981. A general model of comparative fecundity for semelparous and iteroparous life histories. *Am. Nat.* **118:** 27–36.

Chapter 2

SEXUAL SYSTEMS

Many issues in evolutionary ecology revolve around long-standing confusion concerning the evolution of sexual systems. The puzzles are not solved yet, for we have only begun to understand some of the factors involved. The evolution of sex itself is still controversial and, despite a long history of agricultural investigation, so is the evolution of outcrossing and inbreeding. The many variants of sex expression by plants and the evolution of sex ratios are still mystifying in many respects. The goals of this chapter are to present an overview of much of the relevant literature and to attempt a preliminary assessment of possible adaptive values of various sexual attributes.

THE EVOLUTION OF SEX

Sex is a characteristic feature of the life cycle of most eukaryotes. Its basic features include meiosis in the diploid form, producing haploid gametes, and fusion of the gametes to produce a new diploid. The gametes of a few plants are equal in size (isogamous); this is the case, for instance, in some unicellular algae such as *Chlamydomonas*. However, in most plants male and female gametes are different in size (anisogamous), in materials included therein, and in motility; virtually by definition, female gametes are larger, less motile, and usually contain far more included materials than male gametes.

Many plants can reproduce both sexually and asexually, but some appear to have relinquished, largely or altogether, sexual reproduction and rely instead on asexual means. The evolutionary question is what determines the extent of sexual and asexual reproductive modes, both within a species and in the kingdom of plants as a whole. This question can be asked from two perspectives: why is sexual reproduction so common *or* why is asexual reproduction well developed in some cases. From the point of view of vertebrates like ourselves, the second form of the question may seem the most natural. But for several theoretical reasons, the first version has attracted the most attention.

Several authors have noted that sexual reproduction entails various costs. There is the cost of mating: the expense of building special organs or producing attractive substances that bring male and female together (Daly 1978, Muenchow 1978). And there is always a certain risk that no mate, or only an inferior one, will be available. Furthermore, the recombination of genes achieved by sexual reproduction breaks up favorable genetic combinations in each generation, thus creating a continual supply of less fit individuals (Lloyd 1980b, Manning and Jenkins 1980, Maynard Smith 1978, Nyberg 1982). However, the other side of the coin is, of course, that the same process accounts for the assembly of better genotypes, and for varied genotypes, both of which may be very important, as seen later in this discussion.

Sexual reproduction involves the fusion of two haploid gametes, which are the products of meiosis. The evolution of a reduction division presumably came about simultaneously with the origin of the process of sexual fusion as a means of maintaining the standard genome size (but see Williams 1980 for another view). The amount of DNA in a cell can affect not only gene expression but also cell volume and rates of mitosis, all of which may affect the organism's fitness. But the existence of meiosis and sexual fusion mean that additional costs are involved. The precise nature of these costs, however, is somewhat confusing.

Williams (1975, 1980) emphasized that each sexual parent shares, on average, only about half its nuclear genes with its offspring, because each offspring has two parents; unless there are sex chromosomes, the share will be just one-half. We are referring here not to *all* the nuclear genes of the parents, because some alleles are fixed in the populations (e.g., all the specific diagnostic traits), but rather to those loci with at least two alleles and, in particular, to the loci at which the parents bear different alleles. In contrast, asexually produced offspring are identical to their parent (barring mutation or "automixis"—a meiosis whose end-products reunite, and ignoring cytoplasmic genes) and the parent shares all the genes of its offspring. Thus, if fecundities are equal, the sexual parent has passed on half as many genes as the asexual one. Williams called this the cost of meiosis. As presented by Williams, the cost of meiosis is paid by genes rather than individuals, and the perspective is one of gene-level selection rather than individual selection. Willson and Burley (in press), Hartung (1981), and others have argued that gene-level selection probably is less powerful than individual selection. If that is true, then the original concept of the cost of meiosis may not affect individual fitness very greatly. Nevertheless, possible conflict between selection on genes and on individuals, as well as that between elements of the genome (Cosmides and Tooby 1981), need not be resolved entirely in favor of one side or the other, but we don't yet know where the balance lies.

A true meiotic cost should obtain in any organism capable of meiosis. Because meiosis occurs in two successive cell divisions, and mitosis in only one, and because meiosis is followed by syngamy in the production of a zygote but mitosis is not, the formation of a sexual individual may take longer than by asexual means (Lloyd 1980b, Maynard Smith 1978). When rapid reproduction is advantageous, the added time may constitute a cost.

Maynard Smith (1978) suggested that the cost of meiosis in Williams' original sense (the parental sharing of offspring genotypes) might better be viewed as a cost of making males. This is more than a semantic change, because these costs are not exactly equivalent; for instance, the cost of making males is dependent on the sex ratio but that of meiosis itself is not (Lloyd 1980b, Treisman and Dawkins 1976). According to Maynard Smith, an asexual parent who produces all (asexual) daughters can have a contribution to future generations through females twice that of sexuals, if equal numbers of males and females are produced by the sexuals. Although the sexually produced males will mate with sexually produced females, eventually the asexual females will predominate in the population because they turn out a higher proportion of female offspring. His argument is this: if n is the number of asexual females, N the number of sexual females and of sexual males, and k is clutch size per female (assumed to be the same for both sexuals and asexuals), then while n asexual females are producing kn female offspring, the N sexual females are producing $kN/2$ daughters and $kN/2$ sons. Thus asexual females constitute a proportion $n/(2N + n)$ of the adult generation, but this proportion increases to $n/(N + n)$ in the following generation. Clearly, the relative disadvantage to the sexuals depends on the sex ratio of the offspring and, as we shall see, sex ratios can vary widely. Nevertheless, Maynard Smith's algebra seems to be irrefutable. However, Lloyd (1980b,d) argued that the cost of making males disappears when fitness achieved through male offspring is included in the estimation of fitness of sexually reproducing organisms. And there may be advantages to sex that are not reflected by simple measures of relative fecundity. In fact, the assumption of equal fecundity of sexuals and asexuals is risky, in view of the known lowered fecundity of some asexuals (Lamb and Willey 1979, Lynch unpublished manuscript).

Maynard Smith proposed three conditions that might negate the cost of making males. One condition that supposedly eliminates the cost of making males is paternal care. The basic supposition is that extensive care of the young by *both* parents allows the clutch size of a sexual pair to increase enough to match the fecundity of asexuals. Such a proposal applies only to certain rather complex animals. The evidence that paternal care actually doubles the clutch size is minimal (see also deJong 1980). In any case, this condition does not apply to plants and can be neglected here.

The second condition is found in hermaphroditic organisms. The cost will be retrieved if the plant is an obligate self-fertilizer because then the allocation to male function can be greatly reduced and the resources reallocated to production of seeds. However, this situation is more complex than described by Maynard Smith, because resource allocation to male and female function frequently is unequal in regularly outcrossing hermaphrodites (see also Lloyd 1980b) and the postulated twofold differential in effective fecundity may not be found.

The third of these is isogamy. Isogametes are formed by unicellular organisms that simply divide into two equal daughter cells. Regardless of whether a diploid cell undergoes meiosis to form four gametes that fuse in pairs to create two diploid zygotes or the diploid cell forms daughter cells by simple mitosis, the end product is two diploid daughters (Maynard Smith 1978) of equivalent fecundity (see also Charlesworth 1980a). Furthermore, a zygote formed by the fusion of two isogametes could have an initial size twice that of an asexual formed by fission, a difference that could counter the numerical advantage of the asexuals (Manning 1975). On the contrary side, however, Charlesworth (1980b) suggested that under some circumstances (as when zygote fitness does not depend closely on size) this cost may still be relevant. In addition, meiosis and syngamy may have evolved first in isogamous organisms (but see also Alexander and Borgia 1978), which have no cost of making males, and Maynard Smith (1978) then said "It follows that, when considering the origins of sexual reproduction in eukaryotes, one need not allow for a twofold disadvantage." If sex originated without a cost of making males, then what needs to be explained is why eukaryotes took on this cost—why male and female roles became differentiated, and why anisogamy evolved. It is certainly fair to say that there can't be a cost of making males unless male and female are differentiated from each other.

If sex originated in isogamous organisms, the only costs (to individuals) that need to be accounted for are the costs of mating. Gender specialization and anisogamy may have arisen through disruptive selection on gamete size (Parker 1978, Parker et al. 1972): if zygote fitness increases with zygote size, there will be selection to increase gamete size. At some point, however, there will also be selection on other gamete-producers to produce gametes with a high probability of "capturing" and uniting with the large gametes; successful capture is facilitated by the production of numerous smaller (and possibly more mobile) gametes. We call the producers of large gametes females; those producing small gametes are males.

The cost of making males could be mitigated, perhaps, if males sometimes had very high RS (see also Charlesworth 1980b). Obviously, average male RS must equal average female RS in sexually reproducing populations. But if female RS of an hermaphrodite is resource-limited (a common condition), many more genes can still be contributed to future

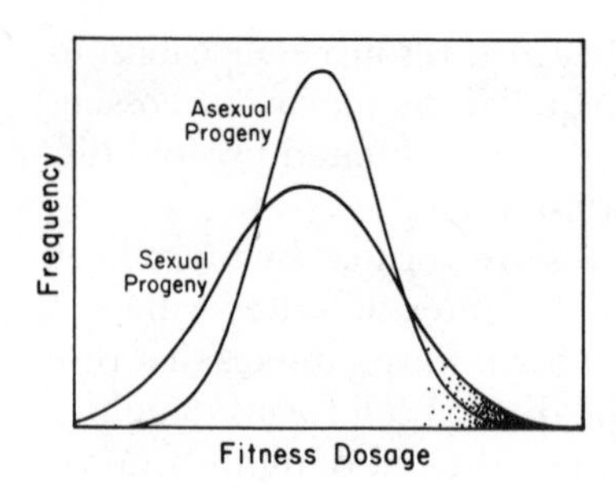

Figure 2.1 A comparison of potential variance in fitness of sexual and asexual progenies. The mean value for sexuals is lower, because fit genotypes are broken up, but the variance is higher, also because of recombination. From George C. Williams, *Sex and Evolution.* (© 1975 by Princeton University Press).

generations through male function. If the sexes occur in separate individuals and a single male can mate many times so that his RS can equal that of numerous females, again the value to a parent of making males may be great. Males—and male function of an hermaphrodite—may be a means of evading strict resource-limitation of RS and potentially enhancing the total parental RS (see also Lloyd 1980b). It is a high-risk means, however, since for every highly successful male there must, on the average, be one that fails to the same degree. This argument is derived from that of Parker et al. (1972, see also Parker 1978) for the evolution of anisogamy, and of Hartung (1981).

On the plus side, sexual reproduction makes possible the assemblage of genes in different combinations having differing fitnesses. Then if selection is so intense that only individuals in the very top categories of "fitness dosage" can survive and reproduce, sexual reproduction may prevail over asexual because it provides a higher likelihood of achieving the very fit combinations (Emlen 1973, Williams 1975). This is illustrated in Figure 2.1. But because these most-fit combinations are disassembled in each generation and because the environment is very changeable, the process of achieving the right combinations is a continual one, never quite completed. Williams (1975) calls it the "sisyphean genotype," after the legendary Greek character named Sisyphus.* The capacity for mate

*Sisyphus had a checkered career, offending the gods on many occasions. Zeus, the chief god, finally sent the messenger of death to claim Sisyphus for the Underworld. Before he died, the crafty Sisyphus advised his wife not to give him the customary funerary honors. Arriving in the Underworld, he went promptly to Hades to complain of his wife's negligence and to ask for permission to go briefly back to Earth to punish her. Permission was granted and Sisyphus, thinking he was very clever, returned to Earth and refused to honor his pledge to stay only briefly. He was eventually dragged back to the Underworld and, for his bad faith, condemned for eternity to roll up a mountain a huge boulder that forever escaped and rolled back down just as it neared the top. Thus "sisyphean" has come to refer to something that must be continually done over and never completely succeeds.

choice makes possible a more rapid assembling of appropriate combinations of genes in the offspring.

Another possible advantage to sexual reproduction may lie in the relative rates of accumulation of beneficial mutations and elimination of deleterious ones (reviewed in Maynard Smith 1978): asexual lineages cannot reduce the number of mutations they carry; but a sexual lineage, with recombination, produces offspring with more and with fewer mutations. Because most mutations are thought to be deleterious, the ability to reduce their frequency is probably advantageous, and the capacity to bring together whatever beneficial mutations there are is also advantageous (Hartung 1981, Lloyd 1980b). There are a variety of circumstances in which these features can favor the evolution of sex, and recent work suggests that conditions are not as restrictive as was previously thought. If males compete among themselves for mates and if females choose the males with which they mate, the elimination of deleterious mutation may even be hastened sufficiently to obviate the cost of making males (Stanton unpublished manuscript). Evidence that offspring fitness improves when sexual selection (intermale competition and female choice) occurs would be useful in this context, and a small amount is available. Mate choice by *Drosophila melanogaster* affects at least one component of offspring fitness (larval competitive ability; Partridge 1980). And, for plants, offspring that result from intense competition among pollen grains on a stigma sometimes exhibit greater size, growth rate, or competitive ability than those from low levels of pollen competition (McKenna and Mulcahy unpublished manuscript, Mulcahy and Mulcahy 1975). Although it is sometimes thought that traits affecting fitness should be fixed (or nearly so) in the population because of their great importance (e.g., Harpending 1979, Maynard Smith 1978, Williams 1975), such a view is basically unreasonable for it implies that evolution has ceased; in a varying environment (both physical and biological), fixation should be less likely.

A third selective advantage of sexual reproduction is thought to lie in the ability to produce a varied progeny; the ecological importance of this aspect of sex is potentially great. The more variation among the offspring, the higher the likelihood that one or some of them will have the right genetic constitution to survive and reproduce in some environment (Williams 1975). This requires, obviously, that the environment itself be variable in time and/or space.

Evidence supporting the hypothesis that a varied progeny is adaptive in a variable environment is meager. A great deal of ink has been spilled in genetic or mathematical models showing that the hypothesis is, at best, improbable, or at least subject to significant constraints (e.g., Moore and

Hines 1981). Nevertheless, sex is clearly with us and the failure of many models may indicate deficiencies in the model-building process and the perceptions of what is indeed possible in the real world more than in the idea itself. One major problem has been that, until recently, biological aspects of the environment have been neglected. Some of the most important features of an organism's environment are other organisms: predators, parasites, and competitors. Every organism interacts with other organisms in at least one of these categories. The presence of sexual reproduction and sisyphean genotypes in any one of the interacting species will create a highly unpredictable environment that selects for sexuality in the rest of the species in the interacting system (Glesener and Tilman 1978, Jaenicke 1978, Levin 1975). For every genetic change in a predator that improves its rate of prey capture, for every change in a competitor that gives it an upper hand, selection will favor a corresponding change in an interacting species that allows it to counter the changes in predator and competitor. One might view it as a coevolutionary "race" in which each participant must continually change just to avoid losing. Van Valen (1973) likened it to the Red Queen in *Through the Looking Glass,* who runs constantly merely to stay in place.*

Two independent surveys suggest that completely asexual organisms generally have different geographic distributions than their sometimes or completely sexual relatives. (Note that these studies do not address the question of the balance of sexual and asexual reproduction within the life history of an organism that does both.) For terrestrial animals, Glesener and Tilman (1978) reported that asexual forms tend to occupy biologically less complex environments, being found especially at high altitudes and latitudes, in xeric rather than mesic habitats, on islands more than mainlands, and in disturbed rather than undisturbed habitats. For plants, Levin (1975) suggested that sexual reproduction is of immediate value to species interacting with herbivores and pathogens. Such interactions may be more intense in climax as opposed to successional communities and in tropical more than temperate areas (Levin 1975), although this common supposition needs verification. That these correlations appear for two such different taxa is heartening, although correlations of course do not always indicate causation. Furthermore, a number of tropical trees are probably apomictic, producing embryos by asexual means (Kaur et al. 1978).

Hamilton (1980, and see also Hamilton et al. 1981) has recently de-

*"A slow sort of country!" said the Queen. "Now, here, you see, it takes all the running you can do to keep in the same place. If you want to get somewhere else, you must run at least twice as fast as that!"

vised a model that bolsters the credibility of the varied-progeny hypothesis. He showed that a short-lived, rapidly evolving, and genotype-specific parasite may provide sufficiently strong selection to raise the fitness of sexual types above that of asexual types, despite costs of making males and so on. Like all models, it is oversimplified and, furthermore, relies on a long-term (geometric) mean fitness rather than immediate advantages. Nevertheless, Hamilton's endeavor is a step forward because it removes many of the previous intellectual constraints stemming from preconceived ideas of what is possible, how variable an environment can be, and what environmental factors are most germane. Other models bolster Hamilton's conclusions (e.g., Bremermann 1980, Glesener 1979).

However, even this skimpy array of evidence is subject to question (Lynch unpublished manuscript). First of all, it is by no means clear that asexual organisms must evolve slowly or that they are evolutionary dead ends. Many asexual lines may have higher mutation rates than related sexual lineages and they have some other means of generating genetic variability as well; phenotypic variance in field populations is often as high as for sexuals (see, e.g., Usberti and Jain 1978), and artificial selection programs often produce changes in gene frequencies. In addition, many plants seem to be somatic chimeras and somatic mutations are readily passed on (Whitham and Slobodchikoff 1981). Therefore, asexual organisms may be fully capable of responding to selection pressures from other organisms. Furthermore, although some asexuals are found at the extremes of generic ranges, others coexist with their sexual relatives in the thick of intense biological pressures. Many asexuals appear to be broadly adapted, being possessed of a "general-purpose genotype," and they often occupy a greater variety of habitats and wider geographic ranges than their sexual relatives. Lynch suggests that asexual reproduction may be rare mainly because of genetic and cytological difficulties such as interbreeding between an ancestral sexual taxon and its asexual offshoot and genomic incompatibility (e.g., odd levels of ploidy) between them.

The theoretical costs of sex continue to trouble evolutionary biologists greatly. Williams (1975) has suggested that sex in many organisms is just a maladaptive trait that became fixed in many populations, which survive despite rather than because of their reproductive mode. However, the persistence of both modes in many organisms strongly suggests that something more than historical factors can account for the maintenance of sexuality. And it is somehow intellectually unsatisfactory to suppose that such apparently successful taxa as insects and vertebrates, not to mention the plants, could have radiated so extensively despite the putative handicap of sexuality. Others have argued that selection at the level of the population or, at the other extreme, the gene may account for the evolution and maintenance of sex. Neither approach is entirely felicitous. There may indeed be long-term consequences of sex to be found in rapid rates of evolution and improved persistence of a species (or a gene; Stanley 1975), although even this is questioned by Lynch (unpublished manuscript). But most evolutionists prefer to consider the usual unit of selection to be the individual, not the group or the gene, and so we must seek explanations of sex in terms of advantages to individuals. However, just how such advantages might outweigh the apparent costs is still a matter of conjecture. More than one selective advantage may obtain; sex and recombination may be important in DNA repair (Bernstein et al. 1981), but there are ecological consequences as well. Moreover, particular population structures and patterns of selection may facilitate the maintenance of sex (Price and Waser 1982). It is even possible that sexuality or asexuality per se are of less evolutionary consequence than other associated life-history traits (Templeton 1982).

It is commonplace to argue that the timing of sex in organisms that reproduce, on a seasonal schedule, either sexually or asexually can be used as evidence that sex is adaptive to environmental uncertainty. Bonner (1958) said that in such organisms sex occurs either before or after a time when the environment undergoes a drastic, usually seasonal, change. This idea has often been uncritically repeated as explanatory gospel. It explains nothing, in fact. First of all, the change referred to is usually one in the physical environment, such as the onset of winter, and we have no real basis for supposing that this change is somehow more critical than an assortment of biological changes throughout the summer. Second, everything is likely to be either before or after something else, strict simultaneity being most improbable, so the question really is *how long* before or after *what* environmental change. Third, and perhaps most important, this is a very one-sided view that neglects the probable significance of the asexually reproducing phase of the cycle, treating this phase as somehow adaptively unimportant. I expect that a more antiquated view (e.g., Coulter 1914) may be appropriate. Asexual reproduction is likely to be adaptive in exploiting ephemeral habitats, capturing resources, and multiplying a genotype (e.g., Rollins 1967); this mode should be employed whenever suitable conditions prevail. Sexual reproduction, on the other hand, should (given the organism is ever sexual) be pushed to whatever season is unsuitable for asexual multiplication, provided that conditions are then suitable for successful sex. In fact, if sexual reproduction is more expensive than asexual due to costs of matefinding or provisioning eggs, it may be constrained to occur principally when density is high and matefinding is thus easier or when resources are particularly abundant. Environmental change, imminent or just past, may have nothing to do with it, except by casual (*not* causal!) association (see Gerritsen 1980, Willson 1981a). Therefore, I think that restatements of Bonner's assertion are generally misguided and misleading; they certainly do not constitute evidence for the adaptiveness of sex in varying environments.

These three possible ecological advantages of sexual reproduction are not mutually exclusive. The constitution of favorable genotypes and elimination of deleterious mutations by means of recombination and the generation of variety per se could all be important. At this point we cannot assign relative weight to each of these factors and additional factors are likely to be proposed as well. Nor is it possible to weigh the relative advantages and disadvantages of sexuality and asexuality in particular circumstances. Some of the possibilities have been discussed here, but the actuality and magnitudes of possible costs, benefits, and constraints have not yet been adequately assessed.

A review of the distribution of sexual and asexual reproduction among plants, and comparing plants with animals, might be very useful. But it will also be very difficult to draw accurate comparisons, in view of the great flexibility of a number of plants and the great gaps in our knowledge about many species. That task is left to the future (and see Bell 1982).

SEX EXPRESSION

Given that an organism is going to reproduce sexually at some point in its life cycle, the next question might be whether male and female genders should be expressed by each organism, either sequentially or simultaneously, or whether the genders should be expressed in separate individuals. We can distinguish several possible conditions of sex expression.

Hermaphroditism. In defiance of the customary botanical usage, I employ this term chiefly in the way it is used for the animal kingdom, meaning individuals that function as both male and female. Some botanists now also use the term this way. In some cases I will refer to hermaphroditic flowers, with both male and female functional parts, but "hermaphrodite" as a noun always refers to individuals.

Simultaneous hermaphrodites. Individuals are functionally male and female at the same time.

1. ***Sexual monomorphism.*** Flowers or sex organs function as both sexes, either simultaneously or sequentially. Note that when a flower is first one sex and then the other, the whole individual is still likely to be functioning as an hermaphrodite. Each flower may be protandrous (first male) or protogynous (first female) or have coincident sex expression. Bawa and Beach (1981) emphasize that temporal separation of sexual functions in hermaphroditic flowers is poorly recorded.
2. ***Monoecy.*** Each flower or sex organ is unisexual, either male or female. Andromonoecy refers to the possession of some hermaphrodite flowers and some male flowers, gynomonoecy to the presence of both hermaphrodite and female flowers on the same individual.

Sequential hermaphrodites. An individual shifts from one sex to the other once or more in its lifetime. Protandry refers to being first male, then female; protogyny is the reverse.

Dioecy. Male and female individuals are separate. In some cases a few individuals in a population are hermaphroditic, whereas, the remainder are unisexual. The rare condition of androdioecy refers to a population in which most individuals are hermaphrodite, but there are some male individuals. In gynodioecy, most individuals are hermaphrodite, but some are strictly female.

What ecological conditions favor each of these states? We will begin with the extremes—monomorphic hermaphroditism and dioecy—and proceed to the intermediate and mixed sexual systems. It is essential to realize that there exist many intergrades among these systems, that normally dioecious populations may contain some hermaphrodites, that normally monoecious species may have some unisexual individuals, and so on.

Hermaphroditism and Dioecy

A graphical representation and preliminary solution to the problem of when to be hermaphroditic and when to be dioecious is provided by Charnov et al. (1976). The general conditions for hermaphroditism to be advantageous are that the reproductive success (RS) of an outcrossing hermaphrodite when functioning as a male must be greater than half of the RS of a male and the RS of an hermaphrodite functioning as a

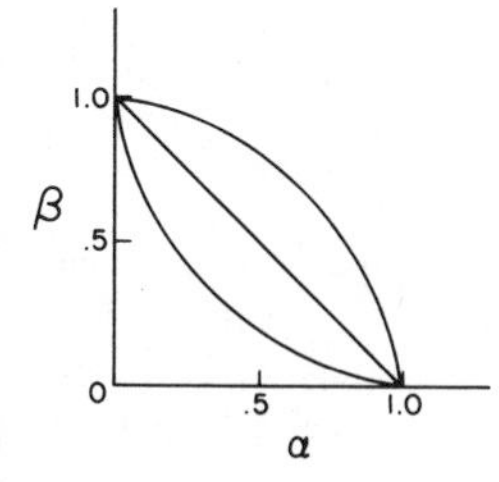

Figure 2.2 Three possible relationships between α (the pollen output of an hermaphrodite relative to that of a male) and β (the egg output of an hermaphrodite relative to that of a female). The shape of the tradeoff curve defines conditions favoring hermaphroditism or dioecy. Modified from Charnov et al. (1976), by permission from *Nature*, vol. 263, No. 5573, p. 125. Copyright 1976 Macmillan Journals Limited.

female must exceed half of the RS of a female. Suppose that β is the number of eggs produced by an hermaphrodite compared to the number produced by a female, and α is the number of sperm produced by an hermaphrodite relative to a male. Also imagine that α and β are not independent of each other but that an increase of one necessitates a decrease of the other. The consequences of this interdependence depend on the shape of the relationship.

The simplest case occurs when egg and sperm production are limited by the same resources, and each change in α brings with it an exactly corresponding change in β, as represented by the diagonal in Figure 2.2. This line represents the equation $\alpha + \beta = 1$ and there is no selection for or against separate sexes. Maximum RS occurs when $\alpha = \beta$ and the product $(\alpha \cdot \beta)$ is maximal. However, if $(\alpha + \beta) < 1$, the curve is concave upward and dioecy should be favored. Here, if α is large, an increase in β necessitates a large drop in α, and if β is large, the reverse is true. In short, an increase in one function necessitates a cost to the other, and once large α or β is reached, evolution away from that situation is difficult. But sperm and egg production may be to some extent controlled by separate resources, perhaps because seed maturation consumes resources after the time of pollen production, or perhaps because different kinds of resources are involved. Then $(\alpha + \beta > 1)$ and the curve is convex upward, and simultaneous hermaphroditism should be advantageous. Here, when α or β is large, the cost to α of a change in β, or vice versa, is relatively small. Of course, the drawing of such curves tells nothing about the biological conditions that shape the curves.

A further consideration is the rate of increase of RS as either α or β increases (Charnov 1979). Suppose that female RS is a function of the resources available for the maturation of seeds but that male RS is limited by the number of eggs fertilized. We can consider the rate of increase of male RS as a function of the amount of resources allocated to male function. According to Charnov, if that gain in male fitness increases only slowly and becomes large only at extremely high values of

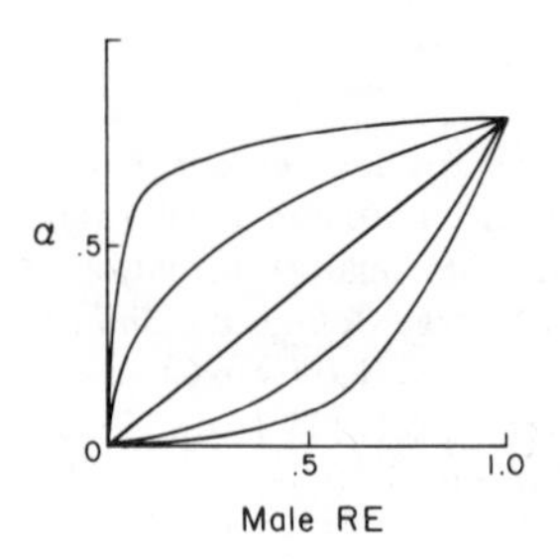

Figure 2.3 Several possible relationships between male fitness (indexed by relative pollen output) and male RE. The rate of fitness-gain determines the advantage of becoming completely male. Modified from Charnov (1979) by permission of E. L. Charnov.

RE, dioecy should be favored because an hermaphrodite would have to allocate most of its RE to maleness to improve its male success, and this would sacrifice too much female RS. However, if male RS increases quickly for small increases in allocation, at least at low values of allocation to maleness, then hermaphroditism may be advantageous (Figure 2.3).

The ability of hermaphrodites to self-fertilize may alter these conclusions (Charlesworth and Charlesworth 1981). Because of the high frequency of hermaphroditic plants that can self (either within or between flowers), the modifications of the model are potentially important. The fitness-gain through male function is restricted when selfing and inbreeding depression occurs, so that fitness of females may exceed that of hermaphrodites even when the ($\alpha + \beta$) curve is convex upward, thus rendering hermaphroditism less favorable. If the selfing rate varies with male RE in an hermaphrodite, the situation is still more complicated and no clear expectations emerge. Moreover, all these modelings depend on the assumption that the initial population is at an evolutionary equilibrium with respect to its sex expression, and it is by no means clear that this need be so (Charlesworth and Charlesworth 1981).

Additional difficulties with simple models appear when one considers the biological factors that might shape the curve of fitness-gain and the relationship of α and β. Low mobility has been suggested as one such factor. Because plants are generally sessile and the distribution of sperm around a male parent is almost always tightly clumped (with few male gametes reaching distant females; e.g., Levin and Kerster 1974), the number of accessible eggs is limited, and an increase of allocation to maleness may bring little increase in RS at any level of RE (Charnov 1979). However, there is a possibility that fitness-gain from breeding with relatively distant females will be greater than from inbreeding with neighboring relatives. Production of enormous amounts of pollen will increase the probability of reaching distant females (Willson and Burley, in press) and might permit males a relatively great gain in fitness for large expenditure of resources. And in any event, we would need to know why sessileness in some plants fostered hermaphroditism but in others allowed dioecy. Another limitation of this suggestion is that most plants so far analyzed breed primarily with neighboring individuals because pollen flow from the source is usually so restricted, and this leaves us with the unanswered question of why only certain plants can nevertheless manage to be dioecious.

The suggestions that the ability to attract animal pollinators (Charnov et al. 1976) or the saturation of pollen vectors (Charnov 1979) may favor hermaphroditism are also not very persuasive because a great many insect-pollinated plants are, in fact, dioecious. There probably are some plants, however, in which the male and female aspects of the floral display work together in attracting pollinators, as when stamens and stigmas determine the visual conformation of the target of the pollinator (Bawa and Beach 1981). In such cases, hermaphroditism is economical for reasons of pollinator attraction. Furthermore, if certain expensive structures, such as the large and showy spathes and spadices in the Araceae (Bawa and Beach 1981), are common costs of both male and female function, selection may favor hermaphroditism in some form over dioecy (Charnov et al. 1976, Heath 1977).

Perhaps the classical explanation for simultaneous hermaphroditism is that it reduces the problem of finding a mate when population densities are low (Baker 1965, Ghiselin 1969). If the individual is capable of self-fertilization, then mate-finding depends only on having a suitable vector to bring the gametes together; in the case of flowers that are both male and female and that can pollinate themselves by means of autogamy, even the need for a vector is eliminated. If the individual must cross-fertilize, in most cases *any* other individual is a possible mate, except when there are distinct mating types (as in heterostyly, see below), genetic incompatibilities, or strong individual "preferences" to mate only with certain other individuals. Just what constitutes low density is not always easy to know; some organisms have special mechanisms that increase their effective density at times; strongly synchronized flowering is one such mechanism. By implication, this argument suggests that simultaneous hermaphroditism is advantageous when female reproductive success is limited by the availability of sperm. And, as we have seen, this condition seems to be less common in plants than one of resource limitation. But it may pertain particularly to colonizing species and a few others.

If female RS is held at some limit, perhaps by the ability of the parent to physically support and to nourish the offspring, an hermaphrodite still has the option of increasing its overall RS through male function

(Heath 1977, 1979). The magnitude of this fitness-gain is affected by many factors, including the rate of self-fertilization and the size of the pollen "shadow," but at least in principle it is clear that flexibility of sexual function may give hermaphrodites an advantage in some circumstances (which deserve further study).

However, even though an individual is morphologically and functionally hermaphroditic, there may be great variation in actual functioning in male or female roles. The ability to function as either sex or as both may permit reproductive success in a variety of circumstances beyond the capacity of a single sexual function. Different genotypes of *Lupinus nanus* achieve very different levels of outcrossed progeny; this is related to pollinator attraction and is not simply a reflection of degrees of self-fertility, because some highly self-fertile genotypes outcross extensively (Horovitz and Harding 1972a). Similar results are known for lima beans and maize (Harding and Tucker 1964, 1969), and for *Vicia faba* (Drayner 1959), *Nicotiana rustica* (Breese 1959), *Lupinus succulentus* (Harding and Barnes 1977), and *Senecio vulgaris* (Campbell and Abbott 1976). Pollen deposition and seed-set varied significantly among corolla-color variants of *Phlox drummondii* (Levin 1976). Not only do different individuals outcross at different rates in terms of their receipt of pollen and female function, the same is true for pollen donation and male function. Lloyd (1972) noted that monoecious species of *Cotula* that appeared to be morphologically adapted for high levels of outcrossing also produced a higher proportion of male flowers. Furthermore, female and male outcrossing in *Lupinus nanus* varied independently, so that the same individuals were not highly successful in both fatherhood and motherhood (Horovitz and Harding 1972a,b). Similar results have been noted for *Lotus corniculatus* (Schaaf and Hill 1979). Male RS was, in general, inversely correlated with female RS in horticultural varieties of *Freesia* (Sparnaaij et al. 1968) and in *Campsis radicans* (Bertin 1982a). Male performance was more variable than female in *L. corniculatus,* although not in all species that have been studied (Schaaf and Hill 1979).

Horovitz (1978) has emphasized that hermaphroditic individuals need not function equally as both sexes. The importance of both male and female function has recently been fostered by other studies (e.g., Lloyd 1980a, Primack and Lloyd 1980b, Sutherland and Delph, unpublished manuscript, Willson 1979, Willson and Rathcke 1974). Hermaphroditic flowers of *Passiflora edulis* sometimes function only as males, when the extended styles fail to reflex down toward the corolla where the bees can touch the stigmas (Purseglove 1968). *Catalpa speciosa* flowers function only as males after about four flowers in the inflorescence have been pollinated (Stephenson 1979). *Sagittaria brevirostris,* a monoecious species, adds mostly male flowers as inflorescence size increases (Kaul 1979). The number of fertile stamens in hermaphroditic flowers of *Spergularia media* varies from 0 to 10, and the frequency distribution of stamens in number differs among populations (Sterk 1969a). Stamen number of *S. marina* is also variable (Sterk 1969b). To the extent that each population outcrosses, the difference in fertile stamen number hints at individual differences in male function.

Other possible examples can be found in the literature, but they require careful scrutiny. Hermaphrodites of *Euonymus europaeus,* a species introduced to New Zealand, achieved a great range of seed production, from a level equivalent to that of females to $< 1\%$ of flowers producing seed. The latter individuals probably functioned largely as males (Webb 1979a). A variety of umbellifer species also exhibit great variability in the female role of hermaphroditic plants, ranging from 0 to 100% fruit set (Webb 1979b). There is, however, a potential problem with these estimates in that actual male function is not measured (for very practical reasons) but is assumed to vary inversely with female function. Actually, an hermaphroditic plant with no seeds might also have contributed no pollen. In short, these estimates assess the potential balance of male and female roles in hermaphrodites (Lloyd 1976) but they do not tell us if hermaphrodites that are poor females are, concomitantly, better males than those that are good females.

Sequential hermaphroditism occurs when the fitness-gain of each sexual role changes with age or size of the individual (Ghiselin 1974, Leigh et al. 1976, Warner 1975, Warner et al. 1975). Protandry should prevail when small individuals can be successful males but female RS increases more rapidly with increases of size or age. Protogyny should be the rule when male RS is more size or age-dependent. Most cases of sequential hermaphroditism among plants seem to be protandrous; even when the adult will be a simultaneous hermaphrodite, young individuals often function principally as males. However, there are exceptions to this generalization; two possible examples (*Thymus vulgaris* and *Impatiens biflora*) are discussed later in this chapter. Sequential development of male and female function also spreads RE through time, reducing peak resource allocation to reproduction (Onyekwelu and Harper 1979); it is unlikely to be merely a means of outcrossing (Onyekwelu and Harper 1979).

So far we can suggest that simultaneous hermaphroditism will be favored when the rate of fitness-gain of one sex does not increase greatly with increasing allocation of resources to the functioning of that sex. The conditions that make this true might include the sharing of high fixed reproductive costs by male and female function and the cuing of animal pollinators on floral sex organs themselves. Mere sessileness, attractiveness to pollinators, or reduced mobility of pollen vectors are insufficient explanations because many dioecious plants share these features. There may be some circumstances that make advantageous the ability to shift sexual roles or to function as either sex as demanded by the occasion; that is, sexual versatility might increase overall RS. But we do not yet know what these circumstances are. Clearly, no particular general conditions for the favoring of hermaphroditism are apparent; indeed, there may not be any single set of conditions that favor it. Let us, therefore, come at the problem from the other direction and ask what are the conditions that select for separate sexes.

The standard explanation for the evolution of dioecy is that it enforces outcrossing and serves, for this purpose, as an evolutionary alternative to self-incompatibility (e.g., Thomson and Barrett 1981a). Genetic models for the evolution of dioecy typically include inbreeding depression as an important element (Charlesworth and Charlesworth 1978a,b). This view is undoubtedly too simplistic (Bawa 1980, 1982; Darwin 1877; Givnish 1982; Lloyd 1979b; Ross 1982; Willson 1979, 1982). It fails to address the question of why obligate outcrossing is of such overwhelming importance, a question particularly germane in view of the large number of plants that do at least some selfing; inbreeding depression is a feature chiefly (but not exclusively) seen in habitual outcrossers (Wright 1977). Nor does it take into account that outcrossing does not necessarily decrease the level of inbreeding among relatives, which are often neighbors and, therefore, also mates (Levin and Kerster 1974). This restricted view also neglects the fact that separation of the sexes has several other consequences of probable ecological and evolutionary importance: role specialization, possible niche differentiation of the sexes, and the possibility of manipulating offspring sex ratios (if sex is genetically determined), to name three. No simple correlations of the frequency of dioecy and habitat or geographic distribution have emerged (e.g., Bawa 1982, Conn et al. 1980, Flores and Schemske unpublished manuscript).

Recently there have appeared suggestions of several other factors besides outcrossing that may select for dioecy. All of these ideas are linked to possible advantages of sexual role specialization and differential fitness-gain (see also Lloyd 1982, Ross 1982). Pollination syndrome, seed-dispersal mode, flower and seed predation, niche segregation, and sexual selection are among the factors considered. Although dioecy occurs in some algae, mosses, and ferns, most ecological suggestions have been developed especially for seed plants.

Bawa (1980) summarized an array of evidence that dioecy is particularly prevalent in species pollinated by insects; Givnish (1982) put it in

Table 2.1
A. Association of pollen vector with frequency of dioecy (from Bawa 1980)

	Insect pollinated	Vertebrate pollinated
Neotropical rain forest	high	low
Neotropical dry forest	high	low
Small, opportunistic bees	80% dioecious	
Other insects	17% dioecious	
Wind	3% dioecious	
Loranthaceae	100% dioecious	0% dioecious
Fuchsia	92% dioecious	0% dioecious
Simaroubaceae	100% dioecious	0% dioecious

B. Association of dispersal mode with dioecy (from Bawa 1980, Givnish 1980, 1982)

	% Dioecious	
	Dispersed in animal interiors	Dispersed by other means
Angiosperms		
Families (without great variability)	27	7
Neotropical rain forest	23	0
Neotropical dry forest	33	10
Meliaceae (genera)	64	0
Gymnosperms	90	5

terms of possessing small, inconspicuous flowers. This is not to say that all dioecious species have small flowers and are insect-pollinated; many in the north temperate zone are wind-pollinated (e.g., Freeman et al. 1980), and a case can be made for the evolution of dioecy in certain vertebrate-pollinated species in which the vector differentially damages male flowers (Cox 1982). Nevertheless, the correlation of dioecy with insect-pollination holds among tropical plants (Table 2.1A). For a tropical dry forest in Costa Rica, small or opportunistically foraging bees are especially common as pollinators of dioecious species; similar results obtain for a Venezuelan cloud forest (Sobrevila and Arroyo 1982). The basis for the association is still a matter of conjecture. When dioecy has evolved from heterostyly (a floral polymorphism—usually one morph has a long style and short stamens and the other the reverse arrange-

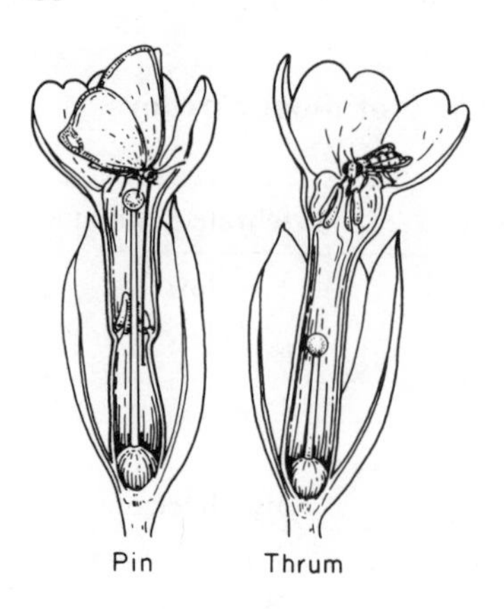

Figure 2.4 A long-tongued lepidopteran reaching the anthers of a pin flower, which the short-tongued fly on the thrum flower could not do.

ment), Beach and Bawa (1980) argued that it reflects a switch from a restricted group of pollinators with long mouthparts that probe deep into the flowers to more general and smaller pollinators that do not have access to both stamens and stigmas in the heterostylous morphs (because they cannot reach whichever parts are lower in the flower, Figure 2.4). As a result, pollen transfer becomes directional, with functional pollen transfer occurring from short-styled (thrum) flowers to long-styled (pin) flowers but not vice versa (see also Lloyd 1979a). Role specialization thus is thought to be related to pollinator behavior. Corroborative evidence for this argument comes from *Coussarea* (Rubiaceae) and *Cordia* (Boraginaceae) in Costa Rica, two genera in which dioecious species have smaller flowers and are visited by small bees, whereas heterostylous or monomorphic species have large flowers and are visited mainly by long-tongued lepidopterans.

Dioecy has also evolved from nonheterostylous insect-pollinated forms. Beach (1981) made the general argument that opportunistic pollinators (small bees and flies especially) may cause pollen flow to be somewhat unidirectional, from the plants with more flowers (and nectar and pollen rewards) to those with fewer. Then some individuals may be more successful as pollen donors and achieve high fitness-gain as males and others as pollen recipients, and selection may then favor increased role specialization to improve still further the reproductive success as whichever gender is most appropriate. Note that it is the larger individuals that are thought to function more as males, which is the reverse of the condition in protandrous hermaphrodites. Suggestive evidence for different functional roles is provided by *Asclepias,* in which pollinia removal increases far faster than pod production as a function of number of flowers per umbel and, therefore, large umbels may have a greater male role than small umbels and may act more as males than as females (Willson and Price 1977). That some genotypes can be better pollen donors than recipients is known for at least five species (Bertin 1982a, Harding and Tucker 1969, Horovitz and Harding 1972b, Schaaf and Hill 1979, Sparnaaij et al. 1968, and others surveyed by Willson unpublished manuscript). However, the extent to which opportunistic pollinators produce the postulated pattern of pollen flow is debated: Givnish (1980) suggested that small insects cause little between-plant pollen transfer (an obvious exception are the fig-wasps, see Chapter 3) and K. Garbutt (pers. comm.) avers that syrphid flies do not behave in the stipulated manner. Size-specific changes in sexual roles and immense environmental variation in size of floral display, which are known in many species, also make improbable any consistent directional patterns in pollen flow. Nevertheless, it seems clear that, for some reason, plants whose flowers are pollinated by relatively specialized pollen vectors such as birds are less commonly dioecious than others (Bawa 1980; Givnish 1980, 1982).

Dioecy also seems to be particularly common in species that have fleshy fruits and seeds dispersed by animals, especially birds, that consume the fruit (Bawa 1980; Givnish 1980, 1982). The association is very strong for gymnosperms and clearcut for tropical angiosperms (Table 2.1B). On a local scale, dioecy is also common among the bird-dispersed species of eastern deciduous forest: at least 32% of the 22 species in Trelease Woods in Illinois are partly or fully dioecious, a much higher representation of dioecy than in the remaining flora of the woods. For the flora of Puerto Rico and the Virgin Islands, dioecy is again prevalent in fleshy-fruited species (Flores and Schemske unpublished manuscript), with the added observation that for trees and treelets the association breaks down and, instead, dioecy is associated with the production of single-seeded (rather than multiseeded) fruits of any dispersal type. (No explanation is presently available for this observation.)

The basis for the association between dioecy and bird-dispersal is uncertain. The cost to the parent plant of building a fleshy fruit to entice birds to eat and carry away the seeds may be high compared to other dispersal modes. Fleshy fruits are sometimes nutrient-rich and divert resources away from other functions. However, the actual cost *per seed* for various dispersal modes has not been measured to my knowledge, and some protective devices such as heavy bracts or husks might be equally expensive. Givnish and Bawa have argued that large fruit crops may attract dispersal agents at a disproportionate rate, resulting in unusually high gains in female fitness as reproductive effort increases. However, the evidence at hand does not support this suggestion; studies of two tropical trees and one north-temperate shrub failed to show disproportionate fruit-removal rates from large fruit crops (Howe and De Steven 1979, Howe and vande Kerckhove 1979, Moore and Willson 1982). It remains possible that large fruit crops might attract a different array of dispersal agents: more large-bodied, far-flying birds visited *Didymopanax morototoni* trees with large fruit crops than those with small crops (T. Martin, pers. comm.).

The role of flower or seed predation in the evolution of dioecy is problematical. Pollinator destruction of the flowers of *Freycinetia* does not greatly impair male function, because the vector carries away a load of pollen. However, female function would be markedly diminished. Cox (1982) suggested that separation of female from male floral organs might prevent destruction of female function. If female flowers remained sufficiently like male flowers to deceive the vectors briefly, females might be visited but not demolished. However, it is not quite clear why these pressures would not lead equally well to monoecy instead of dioecy. Janzen (1971) suggested that females of a dioecious species may be effectively rarer and more widely spaced than would be true for an otherwise equivalent hermaphroditic species and perhaps, therefore, harder to find by a seed (or flower) predator. We have no documentation for this idea yet. And it is not clear that insect pollen vectors might not find such trees equally difficult to find. Distance between obligately outcrossing individuals in Venezuelan secondary and cloud forests did not limit seed-set (Sobrevila and Arroyo 1982, Zapata and Arroyo 1978), but greater distances might be involved in other habitats.

Occasionally, sex changes may be induced pathologically. The feeding of aphid larvae on flowers of *Olinia cymosa* leads to very abnormal floral morphology, with stamens often absent and variable results on female parts (Phillips 1926). A rust reportedly consumes the pollen of dioecious *Silene alba* and thus renders the male flowers infertile; it also transforms females into males—and then consumes the pollen of the transformed flower (Schemske pers. comm.). Whether the effects of such pathogens are simply happenstance consequences or play a role in the evolution of dioecy is a matter for constructive conjecture.

Male and female plants may occupy different microhabitats. Such is the case for a number of arid-country species such as *Thalictrum fendleri, Distichlis spicata, Ephedra viridis, Atriplex confertifolia, Acer negundo* (Freeman et al. 1976), and for a variety of other forms (Cox 1981, Lloyd and Webb 1977, Meagher 1980). A common pattern is to find females in wetter sites than males, although this may reflect the particular species and habitats surveyed more than some general trend. However, other studies have found little or no habitat segregation of males and females (Bawa and Opler 1977, Melampy 1981).

The shrub jojoba (*Simmondsia chinensis*) is sexually dimorphic, desert-grown females having larger leaves and a more open canopy than males; a few other species are also dimorphic in nonfloral characteristics (Conn 1981, Dzhaparidze 1963, Lloyd and Webb 1977, Wallace and Rundel 1979). Sexual differences in extent of vegetative propagation (*Rumex* spp., Putwain and Harper 1972; *Thalictrum dioicum,* Melampy 1981; *Fragaria chiloensis,* Hancock and Bringhurst 1980) are not uncommon and may indicate that males and females exploit their environment in different ways. Seasonal differences in growth and flowering pattern are known in *Rumex hastatulus, Silene alba, Spinacia oleracea* (Conn 1981, Conn and Blum 1981b, Gross and Soule 1981, Onyekwelu and Harper 1979), and suggest niche differentiation (although the differences may have evolved more as a consequence of differing reproductive ecologies than as a means of reducing competition between the sexes). Male and female plants of *Rumex acetosella* exhibit different tolerances to drought stress, the greater tolerance of males being associated not with greater physiological efficiency of water use but with greater biomass allocation to roots and leaves, which allowed them to exploit the available water more rapidly than females (Zimmerman and Lechowicz 1982).

From a parent plant's point of view, niche differentiation (when present) among the offspring may enhance their probability of survival beyond the degree achievable simply by sex and recombination. Furthermore, the variation in family fitness may be reduced if a lineage passes on its genes through either males or females. If males can achieve higher fitnesses in some circumstances but females excel in others, the variance in fitness can be lowered without changing the mean (Lloyd 1980c). A lower variance can result in greater average fitness over many generations (Gillespie 1977) if the cost of lowering the variance is not too high (Ekbohm et al. 1980). One cost of lowered variance is a reduced chance of producing appropriate sisyphean genotypes.

Another ecological-evolutionary explanation for dioecy may lie in sexual selection. Given that most plants (those studied so far, at least) seem to have seed production limited by consumable resources rather than pollen or sperm availability, the stage is set for sexual selection. The idea dates back to Darwin (1871) and has been developed by Bateman (1948) and Trivers (1972). Sexual selection includes competition for mates, usually by males, and mate choice, usually by females. If female RS is limited by resources for producing offspring, selection will favor females that expend those resources on the best possible offspring and thus females that exercise some choice among potential mates. Male RS, on the other hand, is then limited by the number of females obtained or the numbers of eggs fertilized, and males will compete with each other to increase their mating success. The variance in male RS generally exceeds that of female RS because some males are likely to be far more successful than others in the competition whereas females usually will differ less. Differences in the variance of RS may serve as an indication of the intensity of sexual selection (Wade 1979). Adumbrated by Janzen (1975, 1977) and Gilbert (1975), Charnov (1979) and I (1979) have developed the idea that sexual selection may help account for the evolution of dioecy from hermaphroditism (and perhaps monoecy, heterostyly, etc.) in outcrossing plants. Sexual selection cannot operate in complete selfers, in which all mate choice and competition have been eliminated. Basically, the argument is that competition among individuals functioning partly as males selects for individuals of greater male capacity—producing and delivering (perhaps by means of particular vectors) larger quantities of pollen.

One way of becoming successful as a male may be to increase flower production and enhance delivery of pollen. Whether these "extra" flowers would be both male and female or solely male would depend on the relative expense of the female component and whether those costs could be allocated elsewhere, whether there are abortion capacities to eliminate any unwanted fruits that might result from increased pollination, other possible functions of the female parts (as in floral display or in the integration of development), and so on. If an increase in flower number allows a greater increase of male RS than is lost as female RS, selection may favor greater male expression. Individuals producing larger quantities of pollen may be able to capture so many more ovules for fertilization that they outcompete individuals with a smaller male effort. The greater this effect, the greater the tendency toward unisexual phenotypes, because as some individuals obtain ever greater success as males, the remainder will do better to drop out of the male competition and specialize as females. On the other side, high costs of offspring production mean that only certain individuals may be able to marshall enough resources to function successfully as females. If female RE is very high, the resources available for male RE are much reduced and the probability of male success is much less, especially if intermale competition is intense. Female function may then be emphasized.

One problem with the suggestion of ever-greater allocation to maleness in response to intermale competition is that diminishing returns would be expected. If pollen deposition occurs at random (with respect to whether pollen is already present on that stigma), a doubling of pollen output would increase by only 50% the number of stigmas reached. How then can male RS increase sufficiently with greater male RE? Perhaps a 50% gain in pollen delivery for each doubling of RE is sufficient to outcompete other individuals in the intermale rivalry. Possibly an increase in pollen production increases the number of more distant females reached and, because these matings may be less subject to inbreeding effects (references in Levin 1981), there could be a disproportionate gain to be achieved. Or maybe pollen deposition is not quite random with respect to pollen already deposited. The matter has yet to be investigated.

The distance over which a male's genes are dispersed is typically greater than that for a female's genes because genes in pollen move first in the pollen grain and subsequently, with the maternal genes, in the seed (Lloyd 1982). As a result, progeny of a male may encounter more diverse microhabitats and perhaps less sibling competition among themselves than the progeny of a female (Bulmer and Taylor 1980, Lloyd 1982, Taylor and Bulmer 1980). Lloyd suggests that this limits the fitness-gain to be achieved by gender specialization, presumably particularly by females and, therefore, retention of both male and female functions in the same individual would be favored. Left unexplained by this view is the evolution of dioecy by *any* plants, since all plants are thought to share the gene-dispersion differential (Lloyd 1982). On the other hand, using the perspective of Bulmer and Taylor, one could argue that the fitness gain through male function could be sufficiently enhanced to favor a shift toward unisexual maleness, perhaps particularly in certain individuals that, for whatever reason, cannot augment their female RS. The remainder might then find it advantageous to emphasize female function, despite the shorter dispersal distances of their offspring, if competition among the "males" became sufficiently intense. Bulmer and Taylor propose that differential dispersal ability of male and female offspring in animals might select for a shift in the progeny sex ratio to favor the sex dispersing farthest, because reduced sibling competition among offspring of the wide-ranging sex could enhance their prospects of survival and reproduction. In vertebrates, however, male and female young commonly disperse for different distances, for various ecological reasons. Yet there is no conspicuous concomitant trend to bias offspring sex ratios in the manner suggested. There are other ecological pressures that seem to increase the cost of dispersal and favor reduced movement of one sex (Bengtsson 1978, Greenwood 1980, Smith 1979). Fitness-gain through offspring of each sex seems not to vary consistently with the potential for sib-competition. Many of the ecological pressures relevant to vertebrates reside in their social behavior, a feature for which plants are not noted. Perhaps, then, the animal comparison lends no insight into sibling dispersion in plants, although the cautionary message concerning the importance of sib-competition remains.

We know very little about the conditions that control the intensity of sexual selection—the degree of intermale competition and the degree of female choice. What is needed at this point are detailed studies of a variety of species to determine the ability of females to choose among potential fathers, possible countermeasures of the males, and the relative (sexual) competitive ability of hermaphroditic individuals with different degrees of maleness. Sexual selection is not necessarily the sole factor in creating specialization of sexual roles, but its possible importance should not be neglected merely because plants cannot dance, or sing, or battle to prove themselves worthy mates or because of the belief that pollen production is keyed only to the number of ovules available in the population and the efficiency of pollen-transfer. Efficient pollination is one factor permitting an evolutionary reduction of pollen production relative to ovule production (Cruden 1977), although Cruden and Miller-Ward (1981) argue that the relevant factor is not pollinator constancy but the spatial relations between stigma and pollen-bearing surface on the pollinator. These arguments seem to be based on the assumption that seed-set (female RS) is usually pollen-limited. However, when male RS can be increased by delivering more pollen and selection favors increased male RS, I would not expect to see such declines in pollen production.

Most of the proposed ecological pressures favoring dioecy emphasize differential costs and fitness-gains through male and female functions that favor specialization of sexual roles. Outcrossing may affect the rates of fitness-gain, and sexual selection cannot operate unless outcrossing occurs, but the role of outcrossing per se in the evolution of dioecy is clearly far smaller than formerly believed.

Monoecy

Gender specialization can be used to help explain the evolution of monoecy in its various forms. If some flowers serve only a male function, it may be economical to construct them without female parts (and vice versa), particularly if shared costs (e.g., corolla, nectar, etc.) are low. This could be especially true if the prospects of fitness-gain through male or female function are variable but predictable and the individual is capable of adjusting its relative allocation to each sexual role. At the same time, many monoecious species (but not all) are self-compatible and thus are capable of selfing and inbreeding with any relatives.

The ability to shift the floral sex ratio to capitalize on conditions favoring greater RS as one sex or the other could be important to overall fitness of the individual. Several cases of such shifts are known, although their fitness consequences are undescribed. Individuals of two monoecious wind-pollinated species, *Ambrosia trifida* (Abul Fatih and Bazzaz 1979) and *Quercus gambelii* (Freeman et al. 1981), shift toward female expression when growing under a canopy, where pollen receipt is far more likely to be successful than pollen donation. Moreover, predation on *A. trifida* seeds is reduced on the smaller, female individuals. Three wind-pollinated woody species in Utah shift their sex expression in relation to the availability of moisture; male flowers predominate on xeric sites, female flowers on mesic sites. This pattern suggests that female success may be limited by water or access to water-soluble nutrients but that males are not (Freeman et al. 1981).

Given that flowers are unisexual, they can be arranged on the plant to suit the needs of each sex (Heslop-Harrison 1972). Thus many conifers place female cones at the top of trees and male cones below. The usual argument is that this promotes outcrossing in wind-pollinated systems (but see below regarding grasses); it might be better to emphasize that the arrangement probably enhances receipt of pollen from more distant sources and the arrival of pollen from a diversity of males as well. Which aspect of pollen arrival is more important is unknown.

Some grasses share this arrangement, but others have male flowers above female (Connor 1979). Inasmuch as the latter are apparently no more self-incompatible than the former (Connor 1979), one cannot argue that the males-above arrangement is affordable because the plants cannot self anyway. There are at least two other possibilities. One is that some selfing is desirable, either because some inbred offspring are advantageous for some reason or because pollen is limiting to seed-set in these species. The other possible explanation is that, for equally unknown reasons, females of these species obtain a sufficient variety of pollen and selection has favored males that can disperse their windborne pollen for greater distances (which might become possible if male flowers are higher on the plant). The advantage of outcrossing for males is considered doubtful, but the models leading to this conclusion are based on inbreeding depression, random mating, and equal average fitnesses of all genotypes of each sex (Charlesworth and Charlesworth 1978, Lloyd 1975), none of which conditions may actually obtain. Moreover, if we look beyond outcrossing per se, we might find that the intensity of intermale competition in these species has favored males that can reach a variety of females. All such explanations are highly conjectural, of course; at the least it is important to recognize that there are alternatives and that any attempt at explanation needs to account for a variety of floral arrangements (especially among the grasses).

The more complex conditions of mixed unisexual and perfect flowers are even more difficult but, in general, the ecological pressures must be of the same sorts as already discussed for hermaphroditism, dioecy, and monoecy. The mixed systems differ greatly in frequency of occurrence, but we have no very satisfactory explanation for this. These systems have attracted a significant amount of interest, so a brief review of some of this research is appropriate here.

Gynodioecy

Found in quite a variety of angiosperms, gynodioecious systems often show between-population variation in the frequency of fully female individuals (e.g., Brockman and Bocquet 1978, Dommée et al. 1978, Horovitz and Beiles 1980, Philipp 1980, Webb 1979b, Young 1972, and references therein) and in the level of fruit production by hermaphrodites (Webb 1979a,b). Actually, the system represents a complete range of conditions between almost complete hermaphroditism and almost complete dioecy (Lloyd 1976, Webb 1979b).

The maintenance of females in gynodioecious systems has been studied quite extensively. When male sterility is controlled by nuclear genes, a self-compatible, self-fertilizing hermaphrodite may contribute three genes for every one from a female, because it can provide its own pollen and egg nuclei as well as a pollen nucleus to the female (Lewis 1941, Lewis and Crowe 1956, Lloyd 1975). However, this 3:1 advantage cannot apply in cases of self-incompatibility (Charlesworth and Ganders 1979) and dwindles toward 2:1 in populations with few females where hermaphrodites have little chance to father offspring by females. Seed for seed, then, the females must have a greater than twofold selective advantage over the hermaphrodite (when mating is random) to compensate for the difference in gene passage, according to this line of thought. Such advantage may accrue through increased fecundity or through increased offspring fitness. On the other hand, if femaleness is controlled by cytoplasmic genes (and if cytoplasmic inheritance is maternal), then females and hermaphrodites contribute equally to the next generation. Both cytoplasmic and nuclear patterns of inheritance of male sterility are known (e.g., Charlesworth and Ganders 1979, Frankel and Galun 1977), but in no known case is the inheritance mechanism single and unifactorial (Charlesworth 1981). These genetic models seem to be founded in gene-level selection; if the number of progeny bearing parental genes is the appropriate measure of individual fitness, then the relevant measure of RS for an hermaphrodite is simply the number of seeds plus the number of successful pollen grains, and there may not be an inherent 2:1 advantage.

Several studies have shown that female fecundity is often higher than that of hermaphrodites, at least to some degree (Lloyd 1976, Webb 1981), although *Geranium sylvaticum* seems to be an exception (Vaarama and Jääskeläinen 1967). Female flowers are commonly smaller than perfect flowers (e.g., Darwin 1884, Styles 1972, Vaarama and Jääskeläinen 1967, Whitehouse 1959, except in *Cortaderia,* see below), yet no one seems to have reported any lack of pollination of females, despite their small size and lack of pollen rewards. Seed production by female *Rhus ovata* and *R. integrifolia* markedly exceeded that of hermaphrodites; the latter are self-incompatible, but it is not known if the inability to self-fertilize actually contributed to lowered seed production (Young 1972). Females of the self-incompatible *Hirschfeldia incana* achieved slightly higher fecundities than hermaphrodites (Horovitz and Beiles 1980). Females of *Plantago lanceolata* were more fecund than hermaphrodites after the first year of growth, and their seeds were larger (Krohne et al. 1980). *Thymus vulgaris* females produced significantly more seeds and a higher percentage of germinable seeds than did hermaphrodites (Assouad and Valdeyron 1975, Assouad et al. 1978), and female *Stellaria longipes* outproduced hermaphrodites (Philipp 1980). The females of gynodioecious species of the grass *Cortaderia* have larger female sexual parts, a much larger stigma, and more florets per spikelet than hermaphrodites; their seeds are larger and heavier but also more buoyant and capable of dispersing much farther (Connor 1973). *Cortaderia richardii* is gynodioecious and self-compatible; female fecundity is slightly greater than that of hermaphrodites but all the progeny of females are more vigorous than the offspring of hermaphrodites, whether derived from selfing or outcrossing (Connor 1973). However, female fecundities reported for these species were often not twice as great as those of hermaphrodites.

The offspring of females might be genetically superior to those of hermaphrodites if either heterosis or lack of inbreeding depression are more common in the obligately outcrossed offspring of females. If sufficiently superior, the proposed 2:1 fecundity differential may be unnecessary (Charlesworth and Charlesworth 1978a).

Some authors claim that the intensity of selection for outcrossing in self-compatible species governs the frequency of females (e.g., Jain 1968, Valdeyron et al. 1977). In a gynodioecious population of the self-compatible *Nemophila menziesii* that lacked the normal pollinators, females may be maintained because of high levels of selfing and inbreeding depression in the offspring of hermaphrodites (Ganders 1978). However, one wonders, then, why there are still any hermaphrodites in the population at all: if outcrossing is good for some individuals, why not for all? It may be that the advantages of outcrossing do, in fact, vary among individuals. Or perhaps the value of outcrossing is variable, and variable selection intensities might allow maintenance of relatively high frequencies of hermaphrodites in some populations. Most theoretical treatments of the evolution and maintenance of gynodioecy (and dioecy) have relied on either compensations of female fecundity and/or inbreeding depression and heterosis (e.g., Charlesworth 1981, Charlesworth and Charlesworth 1978, Charnov et al. 1976, Lloyd 1975, and earlier studies).

However, it seems likely that other factors must be involved. First, further selection for outcrossing is irrelevant in gynodioecious species that are self-incompatible, in which the offspring of hermaphrodites must be as outcrossed as those of females (Connor 1973, Ganders 1978, Ross 1973). Krohne et al. (1980) argued that outcrossing is very unlikely to explain the evolution of gynodioecy in self-incompatible species. They also found that genetic variation in females and hermaphrodites of *Plantago lanceolata* was similar, being low in both cases, suggesting that outcrossing alone cannot account for the presence of females in this species.

Thymus vulgaris has been studied quite intensively in France (Assouad and Valdeyron 1975, Assouad et al. 1978, Dommée et al. 1978, Valdeyron et al. 1977) and may be used to suggest some additional factors that might be involved in the evolution of gynodioecy. To summarize existing information: seed output and germinability of females exceed that of hermaphrodites. Females are found especially in grassy fields and become rare on rocky sites; they also occur more frequently on non–south-facing slopes. The rocky sites are considered to be normal habitat for the species. The above-ground dry-weight of individuals from predominantly female populations was much smaller than those from mostly hermaphroditic populations, suggesting that female plants are usually smaller than hermaphrodites (but note that the really appropriate comparison would be between sexual morphs from the same populations). The investigators offer an interpretation that the unstable and disturbed environment of the fields favored outcrossing (although the rates varied among individuals, Valdeyron et al. 1977), whereas the relatively undisturbed rocky areas permitted some selfing. Enhanced female fecundity may also contribute to female success in some habitats, and seeds of females are larger (Darwin 1884).

Without denying the possible relevance of these factors at this point, an additional possibility can be suggested. Large plant size may disproportionately improve the prospects of success as a male by attracting more pollinators and enhancing pollen donation. Even though the frequency of selfing can be quite high, the ability to donate to other individuals may raise male RS significantly (see, e.g., Lloyd 1976, Wells 1979), especially when female RS is constrained. Hermaphroditic plants tend to be found at marginally lower conspecific densities. The species is pollinated by bees, whose behavior with respect to plant density and size is apparently not studied, nor do we have information on bee populations and activity patterns in different habitats. The suggestion of selection for male success depends on the activity of the pollen vectors (see also Bawa and Beach 1981).

It has been claimed that the frequency of female plants should be inversely correlated with the seed production of hermaphrodites relative to females (Lloyd 1974c, 1975, 1976; Webb 1979a). Everything else being equal, the success of the male role should increase as the number of available females increases, and hermaphrodites should invest more in male function. Given equivalent resource budgets, this means a relative decrease of female function, hence the prediction. The evidence so far is anything but convincing for most cases. In view of the great differences among gynodioecious species in degrees of compatibility and of inbreeding depression, in addition to our lack of knowledge of the available resource budget and of environmental differences among populations, perhaps the lack of good evidence is not surprising.

Androdioecy

As noted by Lloyd (1975) and Charlesworth and Charlesworth (1978), this system is very rare. Nevertheless, androdioecy does occur in some *Solanum* (Symon 1979) and is reported for *Decaspermum* (Myrtaceae) and *Symplocos* (Symplocaceae; Yampolsky and Yampolsky 1922). Some instances of apparent androdioecy may result when small or young individuals of otherwise andromonoecious species are solely male; *Lomatium grayi* (Apiaceae) is an example (J. N. Thompson, pers. comm.). And some structurally androdioecious plants seem to be functionally dioecious (*Matayba tovariensis,* Sobrevila and Arroyo 1982).

A common explanation for the rarity of this form of sex expression is based on inheritance patterns. Many genes that bring about sterility of males (i.e., femaleness and gynodioecy) are cytoplasmic and transmitted in the egg cytoplasm. Because male cytoplasm was generally believed to be excluded from the egg, some researchers argue that sterility of females (i.e., maleness and androdioecy) cannot be transmitted cytoplasmically by males. However, abundant evidence for some plants shows that male cytoplasm may contribute significantly to the zygote at fertilization (references in Willson and Burley in press), so perhaps this explanation has been invoked too freely. We need to know if maleness is, in fact, genetically controlled at all; perhaps males are produced only in some environmental circumstances. If androdioecy is really as rare as it appears to be, it may actually be maladaptive or transitional to some other form of sex expression. Apparently, we do not know the extent of male function by hermaphrodites or the frequency of males in androdioecious populations. Male RS in an androdioecious system must be sufficient to compensate for the loss of female function if males are to achieve a fitness level equivalent to that of hermaphrodites. This would require, presumably, a very effective pollen-transfer system (Bawa and Beach 1981), but that condition is not sufficient to necessitate androdioecy. Because this sexual system is so uncommon, it is even less studied than the others; speculation about its possible adaptive value will be more profitable when at least a little ecological information is available.

Andromonoecy

This form of sex expression occurs in a variety of species, both wind- and animal-pollinated (Bawa and Beach 1981, Bertin 1982b, Lloyd 1979a, Lovett Doust 1980, Primack and Lloyd 1980a, Thomson and Barrett 1981, Yampolsky and Yampolsky 1922). Although floral sex ratios are very consistent in some umbellifers (Apiaceae; Bell 1971, Lovett Doust 1980), the range of variation of sex expression is often considerable. For instance, some umbellifers exhibit considerable variation in pollen/ovule ratios, notably among umbels, but also among populations (Lindsey 1982). Furthermore, hermaphrodite flowers of red buckeye (*Aesculus pavia*) averaged only 2% of the total number of flowers per tree in wooded sites but up to 40% in open field sites in Missouri. Andromonoecy is not likely to have evolved as an outbreeding mechanism, because the hermaphroditic flowers can still self-pollinate if no other preventative mechanisms exist (Bertin 1982b, Lloyd 1980a). Furthermore, at least some andromonoecious species are self-incompatible (Bawa and Beach 1981), Zapata and Arroyo 1978). Nor is it likely that it is a means of increasing the probability of adequate pollination of pistils on the same plant, in part because hermaphrodite flowers could also do that job. Note that this argument presumes that selfing is not disadvantageous and that pollen availability limits fruit set. At least two andromonoecious species are not limited by pollen availability but by resources for fruit maturation (Bertin 1982b, Primack and Lloyd 1980a).

A third possibility is that male flowers somehow make the female-functioning flowers less conspicuous to predators on ovaries or seeds. At least in red buckeye this was not the case: predation on hermaphroditic flowers was not related to their frequency in an inflorescence, although the proportion of hermaphroditic flowers was inversely correlated with the frequency of fruit and flower predation through the season. Moveover, buckeye has very high fruit-abortion rates (> 90%), indicating that the plants make many more pistillate flowers than can be expected to mature fruits. If ovary predation were an important agent of selection, we would expect the plant in these circumstances to reduce the number of pistillate flowers and make males instead, thus reducing the risk of predation; this may occur through the season, but apparently not generally among inflorescences (Bertin 1982c).

A fourth interpretation is that production of male flowers might increase the conspicuousness of the floral display (at a lower cost than hermaphroditic flowers) and this might increase the rates of visitation by pollinators, which could (1) improve the pollination success of the associated hermaphroditic flowers (apparently not true for red buckeye), (2) increase the variety of pollen donors, permitting greater choice of which fruits to abort (Janzen 1977, Willson and Burley in press), and/or (3) increase the probability of success as a pollen donor to other individuals. Both of the last two possibilities are related to sexual selection, but neither has been tested. Bertin estimates that a red buckeye plant could produce about 500 male flowers for the cost (in biomass) of a single matured fruit (including the cost of those normally aborted); thus if these 500 fathered as little as one mature fruit on another individual, the genetic contribution to future generations would be equivalent to that achieved by mothering a single additional fruit.

Andromonoecy seems to be common (but by no means universal, if the list of andromonoecious species given by Yampolsky and Yampolsky 1922, is to be believed) in species whose fruits are large and expensive (Bertin 1982b, Primack and Lloyd 1980a). This makes fruit maturation by every flower unlikely and permits reduction of the number of pistillate flowers without affecting the total fruit crop. But why not separate male and female functions altogether and become monoecious? Retention of functional stamens by female-functioning flowers is not related to maintaining an ability to self-fertilize, at least in the several self-incompatible andromonoecious species, but it might be important in self-compatible species. Pollen may be an attractant for pollinators in both male and hermaphroditic flowers, but this is not relevant for the wind-pollinated andromonoecious grasses and sedges. If the ability to donate pollen to other individuals is important, stamens do not interfere with pistillate functions, and selfing is not a problem, then one could argue that retention of stamens in pistillate flowers is an economical way to augment male function—by using existing flowers to increase pollen offerings instead of building entire new (and solely male) flowers. Still, this does not explain why only certain species are andromonoecious and others are monoecious. Whether there is a general explanation for andromonoecy remains to be seen.

Gynomonoecy

A less common form of sex expression than andromonoecy, gynomonoecy appears to be especially common among the composites (Asteraceae; Lloyd 1979a, Yampolsky and Yampolsky 1922). Because composite flowers are aggregated into capitula or heads, Lloyd argued that adding male flowers does not much increase the floral display or pollinator rewards, and the only economical way to increase parental success of each head is to add female flowers. Even though the floral head may be the unit of pollinator attraction in composites, I do not think that this necessarily precludes an increase of male RS through pollen donation. An umbel of *Asclepias* is also likely to be a unit of attraction for pollinators, and the addition of—in this case, hermaphroditic—flowers results in a marked increase of pollinia removal (Willson and Price 1977, Willson and Rathcke 1974). The ability to self-fertilize (in self-compatible species) could be advantageous if other pollen sources are inadequate, but how often this occurs is uncertain. The placement of the female flowers on the periphery of a composite inflorescence not only gives them access to a better vascular supply than more central flowers but also improves the probability that they will receive outcross pollen when pollinators arrive from other plants (because they usually land at the edge of the inflorescence; D. A. Goldman, pers. comm.).

Addition of purely female flowers implies that (1) resources are available for seed maturation and (2) the ability to increase female RS exceeds the ability to increase male RS by increasing pollen production. But we do not seem to know what environmental factors are correlated with increasing proportions of female flowers or what conditions favor an increase of female RS.

Silene vulgaris (Caryophyllaceae) is commonly gynodioecious, but a small percentage of individuals are gynomonoecious, the form of sex expression varying with habitat (Brockman and Bocquet 1978). In *Baltimora recta,* a composite weed of the neotropics, the ratio of female to hermaphroditic or functionally male flowers in a head is quite variable (unpublished observations). Gynomonoecy may be seen to grade into gynodioecy, but the ecology of the situation is little explored.

Explanations for the evolution of sexual systems are summarized in Table 2.2. It is likely that some of these act synergistically to make a

Table 2.2 Summary of possible adaptive values for various forms of sex expression[a]

Hermaphroditism—Simultaneous
  Facilitates or eliminates mate-finding
  Increased RE increases both ♂ and ♀ RS
    High common costs of ♂ and ♀ function
    Cuing of pollinators on sexual parts
  Sexual role versatility in unpredictable circumstances

Monoecy
  Enhances outcrossing, but usually permits selfing also, and thus reduces problem of mate finding
  Contributes to economy of flower production
  Sexual role versatility in unpredictable circumstances
    Allows shifting of floral sex ratios
  Permits placement of unisexual flowers or sex organs for best functioning in ♂ and ♀ roles

Andromonoecy
  Enhances outcrossing (but allows selfing)
  Reduces predation on female parts
  Permits economical ♂ role specialization, favored by intermale competition

Gynomonoecy
  Enhances outcrossing (but allows selfing)
  Permits economical ♀ role specialization (in what conditions?)

Hermaphroditism—Sequential
  Enhances outcrossing
  Fitness gain as ♂ or ♀ is age or size-dependent

Dioecy
  Enforces outcrossing
  High RE to each sexual role results in large fitness gain and role specialization
    Male–male competition
    Directional patterns of pollen flow
    High cost of ♀ RS
  Niche differentiation of offspring
  Adjustment of progeny sex ratios

Gynodioecy
  Enhances outcrossing
  Some individuals achieve high fitness as ♀♀

Androdioecy
  ?

[a]Suggestions are not mutually exclusive, but some are less likely than others, as discussed in the text.

particular form of sex expression adaptive. However, assigning relative importance to different but interacting factors is difficult. Although one can ascertain the effects of outcrossing on offspring fitness and document changes in fitness-gain with changing budgets of resource allocation, for instance, this does not give us the relative contribution of each factor, unless one of them is shown to be altogether inapplicable. An experiment that would demonstrate the relative importance of each factor is a major undertaking and has yet to be accomplished.

DETERMINATION OF SEX EXPRESSION

When sexes are separate, sex determination by genetic means is possible, and many dioecious plants have such genetic mechanisms (Grant 1975). Sometimes there are specific sex-determining genes, and several monoecious or hermaphroditic species are known to have genes that suppress either maleness or femaleness. Possession of two such opposing alleles can lead to separation of sexes on different individuals. Thus the raspberry *Rubus idaeus* is normally hermaphroditic, but an f gene suppresses femaleness and an m gene suppresses maleness. Cross-breeding and recombination yields plants that are all female (F_mm) and all male (ffM_); however, these artificially synthesized combinations do not breed true and the system is incomplete (Lewis 1942). Conversely, *Vitis vinifera* and *Cannabis sativa* are dioecious in wild populations but hermaphroditic in cultivated varieties. A number of species of both angiosperms and bryophytes have sex chromosomes, but chromosomal mechanisms of sex determination are varied (Allen 1940, 1945, Frankel and Galun 1977, Martin 1966, Westergaard 1958, Wiens and Barlow 1975). Apparently the most common system in angiosperms and bryophytes is one of XY, in which females are XX and males are XY (and sometimes also YY). Sometimes there are multiple X's and Y's, and sometimes sex is determined by the balance of X's and Y's or of X's and autosomes. Typically, the male is the heterogametic sex (except in *Fragaria* and a few others, e.g., Lloyd 1974a; this is like mammals but unlike birds and lepidopterans), regardless of the exact mechanism. There are, however, numerous cases of dioecy without known genetic means of sex determination (Grant 1975, Westergaard 1958).

Cytoplasmic sex determination is well-known in gynodioecious systems, in which male-sterility or femaleness is often carried by cytoplasmic genes. In addition, there are cases of nuclearly controlled gynodioecious systems and possibly of systems controlled by a nucleo-cytoplasmic interaction. The intriguing possibility that sex-determination mechanisms may have evolved in part as a result of conflict among different elements of the genome (nuclear versus cytoplasmic genes, for instance, or autosomes versus sex chromosomes) has been discussed by Cosmides and Tooby (1981). Even if individual-level selection commonly outweighs gene-level selection, as I have briefly mentioned elsewhere in this chapter, conflict between two types of selection might yield a variety of sex-determining mechanisms.

Westergaard (1958) says that when sex is determined by an X/autosome balance, sex expression is little influenced by the environment. But when other systems of nuclear control prevail, environmental influences may, in fact, override the genetic factors. Unfortunately, Westergaard does not seem to provide evidence for his assertion of differential sensitivity to environmental factors. Nevertheless, the phenotypic plasticity of many plants with respect to sex expression is well known, and the ability of environmental factors to overwhelm genetic control of sex is also well established.

Environmental determination of sex expression would seem to be advantageous whenever an individual's fitness as either male or female is greatly affected by environmental conditions and when an individual (or its parent) cannot accurately predict or control the environment the individual will inhabit (Bull 1981, Charnov and Bull 1977). These conditions seem to be so generally true that the evolutionary basis for genetic control of plant sex is somewhat mystifying. Important and potentially variable environmental factors influencing reproductive success of either sex might include (1) the sex ratio of neighbors and probable mates (male success might be high, for instance, if the neighbors are mostly females; e.g., Thomson and Barrett 1981), (2) availability of resources (e.g., if resources limit RS of either sex), (3) access to pollen vectors (e.g., if microhabitat or other factors reduce the probability of pollen donation to nil), and (4) susceptibility to assorted mortality factors that might be sex-specific. These kinds of factors might affect not only the sex expression in dioecious (including gynodioecious) species, but also the relative male:female balance in all varieties of monoecy.

I do not know of any documentation for sex expression in seed plants being influenced by the sex ratio of its potential mates. But the potential is there—for instance, an individual functioning initially as a female but receiving very little pollen might switch to become a pollen donor. Such a response would not be expected in a species where female RS is consistently pollen-limited (because then all females would become males) but rather in cases where pollen-limitation is unexpected yet frequent enough to provide selection for sex-switching ability.

For ferns there is considerable evidence of sex-modification in reponse to neighbors (review in Willson, 1981b). Many types of fern gametophytes bear both antheridia and archegonia and thus produce both sperm and eggs. Some of these hermaphroditic species produce hormonal substances called antheridogens that induce the formation of antheridia; ten species in eight genera have been found, so far, to secrete these hormones. Antheridogens are produced by large gametophytes that bear archegonia and that are, themselves, no longer sensitive to antheridogen. Other, smaller, gametophytes growing near these larger individuals respond to the presence of antheridogen by precociously producing antheridia, often in great abundance. These induced males can function later as females, provided they survive.

The production of antheridogen is usually interpreted as a means of increasing levels of outcrossing by providing nonself sources of sperm for the eggs of the female portion of the gametophyte. Even it true, there is no reason to believe this is the sole function of antheridogen production. It is conceivable that the main function of antheridogen is as yet undiscovered and that it leaks out of the females more or less incidentally. The response of small neighboring gametophytes might still be adaptive, however, in that producing male gametes may allow opportunistic mating by a responsive gametophyte. If mortality rates of gametophytes are high, the ability to respond may give the gametophytes a much better chance to contribute genes to following generations. Even if mortality is low, and the gametophytes have good prospects for later reproduction as a female, early reproduction as a male would increase its RS significantly.

However, it is also possible that antheridogen production is adaptive to its producer for reasons other than or in addition to outcrossing. This could work in at least two ways. Imagine a group of neighboring gametophytes that are not siblings. One gametophyte develops faster than the rest, begins production of eggs, and induces maleness in the rest of the group. If there are several precocious males, the female may be able to choose among the potential fathers for her sporophytic offspring. The number of sporophytes produced per gametophyte is usually low, often only one, so most of the induced sperm, and most of the males that produced them, are doomed to failure. Thus competition among the males is presumably intense. In addition, precocious allocation of resources to sex instead of growth may prolong the development of the male-functioning gametophytes such that their attainment of femaleness is delayed. Retarded individuals also may be subjected to higher mortality (as is true for the rosettes of some semelparous angiosperms, for instance). Retarding development of neighbors thus might increase the availability of resources for the female as well as reduce the fitness of the neighbors. In this view, antheridogen production is seen as a kind of allelopathy. However, there should be strong selection for genotypes that are not susceptible to antheridogen, and one must wonder why females don't suppress their neighbors altogether instead of inducing maleness. Perhaps sperm supplies then become limiting or perhaps the neighbors are, in fact, genetically related.

In the second way, if neighbors and mates are usually sibs, a female's inclusive fitness might be reduced significantly if she suppressed her neighbors entirely. Moreover, responsive males provide sperm to the female when she is ready, which may thus enhance the reproductive output of the family as a whole. It is even possible that the parental sporophyte manipulates sex expression in its gametophytic offspring by adjusting spore size, upon which gametophytic growth may depend. From a parent's point of view, antheridogen production and responsiveness might increase RS by ensuring fertilization of eggs (*if* zygote production is sperm-limited), by staggering the times at which the gametophytes produce their sporophytes (if the environment is so variable that successful sporophyte maturation is unpredictable, or if temporal differences somehow increase the total number of sporophytes that can be produced), or by increasing the probability that the sporophyte with the head start will not suffer extensive competitive pressures from its developing neighbors (especially in crowded areas, dense populations of young sporophytes all about the same size might result in diminished growth and survival of all of them).

At this point, we know so little about the ecology of ferns it is impossible to sort out these various possibilities. Nevertheless, it is intriguing to contemplate the alternatives and to consider the appropriate experimental and observational studies.

Induction of sexual differentiation and activity by hormonelike secretions is known in algae and fungi, at least (van den Ende 1976). But induction of a particular sex expression in other individuals is much less well documented. Males of dioecious species of the colonial green alga *Volvox* apparently can trigger male differentiation in neighboring vegetative colonies, but they may also turn on female development, at least in some species (van den Ende 1976). The ecological importance of this behavior seems to be unstudied.

The availability of resources is known to have profound effects on sex expression in a number of plants (e.g., Dzhaparidze 1963). *Catasetum* and *Cycnoches* orchids grown in good light are large and female, whereas those in shade are small and male (Dodson 1962, Gregg 1975). *Arisaema triphyllum* and *A. japonicum* with large corms and, presumably, stored resources tend to be female; individuals with smaller corms are either male or don't flower at all (summary in Bierzychudek 1981, Heslop-Harrison 1957, Policansky 1981), although such size differences are not universal (Lloyd 1981). The monoecious oil palm (*Elaeis guineensis*) produces more male inflorescences if its old leaves are pruned off but more females if its neighbors are pruned to reduce the surrounding shade. If fruit production by young trees is very high in one season, subsequent growth rates are reduced and the tree shifts toward production of male flowers (Hartley 1970). Heslop-Harrison surveyed a number of experiments showing that increased soil nutrients and moisture often favor female expression, as expected; however, in some cases maleness is favored. These experiments all suggest resource limitation but are hard to interpret in detail inasmuch as we do not know what resources might be limiting to which sexual role. For instance, if phosphorus were limiting to pollen formation, augmentation of available phosphorus might favor an increase of maleness, but an increase in nitrogen, for example, might favor femaleness in the same individual. Severe winter weather induced both higher mortality of females and sex switching of some individuals (toward increased male expression) in *Atriplex canescens* (McArthur 1977). Of great interest is McArthur's observation that sex-switchers produced fewer gametes than those not changing sex expression. This indicates that there may be a cost to switching; unfortunately, we have no data with which to assess the relative costs of *not* switching. Miscellaneous factors (temperature, hormones, etc.) are known to influence sex expression (e.g., Frankel and Galun 1977, Heslop-Harrison 1972), but their ecological basis is apparently unstudied. Hormonal balance is known to be adjusted to environmental condition (Freeman et al. 1981) and hormones provide a proximate means of responding in a potentially adaptive way to variable environments.

That access to pollen vectors may influence sex expression is illustrated by two monoecious species already discussed: *Ambrosia trifida,* the giant ragweed (Abul Fatih and Bazzaz 1979) and gambel's oak (*Quercus gambelii;* Freeman et al. 1981), both of which are predominantly female under a canopy of other plants. These species are wind-pollinated and the prospects of achieving successful pollen donation underneath a heavy canopy are surely very low. Because both studies are correlative, we cannot know yet that this shift is really due to differences in expected success in a particular sexual role; some detailed experimentation would be needed to ascertain causation.

Some cases of sex-specific predation are summarized by Bawa (1980). However, we do not yet know how variable and predictable the predation patterns might be. In some species, individuals of different sex expressions produce seeds of different sizes, but seed size does not determine sex expression (e.g., Darwin 1884).

SEX RATIOS

We can consider both the sex ratios within progenies produced by a set of parents and the sex ratios of flowers on a single individual. In terms of theory, these two problems are sometimes treated similarly, at least up to a point (Charnov 1979, Maynard Smith 1978). In this section we are concerned with the possible adaptive value of particular sex ratios produced by individuals, although a considerable fraction of the literature on the evolution of sex ratios focuses on such nonindividual traits as the efficiency of seed production by the population.

Sex ratios can be considered at three points in time. The primary sex ratio is that at the time of fertilization. We seldom know this ratio, although it is of prime interest for theoretical genetic models on the evolution of sex ratios. The secondary sex ratio is that at birth or hatching in animals; the corresponding time in seed plants might be when seeds germinate. The problem is more difficult in the "lower" plants whose haploid males and females are generated without fertilization; here the primary ratio must be that following meiosis in the sporophyte and the secondary ratio that at germination of the spores. The tertiary sex ratio includes all postjuvenile stages, of any age. This is the sex ratio that usually is easiest to determine and for which the most information is available.

Most of the individual-based theory regarding the evolution of sex ratios stems from R. A. Fisher (1929), who, himself, used both population-level and individual-level arguments. Fisher noted that, in sexually reproducing organisms, every individual has a parent of each sex and derives half of its genes from each parent. From a grandparent's point of view, then, all sons, collectively, are equal in value to all daughters, collectively, because the grandparent's nuclear genes are passed on equally by offspring of each sex when mating is random. Because all sons together are as valuable as all daughters, then selection should favor equal parental expenditure on sons and daughters. This fundamental proposition excludes sex-linked and cytoplasmic genes and parthenogenetically produced offspring and assumes that all females are equivalent to each other, as are all males. The model assumes that there is a given total parental expenditure to be allocated to production of offspring and that the population is in equilibrium with respect to sex ratio. When the cost of producing each male offspring is the same as for a female offspring, the sex ratio in the progeny will be 1 : 1 at the time of fertilization. Then if males are rarer than females in the population, sons are more valuable than daughters and selection favors parents with male-biased families until the sex ratio and expenditures are again equal. In short, any deviations from equal expenditure will tend to be rectified, although slight deviations have little effect. For Fisher's argument it is the parental expenditures before the end of parental care that control the system, and the sex ratio is the result. If each male costs twice as much to produce as a female because males require more nutrition, when total expenditures on sons and daughters are equal, the optimum primary sex ratio will be 1 : 2 (female-biased). At this point, each daughter is worth less than each son but the total value of the set of daughters still equals that of all sons. Notice that, for the same total parental expenditure, the clutch size will necessarily be smaller than when all offspring cost the same (unless the greater costs of males is offset by a compensating decrease in the cost of females). Emlen (1968a,b) provided a modification of the basic theory by suggesting that it is not merely the number of genes contributed to future generations but also the rate at which they are passed on. Then differential mortality of the sexes at any time, not just prior to independence from the parent, and the age at maturity may affect the relative worth of sons and daughters, although this conclusion is not popular in the literature.

Verner (1965), following Fisher, showed that the sex ratio unbeatable by mutant individuals would be based, on average, on a 50:50 expenditure on sons and daughters and a 1:1 sex ratio (with equal per capita expenditures). However, if the population sex ratio was shifted toward, say, males, an individual parent could improve its fitness by producing mostly females. In short, the selection for "sex ratio homeostasis" is frequency dependent (Taylor and Sauer 1980).

The problem is far more complex even than might be thought from the genetical models (of which I have only skimmed the surface). Not only is there environmental sex determination, as we have seen, but also in some cases parents (of species with genetic sex determination) are able to modify at least the secondary, if not the primary, sex ratio of their progeny. Many instances of phenotypically controlled shifts of sex ratio are now known for animals; they are thought to derive selective advantage from a variety of ecological conditions, including inbreeding and local mate competition. Hamilton (1967) initially viewed local mate competition as occurring in very localized populations, where mates are likely to be siblings and all the daughters of a parent can be successfully fertilized by just a few sons. Then a parent producing mostly daughters will have more grandchildren than one producing equal numbers of daughters and sons. The operant factor was considered to be the degree of sib-competition for mates—male–male sibling competition favoring female-biased sex ratios and female–female competition favoring males (Bulmer and Taylor 1980, Taylor and Bulmer 1980). Hamilton dealt primarily with insects that are parasitic on other insects: female parents produce only enough sons to fertilize the daughters that occur in the same host. Bulmer and Taylor (1980) applied similar reasoning to plants. The sex ratio should favor the sex that disperses more widely or evenly (i.e., as pollen or as seeds) because members of that sex are less likely to encounter each other and compete for mates. At some point, this will be balanced by the disadvantage of producing still more young of the sex that is already more common. As we have seen, however, dispersal differences between male and female vertebrates are not known to produce consistent differences in offspring sex ratios, although a few cases can be found (Maynard Smith 1980).

More recent work has indicated that the effects of local mate competition do not depend on sib-mating (e.g., Wilson and Colwell 1981). In one species of wasp, a second female parasitizing the same host makes fewer eggs than the first, and most of the eggs produce sons, which have high mating success with the daughters of the first female (Werren 1980). Sex ratio in many parasitoid hymenopterans shift with size of the host, probably because the fitnesses of sons and daughters are unequally affected by size of host (Charnov et al. 1981). Similar shifts occur in the progeny of trap-nesting bees and wasps, as a function of the size of the nest-tunnel (Charnov et al. 1981). Biased progeny sex ratios are known to be common in other taxa as well (Wilson and Colwell 1981).

Several other studies with animals have also suggested or demonstrated manipulation of primary or secondary offspring sex ratios (Werren and Charnov 1978) in response to the expectation of success of either sex and/or the relative costs of producing each sex (e.g., Altmann 1980, Burley 1981, Maynard Smith 1978, Myers 1978, Trivers and Willard 1973). We know, for instance, that female baboons produce sex ratios that vary with their social rank, high-ranking females producing more daughters than expected, presumably because daughters tend to inherit their mother's rank but sons do not (Altmann 1980). The sex ratio in zebra finch broods favors the sex of the more attractive parent (Burley 1981), and specific patterns of skewed family sex ratios are recorded for a number of other vertebrates (Willson in press).

That parents should produce higher proportions of the sex that will achieve the greatest fitness gain in a particular set of circumstances is intuitively appealing, and it is essentially outside the original Fisherian rules (Maynard Smith 1978). Wilson and Colwell (1981) and Colwell

(1981) argued that only a sort of group selection can explain the evolution of biased sex ratios, but I do not think they account for the possibility of differential fitness-gains.

On the other hand, Williams (1979) despaired of finding adaptiveness in parental sex ratios of outcrossed vertebrates because most data sets show a very large variance of family sex ratios in the population, which is not to be expected if selection is favoring a particular family ratio. In addition, the sex-ratio variance did not differ from that expected by random sex determination. As a result, he argued for a nonadaptive, random determination of sex ratios simply in accordance with Mendelian genetics. Maynard Smith (1980) also remarked on the scant evidence for genetic variance in progeny sex ratios, although some cases are known (see also Wright 1977). Williams' argument may have merit; it certainly underlines the great difficulties of understanding sex ratios. Nevertheless, I am inclined to suspect that very close study of specific cases may expose a variety of conditions that produce adaptive changes in family sex ratios; perhaps the same ratio of sons and daughters is not adaptive for all families.

With this background in mind we can examine what is known about sex ratios in plants. A glance at Table 2.3 shows that sex ratios of mature plants vary tremendously. A preponderance of males is common, but in a number of species females predominate or the sex ratio is undistinguishable from unity. Where multiple samples are available for a single species, sometimes the sex bias is consistent but other times it is not. There may be geographical or habitat patterns. For instance, *Rumex acetosella* in Europe has about equal numbers of males and females, but in New Zealand males predominate in established stands and sex ratios are about 1:1 in new stands (Harris 1968, Putwain and Harper 1972). The quaking aspen *(Populus tremuloides)* varies from female-biased at low elevations to male-biased at high altitudes in Colorado (Grant and Mitton 1979). *Aralia nudicaulis* in New Brunswick has even or female-biased ratios of flowering stems in roadside habitats but commonly has male-biased ratios in the forest (Barrett and Helenurm 1981). The degree of male-bias in *Myrica gale* varied among samples (Lloyd 1981). Can we make any sense out of this welter of variation? Very little, as it happens, at least at present. The variation apparent in Table 2.3 may emerge for diverse reasons, which will require close analysis by future work.

Several factors may contribute to apparently skewed adult sex ratios: differential mortality, length of reproductive life, differential thresholds for flowering, different frequencies of vegetative propagation, and possibly parental manipulation of the offspring sex ratio. Sexual differences in vegetative propagation and different flowering thresholds can alter the number of stems counted in each gender without affecting the actual number of genetic individuals of each sex present in the population. Allocation of greater amounts of energy or nutrients to reproduction by females may reduce vegetative growth and thus diminish the prospects of survival (Harris 1968) or reduce vegetative propagation (Putwain and Harper 1972). Such explanations do not appear to apply to all cases of skewed sex ratios, either because the species in question do not propagate vegetatively or because females allocate more than males to both sexual and vegetative budgets (e.g., *Populus tremuloides,* Grant and Mitton 1979). Sexual differences in length of reproductive life are not well documented; Bullock and Bawa (1981), Clark and Orton (1967), Meagher (1981), Opler and Bawa (1978), Valentine (1975), and others record some cases of earlier maturation by males, but we do not know if such individuals also survive well enough to bias the entire population of adults toward males. Certain orchids (e.g., *Cycnoches lehmannii*) have male flowers lasting three to five days but females lasting six to eight weeks (Dodson 1962, Dodson and Frymire 1961); functionally, the difference may be even greater (Janzen 1981). *Anguria* males may flower for as long as three years, but females cease when fruits begin to develop (Gilbert 1975). Male flowers of *Jacaratia dolichaula* last longer than females (Bullock and Bawa 1981). However, there may be differences in frequency of flowering, as reported for *Aciphylla scott-thompsonii, Jacaratia dolichaula, Trichilia monodelpha,* and *Chamaelirium luteum* in which males flower more often than females (Bullock and Bawa 1981, Meagher 1981, Styles 1972, Webb and Lloyd 1980). Note that in none of these cases is it likely that biased adult sex ratios are favored directly by selection (Lloyd 1973) unless one argues that, for example, vegetative propagation and growth are favored in part because of their effects on sex ratio.

It is sometimes argued that the sex ratio will be adjusted so as to maximize seed production of the population (e.g., Kaplan 1972, Mulcahy 1967) and, in general, the conclusion is that females are likely to predominate in most situations. Since there is no basic tendency for females to outnumber males, this suggestion is, at best, rather limited. It presents the further difficulty of supposing that the fecundity of the population is somehow subject to selection that overrides whatever natural selection is occurring at the individual level (see Mulcahy 1967). Lloyd (1974b) also doubts the validity of arguments for maximization of seed production by the population.

Information on the sex ratios of progenies is much sparser than that for adults, at least in part because offspring usually must be grown until they flower to ascertain their gender. *Rumex acetosa, R. acetosella,* and *R.*

Table 2.3 Selected data on sex ratios (male/female) in adult, dioecious seed plants; the first two species are gymnosperms, the rest are angiosperms

Taxon	Ratio (male/female)	Source	Location
Podocarpaceae			
Dacrydium cupressinum	1.04	Godley 1964	New Zealand
D. fonckii	1.68,[a] 1.52[a]	Godley 1964	New Zealand
Anacardiaceae			
Spondias nigrescens	.98	Opler and Bawa 1978	Costa Rica
Araliaceae			
Aralia nudicaulis	.72[a] → 190.87[b]	Barrett and Helenurm 1981	New Brunswick
	.45[a] → 3.07[a]	Bawa et al. 1982	Massachusetts
Boraginaceae			
Cordia collococca	1.50[a]	Opler and Bawa 1978	Costa Rica
C. panamensis	2.11[b]	Opler and Bawa 1978	Costa Rica
Burseraceae			
Bursera simarouba	1.20	Opler and Bawa 1978	Costa Rica
B. tomentosa	1.23	Opler and Bawa 1978	Costa Rica
Caryophyllaceae			
Silene alba	.80	Gross and Soule 1981	North America
	.79	Lloyd 1973	Europe
	.32 → .68 (1.03)	Mulcahy 1967	North America
	.3 → 3.0[b] (mostly < 1.0)	Willson unpublished	Minnesota
S. otites	About 1.56	Godley 1964	Europe
Chenopodiaceae			
Spinacia oleracea	Low density 1:1; High density > 1.1[a]	Onyekwelu and Harper 1979	Europe

Cornaceae			
Griselinia littoralis	2.18[a]	Godley 1964	New Zealand
Ebenaceae			
Diospyros melanoxylon	1.4 → 4.3[b]	Rathore 1969	India
Elaeocarpaceae			
Aristotelia serrata	1.9,[a] 3.16,[a] 4.07,[a] 1.19	Godley 1964	New Zealand
Epacridaceae			
Cyathodes colensoi	4.45,[a] 2.14,[a] 3.15[a]	Godley 1964	New Zealand
Ericaceae			
Epigaea repens	< .04 → 2.2	Clay and Ellstrand 1981	North America
Erythroxylaceae			
Erythroxylon rotundifolium	2.14[a]	Opler and Bawa 1978	Costa Rica
Euphorbiaceae			
Bernardia nicaraguensis	1.04	Opler and Bawa 1978	Costa Rica
Mercurialis annua	(female > male)	Lloyd 1973	Europe
Liliaceae			
Chamaelirium luteum	1.7–3.4	Meagher 1981	North America
Malvaceae			
Plagianthus betulinus	1.05	Godley 1964	New Zealand
P. divaricatus	.95	Godley 1964	New Zealand
Meliaceae			
Trichilia cuneata	1.24	Opler and Bawa 1978	Costa Rica
T. anisopleura	1.38	Opler and Bawa 1978	Costa Rica

Table 2.3 (*Continued*)

Taxon	Ratio (male/female)	Source	Location
Moraceae			
Cecropia peltata	1.04	Opler and Bawa 1978	Costa Rica
Chlorophora tinctoria	1.23	Opler and Bawa 1978	Costa Rica
Cannabis sativa	$.82 \rightarrow .95$	Dzhaparidze 1963	Europe
Humulus lupulus	$.11 \rightarrow .29$	Dzhaparidze 1963	Europe
H. japonicus	(female > male)	Lloyd 1973, Godley 1964	—
Nyctaginaceae			
Pisonia macranthocarpa	1.36	Opler and Bawa 1978	Costa Rica
Orchidaceae			
Catasetum macroglossum	21.0, 46.0	Dodson 1962	South America
Cycnoches lehmannii	27.0	Dodson 1962	South America
Piperaceae			
Macropiper excelsum	3.50,[a] 7.1,[a] 3.25[a]	Godley 1964	New Zealand
Polygonaceae			
Coccoloba caracasana	.41[b]	Opler and Bawa 1978	Costa Rica
C. floribunda	.85	Opler and Bawa 1978	Costa Rica
Ruprechtia costata	.70	Opler and Bawa 1978	Costa Rica
Triplaris americana	$.26 \rightarrow .79$[b]	Opler and Bawa 1978, Melampy and Howe 1977	Costa Rica
Rumex hastatulus	.75	Smith 1963, 1968	North America
	$.66 \rightarrow 1.02$; $\bar{x} = .92$[a]	Conn and Blum 1981a	North America
R. thyrsiflorus	$.08 \rightarrow .63$[b]	Zarzycki and Rychlewski 1972	Poland
R. acetosa	$.07 \rightarrow .91$[b]	Zarzycki and Rychlewski 1972, Putwain and Harper 1972	Poland, Britain
R. paucifolius	1.21	Smith 1968	North America
R. acetosella	About 1.00	Putwain and Harper 1972	Britain
	$1.00 \rightarrow 1.67$[b]	Harris 1968	New Zealand

Ranunculaceae			
Clematis afoliata	2.16[a]	Godley 1964	New Zealand
C. paniculata	2.00[a]	Godley 1964	New Zealand
Thalictrum dioicum	.84 → 10.48[b]	Melampy 1979	North America
T. polygamum	.65 → 1.26	Melampy 1979	North America
Rosaceae			
Fragaria elatior	.97 → 1.03	Dzhaparidze 1963	Europe
Potentilla fruticosa	.56 → 2.44	Elkington and Woodell 1963, Grewal and Ellis 1972	Europe
Rubiaceae			
Alibertia edulis	1.69[a]	Opler and Bawa 1978	Costa Rica
Genipa caruto	.99	Opler and Bawa 1978	Costa Rica
Randia spinosa	2.70[b]	Opler and Bawa 1978	Costa Rica
R. subcordata	2.04[a]	Opler and Bawa 1978	Costa Rica
Coprosma australis	1.09	Godley 1964	New Zealand
C. rhamnoides	2.03[a]	Godley 1964	New Zealand
C. repens	1.40[a]	Godley 1964	New Zealand
Salicaceae			
Populus deltoides	1.17	Farmer 1964	North America
P. tremula	1.0 → 4.5[b]	Valentine 1975	Europe
P. tremuloides	1.12 → 2.55[a]	Lester 1963, Valentine 1975	Eastern North America
	.79 → 1.77[b]	Grant and Mitton 1979	Colorado
Sapindaceae			
Allophyllus occidentalis	1.48[b]	Opler and Bawa 1978	Costa Rica
Simaroubaceae			
Simarouba glauca	1.08	Opler and Bawa 1978	Costa Rica

Table 2.3 (*Continued*)

Taxon	Ratio (male/female)	Source	Location
Umbelliferae (Apiaceae)			
Aciphylla pinnatifida	.98	Webb and Lloyd 1980	New Zealand
A. scott-thompsonii	1.29, .57, 1.62[a]	Lloyd 1973, Godley 1964	New Zealand
A. subflabellatum	4.21[a]	Godley 1964	New Zealand
A. monroi	2.42,[a] 3.03[a]	Lloyd 1973, Webb and Lloyd 1980	New Zealand
A. poppelwellii	1.24	Webb and Lloyd 1980	New Zealand
A. aurea	.60 → 1.45[b]	Lloyd 1973, Webb and Lloyd 1980, Godley 1964	New Zealand
Anisotome aromatica	3.42,[a] 4.65,[a] 7.42[a]	Lloyd 1973, Webb and Lloyd 1980	New Zealand
A. filifolia	4.00[a]	Webb and Lloyd 1980	New Zealand
A. deltoidea	16.17	Webb and Lloyd 1980	New Zealand
A. flexuosa	1.57 → 5.61[a]	Webb and Lloyd 1980	New Zealand
A. hastii	2.81,[a] 5.80[a]	Webb and Lloyd 1980	New Zealand
Lignocarpa diversifolia	1.40, 1.48, 1.78[a]	Lloyd 1973, Webb and Lloyd 1980	New Zealand
Valerianaceae			
Valeriana dioica	.2 → 9.0	Godley 1964	Europe
Valeriana edulis	.3[a] → 1.0	Soule 1981	North America
Zanthoxylaceae			
Zanthoxylum setulosum	1.92[a]	Opler and Bawa 1978	Costa Rica

[a]Indicates the ratio is statistically significantly different from 1:1.
[b]Indicates that significant between-population differences were tested and found.

thyrsiflorus seeds yielded a 1:1 sex ratio as did some samples of *R. hastatulus,* indicating a lack of correspondence between seed and adult sex ratios (Putwain and Harper 1972, Smith 1963, Zarzycki and Rychlewski 1972). *Ilex opaca* also exhibited a 1:1 sex ratio of seedlings (Clark and Orton 1967), as did *Chamaelirium luteum* (Meagher 1981). Seeds of *Clematis gentianoides* and of four of five species of *Dioscorea* yielded a significantly male-biased sex ratio (Godley 1976, Martin 1966). Progeny sex ratios in artificial population of *Silene alba* ranged from .77 (♂/♀), significantly female-biased, to 1.25, which was not significantly different from 1.0 (Mulcahy 1967).

Early experiments by Correns (1928, cited in Allen 1940, Conn and Blum 1981, Jones 1928) with *Silene alba* demonstrated that light pollen loads on a stigma result in a progeny sex ratio of about 1:1 (slightly female-biased), but that heavy loads of pollen resulted in a progeny strongly skewed toward females. He suggested that pollen tubes with an X chromosome grew faster than those with Y's; when X pollen was insufficiently abundant to fertilize all the ovules, the slower Y pollen eventually claimed the remaining ones, but with abundant pollen arriving, the fast-growing X pollen reached most of the ovules ahead of the Y pollen.

Differential pollen tube growth and skewed offspring sex ratios with abundant pollen loads have been reported also for *Silene dioica, Humulus japonicus, Cannabis sativa,* and *Rumex acetosa* (references in Allen 1940, Conn and Blum 1981, Lloyd 1974b, Mulcahy 1967, Rychlewski and Zarzycki 1975). Pure pollen loads produced a progeny sex ratio slightly favoring males in *Rumex hastatulus,* but diluted pollen loads produced ratios slightly favoring females (Conn and Blum 1981). None of the reported ratios are significantly different from 1:1, but the shift is a significant one. Increasing dilutions did not, however, yield increasingly female-biased ratios (Table 2.4). The apparent effect of pollen competition on the sex ratio was not observed in the field using the progeny of females growing at increasing distances from a single pollen source, perhaps because abundant supplies of air-borne pollen reached all females (Conn and Blum 1981). Thus it remains to be seen if these intriguing effects of pollen competition are of general importance in natural conditions (see also Rychlewski and Zarzycki 1975).

Mulcahy's (1967) data (recalculated from his Tables 2 and 3) indicate a tendency for seed production per female plants of *Silene alba* to decline with decreasing numbers of males in his synthetic populations, suggesting that pollen was commonly limiting to female RS under conditions of his experiment. There was no significant correlation between number of males and the tendency for males to predominate in the progenies,

Table 2.4 Effect of pollen load on sex ratio in *Rumex hastatulus* (Conn and Blum 1981). *Rumex* pollen was diluted with dead pine pollen in the indicated ratios. The sex ratio for pure *Rumex* pollen is significantly different from all the others.

Pollen dilution ratio (*Rumex*:Pine)	*Rumex* pollen density on receptive surface	Overall progeny sex ratio (male/female)
1:0	1299	.70
1:10	203	1.50
1:100	39	1.20
1:1000	13	1.06

although it is interesting to note that females in the population with the most males produced progenies significantly biased in favor of females and in the population with fewest males, the progeny sex ratios were slightly (insignificantly) skewed in favor of males. Because Mulcahy did not use controlled hand-pollinations but relied on natural pollinators (chiefly moths) and because the number of synthetic populations to be compared was small, both very local differences in pollinator behavior and chance variability could have obscured any real trends. Clearly, if females can adjust their progeny sex ratios in response to the abundance of pollen arriving on their stigmas and the number of available males in a population, this would be a matter of considerable interest.

Note that the X–Y differential suggested by Correns can only adjust the sex ratio in the direction of female-bias; there is no way to get more than 50% males. The arrival of little pollen may signal a paucity of effective male individuals and indicate conditions in which production of male offspring would be profitable, but the mechanism of adjustment does not allow the parent to capitalize fully by making mostly males. Several questions remain unanswered: Why is a slow-growing Y chromosome maintained? Do all species with genetic sex determination have some prezygotic means of manipulating the sex ratios? Are there postzygotic means of adjusting the sex ratio, such as differential abortion? It remains to be seen if individual plants adjust their progeny sex ratios in adaptive ways.

In these species, as in most dioecious species with genetic sex determination, males are heterogametic, although a few species with female heterogamety are known. Mulcahy (1967), following Lewis (1942), proposed that male heterogamety might be advantageous (hence its commonness) because it is a mechanism that allows adjustment of offspring sex ratios through differential pollen tube growth. However, I might note that in mammals it is the male who is heterogametic but females, nevertheless, can manipulate the sex ratio; perhaps female plants can too, in other ways.

The application of sex-ratio theory to the allocation of resources to sex expression of an hermaphroditic plant has found some support (Charnov et al. 1981, Maynard Smith 1978, Smith 1981, Willson 1979). However, the production of gametes is not really biologically equivalent to production of offspring, so that theory invented to explain the evolution of offspring sex ratios may not be applicable. Furthermore, Charlesworth and Charlesworth (1981) suggest that, at least with selfing and inbreeding depression, equal allocation to male and female function is not to be expected. It also is hard to say, at this point, whether allocational patterns do conform to the expectations of sex-ratio theory. If we consider biomass allocation, we can find great variation in expenditures on male and female functions; very commonly female costs outweigh male, at least when the materials allocated to seeds are included. But a great deal of information, some already presented, documents enormous variation among individuals in sexual biomass allocation. The basis and possible adaptive nature of such variation is still controversial. We are not even sure that biomass is the appropriate currency for assessing costs.

SELFING VERSUS OUTCROSSING

Considering the number of things that have been "explained" by the putative importance of outcrossing, the actual evidence for the adaptive values of outcrossing and selfing is remarkably scant (Lloyd 1980c). Most rationales for the advantages of selfing depend on (1) selection to ensure or increase seed-set where and when pollination is unreliable, (2) selection to avoid disruption of a successful genotype in noncompetitive situations (this apparently disregards the importance of predators and pathogens), or (3) selection to rapidly exploit an open environment by reducing the costs of mating and allocating the savings to increased fecundity. Self-fertilizing species are sometimes common as colonizers in environments frequently unsuitable for insect pollinators, and sometimes where density or population size is low (e.g., Baker 1965). Selfing also means that a parent effectively avoids the cost of making males and passes on two haploid genomes in every zygote, and facultative selfers may have an advantage by virtue of their ability both to self and to donate pollen to other individuals (Lloyd 1979c, Wells 1979). On the other hand, the advantages of outcrossing are almost universally thought to reside in the heterotic superiority of offspring produced and the avoidance of inbreeding depression. Indeed, such effects have been known for a very long time (Darwin 1877), and offspring resulting from outcrossed matings can be expected to outcompete selfed ones in limited environments (Lloyd 1980c, Schemske in press). However, in certain cases (e.g., *Cirsium* spp.) inbred offspring may begin life with compensating advantages such as larger seed size and greater success in establishment (van Leeuwen 1981). Outcrossing also increases the variation among the offspring and contributes to achieving the possible advantages of sex. Furthermore, facultatively outcrossed plants may have different RS as male and as female and, as we have seen, the two functions can vary independently even in hermaphroditic flowers. It is conceivable that variability of sexual roles provides another potential advantage to outcrossing systems.

At least some of these rationales are very plausible and some are substantiated by correlative evidence. Yet our understanding of the evolution of either extreme of selfing or outcrossing, much less various combinations of them, is paltry indeed. There are many exceptions to the correlation between selfing and colonizing, and selfing does not always bring with it a great reduction of progeny variance because there exist mechanisms for the maintenance of variation in the face of inbreeding (e.g., Allard et al. 1968, Usberti and Jain 1978). Inbreeding depression is a condition found in species adapted to outcrossing and is not necessarily a problem in those *adapted* to inbreeding (Wright 1977, but see Cartier in press, Charlesworth and Charlesworth 1979d, Schemske in press). Moreover, even obligate outcrossing seldom prevents inbreeding with close relatives, which may be the most likely mates in many populations, and inbreeding with relatives ultimately has the same effect on levels of heterozygosity that selfing does. The value of outcrossing may depend sometimes on what individual mates are involved and how many there are. In short, not all outcrossing has the same consequences. There is the further, practical difficulty that levels of outcrossing achieved in natural populations may reflect not some selectively optimal balance but rather what is possible under the circumstances. The amount of work required to establish the evolutionary and ecological conditions for a given breeding system is phenomenal and will, no doubt, be slow in coming. However, we are ahead of our former situation if we realize the variety of factors and problems involved.

Cleistogamy Versus Chasmogamy

A special case of the selfing/outcrossing problem is represented by plants that produce both kinds of flowers. Chasmogamous flowers are what one normally thinks of—open flowers with sex organs accessible to potential pollen vectors. Cleistogamous flowers are closed; they are obligate self-pollinators and commonly produce less pollen than chasmogamous flowers. Cleistogamous flowers are sometimes borne on above-ground shoots (*Viola, Impatiens*), sometimes they are subterranean (*Amphicarpum, Emex, Gymnarrhena*), and sometimes they occur in both places (Uphof 1938). Sometimes cleistogamous and chasmogamous flowers are reported to occur on different individuals, as in *Lithospermum carolinense,* in which the chasmogamous flowers are heterostylous and largely self-incompatible but cleistogamous flowers are obviously self-compatible and are produced especially at high densities (Levin 1972). *Ruellia caroliniensis* has several floral morphs, ranging from chasmogamy to cleistogamy, and all are self-compatible (Long 1971). Apparently, the ecology of this curious system is unexplored.

The proportions of cleistogamous and chasmogamous flowers borne by an amphicarpic individual can vary enormously. At least in the grass *Danthonia spicata,* variation in the proportions of cleistogamous and chasmogamous flowers was partially controlled genetically (Clay 1982). Seed production by chasmogamy prevailed in plants of *Amphicarpum purshii,* an annual grass, growing in more disturbed habitats but was accomplished entirely by subterranean cleistogamy in plants in a less disturbed site (McNamara and Quinn 1977). Although resource allocation to cleistogamy was greater in the least disturbed site, where total plant size was smallest, the average number of seeds produced by cleistogamous means differed very little in different sites; the major difference was found in the almost complete elimination of chasmogamy in the relatively undisturbed site.

Emex spinosa is another amphicarpic annual, in the Polygonaceae. Subterranean fruit production was less affected by plant density, nitrogen availability, and other habitat conditions than was aerial production of fruits. The larger the plant, the more aerial flowers it bore (Weiss 1980). Fruit production by means of chasmogamy in *Gymnarrhena micrantha,* an annual composite, increased with increasing soil moisture; these fruits are less drought-tolerant than those from the cleistogamous flowers, and it is probably no accident that they are produced more frequently when growing conditions are more suitable (Koller and Roth 1964). Furthermore, biomass allocation and seed production in chasmogamous flowers increased linearly with plant size, whereas seed production by cleistogamous flowers stayed constant (Zeide 1978). Whether plant size varies with moisture conditions is not reported but seems likely. Notice that, in both these cases, it is the chasmogamous flower production that varies.

Impatiens capensis (biflora) is an annual herb of forest understory and the shores of lakes and rivers; it has been studied rather intensively by several investigators. Estimates of the percent biomass allocation to reproduction remained quite constant, despite huge differences in plant size (Abrahamson and Hershey 1977, Waller 1980). However, the relative allocation of biomass to cleistogamous and chasmogamous flowers varied greatly—the larger the plant the higher the proportion of chasmogamous flowers, whereas the relative allocation to cleistogamous flowers remained the same over a range of plant sizes (Waller 1980). The proportion of chasmogamous flowers and fruits also increased with light intensity (Schemske 1978, Waller 1980), even when plant size was the same (and density effects did not mask the effect of light; Waller 1980).

Chasmogamous flowers of *Impatiens* are strongly protandrous: the androecium covers the stigma until pollinators remove it or it dries up and falls off. When the stigma is exposed, the flower enters the female phase. Therefore, although the plant is self-compatible, within-flower self-pollination is improbable, but between-flower pollinations on the same plant could be effective (Schemske 1978). The major pollinators are bees, wasps, and hummingbirds (Rust 1977, Schemske 1978). Each flower is short-lived (one to two days), but as a rule the male phase of each flower lasts five times as long as the female phase (Schemske 1978). Chasmogamous flowers are larger and showier, they produce nectar containing both sugar and amino acids, and (for some unknown reason) their seeds are larger and take longer to develop (Rust 1977, Schemske 1978, Waller 1979). As a result, the production of a seed by a chasmogamous flower is estimated to be about two to three times as costly as the production of a seed by a cleistogamous flower (Schemske 1978, Waller 1979). The greater costs of seeds by means of chasmogamy may be one reason why only large and well-lighted plants produce many seeds this way. Having two means of seed production (especially cleistogamy) may permit some successful reproduction in a greater range of habitats than would otherwise be possible (Schemske 1978).

The long male phase of each flower and the improbability of within-flower selfing may also enhance outcrossing (Schemske 1978). Since chasmogamous flowers can be outcrossed, Waller (1980) asked why outcrossing should be favored in large plants. He suggests two possible answers. First, when clutch size (of the whole plant) is large, the plant can afford more genetic variation among its offspring and risk having some of them fail; there should be less effect on the variance of clutch size than when clutch size is small. And second, increased competition from sibs or unrelated neighbors may lead to selection for increased variation, in order to get a winning genotype. Note that these arguments assume that outcrossing is indeed accomplished to a greater degree by plants with many chasmogamous flowers and that within-plant pollen movement does not significantly reduce the level of outcrossing achieved. The second suggestion further supposes that competition is greater for larger individuals or, at least, that larger plants are the only ones that can produce enough seeds to elude such competition.

There is an additional possibility. Production of fruits by cleistogamous flowers can be seen as a low-cost, low-risk form of reproductive effort, because the per-seed cost is small and pollination is guaranteed. Functional sex expression is solely female. Plants growing in good conditions may be able to add not only the more expensive form of seed production but also effective male function. In other words, they don't switch sexes, they add a sex. Male function may be a high-risk form of sex expression because success as a pollen donor depends entirely on the behavior of pollen vectors. Plants with chasmogamous flowers commonly are already committed to rearing a large crop of seeds; pollen donation may be a way to increase overall RS without incurring still greater expense in building seeds. And, as Schemske noted, a chasmogamous flower that fails as a female (because of lack of pollination or because of predation) may be highly successful as a male. That the male phase of chasmogamous flowers so greatly exceeds the female phase hints that male function may be a very important aspect of chasmogamy. This suggestion regarding the importance of male function in chasmogamy may pertain to other amphicarpic species as well—the expression of chasmogamy seems to be generally more variable than cleistogamy.

Both pollen donation and the production of outcrossed seeds may have been more important in the evolution of chasmogamy by plants that also reproduce cleistogamously. It may not be possible to ascertain their relative importance; at least a large research effort would be needed. One would need to know the value of variable offspring across the range of conditions under which the species grows, the juvenile survival and eventual RS of seeds produced by cleistogamy and chasmogamy in assorted conditions, and the relative male and female RS (both mean and variance) of chasmogamous flowers, to even begin to answer the question.

Another important question is why produce seeds cleistogamously at all. Do they have a biological function different from vegetative reproduction? Cleistogamously produced seeds on aerial shoots can be dispersed by normal means of seed dispersal and thus have a higher probability of reaching new sites than most vegetatively produced progeny. But underground seeds lack this dispersal capacity, and one could argue that they are a substitute for vegetative reproduction. However, because meiosis and syngamy even within a flower permit a modicum of recombination, underground seeds produced cleistogamously are not quite equivalent to vegetative reproduction. Furthermore, they may be endowed with the same capacities for dormancy as chasmogamously produced seeds and thus retain the option of delaying activity until conditions are suitable. Whether such differences provide some adaptive basis for cleistogamously produced seeds is a matter for discussion.

INCOMPATIBILITY SYSTEMS

Incompatibility systems have been reported from most major taxa but are best known in angiosperms (and fungi, which are now often segregated into a separate kingdom all their own). For some reason, they are very poorly developed among the gymnosperms and apparently rare, or at least little recorded, in the lower plants. As a result, the emphasis here is on the angiosperms.

It is customary to distinguish true incompatibility, referring to prezygotic events that prevent fertilization, from other events following fertilization that result in embryo failure (de Nettancourt 1977). Postzygotic events may be due to genetic problems in the embryo itself—the expression of lethal genes, perhaps as a result of inbreeding, and mismatches between parental genomes that cause developmental errors, or they may result from rejection and abortion by the maternal sporophyte without regard to embryonic defects. In this section we are concerned chiefly with prezygotic failure of sperm. Prezygotic failure can take place at any point before fertilization; it sometimes occurs on the stigma, sometimes in the style, or in the ovary and ovule. The strength of the incompatibility reaction can be varied; failure may be complete or partial, as in the case of *Amsinckia grandiflora,* in which some pollen tubes are usually outcompeted by others but are functional in the absence of such competition (Weller and Ornduff 1977).

There are two related aspects of incompatibility. The one that dominates most of the literature on the subject is *self*-incompatibility (SI), by which sperm from the same plant (or others carrying the same allele) are prevented from fusing with the eggs. Sometimes incompatibility systems

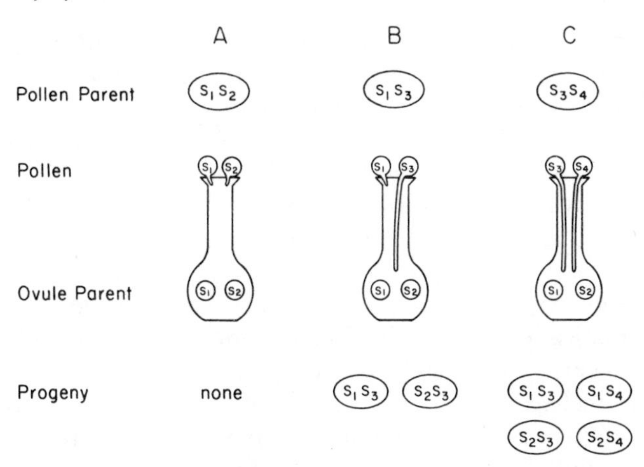

Figure 2.5 The consequences of a gametophytic incompatibility system on the production of progeny and the variability of the progeny. Redrawn from Lewis (1949) by permission.

extend beyond self to a varied array of "mating types" or even individual mating "preferences."

Two major mechanisms of SI are known. Gametophytic SI results from the expression of genes in the haploid male gametophyte, the pollen grain. Gametophytic SI is widely distributed among angiosperms in both dicots and monocots (Pandey 1960), though its presence is more often inferred than tested. Typically, a single locus S has many alleles, but in some cases as many as four multiallelic loci are involved (de Nettancourt 1977, Lewis 1979). At least in some species (e.g., *Nicotiana sanderae*), S alleles differ in their ability to prevent self-fertilization (in Lewis 1954) and modifier genes may change the strength of the incompatibility reaction (Charlesworth and Charlesworth 1979c), but little theoretical attention seems to have been paid to such observations. If the pollen contains the same allele as the diploid maternal tissue, fertilization is inhibited. When the pollen parent is genetically like the seed parent (at this locus), no pollinations will be successful, but when the parents share only one of the S alleles (in a single-locus system), half the pollinations may succeed (Figure 2.5A,B). When parents are unlike in both their S alleles, all pollinations may succeed (Figure 2.5C).

Interestingly, the ability of the ensuing progeny to mate among themselves is different in cases B and C. The prospect of sib-mating, or mating with other relatives, is a real one in most plants, because both pollen and seeds are often dispersed in a tightly clustered pattern around the parent (e.g., Levin and Kerster 1974, but see Levin 1981), and most breeding may take place among relatives. The offspring of B are like their parents and again only half the pollinations among them can be successful. However, in C, some matings among the offspring are entirely cross-fertile although others are not, resembling case B. Although female RS will be affected only if pollen is limiting to seed production, male RS may be strongly dependent on the relative frequencies of siblings with appropriate alleles. An S_1S_3 male, for instance, perhaps can donate more pollen if growing near S_2S_4 females than if S_1S_2 or S_2S_3 females are nearby. When all pollen is potentially capable of success and is not automatically excluded for lack of compatibility, any other factors tending to improve male success necessarily achieve greater results, on an absolute scale, than when a fraction of the pollen is not potentially successful. Fitness gain for males may then be more effective when more pollen is potentially functional. An SI male would have to produce greater total amounts of pollen in order to reach as many accepting females with the same probability as a male unrestricted by an SI system. This discrepancy may contribute, perhaps, to the putative association of dioecy with self-compatibility (Baker 1959, 1967; Thomson and Barrett 1981a): the more relatives suitable as recipients for pollen donation, the more successful may be role specialization that culminates in separation of the sexes. However, the strength of this association has been questioned (Bawa 1982, Willson 1982). Furthermore, if dioecy were attained by this route, the original problem would recur: only a fraction of accessible individuals are potential mates. It may not be plausible to propose a reversal of rates of fitness-gain along a proposed evolutionary pathway.

The second SI mechanism is sporophytic, because pollen failure is determined by the sporophytic producer of the pollen. All the pollen grains of a given male are rejected by the female organs of that individual; the genetic constitution of the haploid pollen grain itself is irrelevant. Sporophytic SI systems are known to prevail in the Asteraceae and Cruciferae and thought to be present in several other dicot families and some lilies (Pandey 1960). As many as four loci are involved, but one multiallelic locus is more common (Lewis 1979). In contrast to gametophytic SI, however, alleles may interact and, therefore, a great variety of outcomes are possible depending on the allelic combination and whether they work the same way in both pollen and style (Lewis 1954). In general, the more alleles and loci involved, the less the effect in reducing sib-matings but, as in gametophytic systems, the effect may still be considerable (Lewis 1979).

The common and obvious explanation for the evolution of SI is the avoidance of self-fertilization and any disabilities attendant upon this closest form of inbreeding. Any SI system brings with it cross-incompatibility with genetically similar individuals, but cross-incompatibility is an outbreeding mechanism only when it reduces mating among relatives more than among nonrelatives (Bateman 1952). The more S alleles and loci there are, the less likely are incompatible responses between nonrelatives and, in addition, the less the reduction of inbreeding among relatives (Lewis 1954, 1979). Therefore, several authors have argued that the selective value of incompatibility systems does not lie primarily in its effects on sib-mating (Bateman 1952, Lewis 1979). However, Lewis' (1979) estimates of the relative frequency of compatible matings among sibs and in the population at large indicate that the percentage of effective pollen and the relative compatibility of sibs can be much reduced (to less than half that of the general population). This suggests that the influence of selection through reduction of sib-matings may have been undervalued, at least in multigenic SI systems. However, the evolution of multiallelic, multilocus SI systems does seem to improve the discrimination possible between relatives and nonrelatives, with the result that fewer potentially suitable mates are excluded.

There seems to be some disagreement over the phylogenetic history of SI systems (de Nettancourt 1977). One school maintains that it originated only once in angiosperms and all existing systems are descendants from the primitive condition of SI (Stebbins, Whitehouse in de Nettancourt 1977). The other school contends that SI must have originated more than once (Bateman 1952, Charlesworth and Charlesworth 1979c, Ganders 1979). Both sides point to the widespread occurrence of SI across the angiosperms as supporting evidence, and the issue may remain moot for some time. However, agreement is more general concerning the phylogeny of the two types of SI: the gametophytic system is thought to be the more primitive and the sporophytic system derived from it (Beach and Kress 1980). The beginning of a rationale based on natural selection is provided by Beach and Kress. They note that there can be a conflict of interest between gametophyte and sporophyte, and that selection may operate upon each generation separately. Once a pollen grain has been deposited on a stigma, its sperm must fertilize an egg or none of that gametophyte's genes will enter the next generation. Selection should favor all possible means of ensuring success by the male gametophyte, whether by physiologically overcoming resistance by the female organs or by avoiding recognition as "self." The quality of the resultant zygote is not an issue, from the point of view of the male gametophyte, because it is already committed and any zygote is better than none. The female-functioning sporophyte, on the other hand, may gain by rejecting some male gametophytes (self-pollen) and accepting those from other sporophytic individuals. Basically, sporophytic SI is viewed as more effective in discriminating against committed male gametophytes, because it operates before the haploid gametophytic genome is expressed. One could argue, perhaps, that the hermaphroditic sporophyte should manipulate its male gametophytes in some way to reduce their aggressiveness upon landing on a self-stigma, but it may be difficult to evolve pollen that is unaggressive on self-stigmas but that maintains its competitive ability on all other stigmas. The more efficient solution thus may be to shut down the activity of all self-pollen through the female side of the interaction. The recognition substance in self-pollen is derived from sporophytic tissue in sporophytic SI systems, but I cannot find out if gametophytic SI systems lack these deposits in the wall of the pollen grain or if there is simply no reaction to such materials by the female organs.

I suggest that a strict SI system contains some elements of conflict between male and female function (i.e., not just between gametophyte and sporophyte), especially with respect to mating among relatives. If female RS is not pollen-limited, the cost of rejecting a neighboring relative's pollen is small, even though the variety of offspring may be slightly reduced, and the cost of raising inferior offspring may be rather high (seeds can be expensive). However, the cost of rejection to potential male RS may be greater, because more pollen is doomed to fail. Even inbred offspring are probably of greater value, once a pollen grain has landed, than none at all; even inbreds can survive and reproduce in many circumstances and the male function, unlike female, invests no more than the pollen grain in rearing the young. Thus even if selfing is equally disadvantageous to both male and female function in terms of zygote quality, the effect of a strict SI system on mating with relatives may provide conflicting selection pressures. Multigene and partial SI systems may contribute to resolving such conflict by permitting gradations of distinctions among potential mates. If the female function has other ways (such as abortion) of eliminating inferior zygotes when occasion demands it, selection on male function could favor production of pollen that functions on the stigmas of many genotypes.

I would have liked to provide an alternative hypothesis for the evolution of self-incompatibility, for the sake of argument if nothing else. Outcrossing has been an unchallenged standard rationale for so long that an alternative would have some heuristic value at least. However, none comes to mind, at least for complete SI. A major puzzle lies in determining the factors that necessitated and facilitated SI in certain species and not in others. Is SI favored chiefly when other mechanisms (sequential flowering, etc.) fail to reduce inbreeding sufficiently? Are the consequences of occasional inbreeding worse for some mainly outcrossing plants than for others? What allows strict outcrossers to sacrifice the flexibility that comes with a variable breeding system? Questions such as these really have not been approached analytically.

Heterostyly and Compatibility Systems

Heterostyly is widespread in the angiosperms, occurring in 24 families; the condition seems to be unusually common in the Rubiaceae. By far the commonest form is distyly with two floral morphs; tristyly, with three morphs, is confirmed in only eight genera in three families (Lythraceae, Oxalidaceae, Pontederiaceae; Ganders 1979). Almost all heterostylous plants are self-incompatible, although SI is only weakly developed in the Pontederiaceae and a few heterostyled species have apparently reverted to self-compatibility. Furthermore, the degree of SI sometimes differs between the morphs, but either the pin or the thrum form may have the stronger SI, depending on the species (Ganders 1979). The relative frequencies of pins and thrums sometimes differs markedly among years, sites, and species (Philipp and Schou 1981).

Because heterostylous flower morphology is commonly associated with sporophytic SI, the morphological feature is sometimes used to create a special category of self-incompatibility such that heteromorphic SI is contrasted with homomorphic SI (e.g., de Nettancourt 1977; Lewis 1949, 1954). However, there is good reason to view floral morphology as a separate matter (Beach and Kress 1980; Charlesworth and Charlesworth 1979a,b; Ganders 1979; Vuilleumier 1967). The advantages of sporophytic SI are derived (at least partly) from the ensurance of cross-fertilization and not from the prevention of sib-mating, because the proportion of the two mating types (only two alleles in distyly) is the same among sibs as it is in the population at large (Bateman 1952, Lewis 1979). When there are only two mating types in the population, cross-compatibility among individuals is much reduced, and incoming pollen has a high likelihood of being rejected. Even if the remaining pollen supplies are sufficient for full seed-set and female RS is unimpaired, male failures must be frequent. The floral dimorphism may have arisen, after SI was established, as a means of improving cross pollination between compatible types. Note that this explanation is predicated on male fitness-gain rather than outcrossing per se.

Testing the hypothesis of improved cross-pollination in a proper way is difficult and almost never done (Ganders 1979). The desired comparison is between the levels of disassortative mating (between unlike morphs) in heterostylous populations and in otherwise identical homostylous populations. Only some heterostylous species possess this range of variation. Another problem is that pin flowers usually produce more, and smaller, pollen grains than thrums and the morphs do not necessarily occur in equal frequencies in a population, so that the expected frequencies of various crosses must be calibrated accordingly. Furthermore, the extent of intraplant movement of pollen must be distinguished from the arrival of own-form pollen from other plants.

Despite these hurdles, there is some evidence of disassortative pollination (not necessarily fertilization!) in several species (Ganders 1979). Taking into account the greater production of pin pollen, thrum stigmas sometimes capture a greater-than-expected load of pin pollen, but sometimes the pollen load favors thrum pollen (self and nonself not distinguished) or is undistinguishable from a random distribution. In contrast, most pin stigmas capture an excess of pin pollen (again, self and nonself not distinguished) and no cases of greater-than-expected thrum pollen loads are recorded. In virtually every case, pin stigmas receive more pollen grains than thrum stigmas. The evidence indicates that pollen flow patterns are often nonrandom and asymmetrical: usually pin flowers are better pollen donors than thrums. At least in some, perhaps most, cases, seed production of the two morphs is equal (e.g., Ornduff 1980) but in others seed production by thrums is greater (Barrett 1980). Thus some heterostylous plants show a degree of sexual role differentiation, with thrums apparently functioning more as females, but it is noteworthy that when heterostyly is a precursor to dioecy, it is normally the pin flowers that assume the female role (Lloyd 1979b). The smaller size of pin pollen, long thought to be associated with the shorter styles of thrum plants on which such pollen is compatible, may need an alternative explanation inasmuch as there is no correlation between style length and pollen size in cross-species comparisons among heterostylous species (Ganders 1979; although in other cases, a correlation of pollen size and style length is significant; e.g., Plitmann and Levin in press). Perhaps selection for large numbers of pollen grains in pin plants has led to a concomitant reduction in pollen size.

LITERATURE CITED

Abrahamson, W. G. and B. J. Hershey. 1977. Resource allocation and growth of *Impatiens capensis* (Balsaminaceae) in two habitats. *Bull. Torrey Bot. Club* **104:** 160–164.

Abul Fatih, H. A. and F. A. Bazzaz. 1979. The biology of *Ambrosia trifida* L. III: Growth and biomass allocation. *New Phytol.* **83:** 829–838.

Alexander, R. D. and G. Borgia. 1978. On the origin and basis of the male–female phenomenon. *In* M. S. Blum and N. A. Blum (eds.), *Sexual Selection and Reproductive Competition in Insects.* Academic, New York.

Allard, R. W., S. K. Jain, and P. L. Workman. 1968. The genetics of inbreeding populations. *Adv. Genet.* **14:** 55–131.

Allen, C. E. 1940. The genotypic basis of sex-expression in angiosperms. *Bot. Rev.* **6:** 227–300.

Allen, C. E. 1945. The genetics of bryophytes, II: *Bot. Rev.* **11:** 260–287.

Altmann, J. 1980. *Baboon Mothers and Infants.* Harvard University Press, Cambridge, Mass.

Assouad, M. W., B. Dommée, R. Lumaret, and G. Valdeyron. 1978. Reproductive capacities in the sexual form of the gynodioecious species *Thymus vulgaris* L. *Bot. J. Linn. Soc.* **77:** 29–39.

Assouad, M. W. and G. Valdeyron. 1975. Remarques sur la biologie du thym *Thymus vulgaris* L. *Bull. Bot. Soc. Fr.* **122:** 21–34.

Baker, H. G. 1959. Reproductive methods as factors in speciation in flowering plants. *Cold Spring Harbor Symp. Quant. Biol.* **24:** 177–191.

Baker, H. G. 1965. Characteristics and modes of origin of weeds. *In* H. G. Baker and G. L. Stebbins (eds.), *The Genetics of Colonizing Species.* Academic, New York.

Baker, H. G. 1967. Support for Baker's law—as a rule. *Evolution* **21:** 853–856.

Barrett, S. C. 1980. Dimorphic incompatibility and gender in *Nymphoides indica* (Menyanthaceae). *Can. J. Bot.* **58:** 1938–1942.

Barrett, S. C. H. and K. Helenurm. 1981. Floral sex ratios and life history in *Aralia nudicaulis* (Araliaceae). *Evolution* **35:** 752–762.

Bateman, A. J. 1948. Intra-sexual selection in *Drosophila. Heredity* **2:** 349–368.

Bateman, A. J. 1952. Self-incompatibility systems in angiosperms, I: Theory. *Heredity* **6:**285–310.

Bawa, K. S. 1980. Evolution of dioecy in flowering plants. *ARES* **11:** 15–39.

Bawa, K. S. 1982. Outcrossing and the incidence of dioecism in island floras. *Am. Nat.* **119:** 866–871.

Bawa, K. S. and J. H. Beach. 1981. Evolution of sexual systems in flowering plants. *Ann. Missouri Bot. Gard.* **68:** 254–274.

Bawa, K. S. and P. A. Opler. 1977. Spatial relationships between staminate and pistillate plants of dioecious tropical forest trees. *Evolution* **31:** 64–68.

Bawa, K. S., C. R. Keegan, and R. H. Voss. 1982. Sexual dimorphism in *Aralia nudicaulis* L. (Araliaceae). *Evolution* **36:** 371–378.

Beach, J. H. 1981. Pollinator foraging and the evolution of dioecy. *Am. Nat.* **118:** 572–577.

Beach, J. H. and K. S. Bawa. 1980. Role of pollinators in the evolution of dioecy from heterostyly. *Evolution* **34:** 1138–1142.

Beach, J. H. and W. J. Kress. 1980. Sporophyte versus gametophyte: A note on the origin of self-incompatibility in flowering plants. *Syst. Bot.* **5:** 1–5.

Bell, C. R. 1971. Breeding systems and floral biology of the Umbelliferae or Evidence for specialization of unspecialized flowers. *Bot. J. Linn. Soc.* **61** (Suppl. 1): 93–108.

Bell, G. 1982. *The Masterpiece of Nature: The Evolution and Genetics of Sexuality.* University of California Press, Berkeley, Calif.

Bengtsson, B. O. 1978. Avoiding inbreeding: At what cost? *J. Theor. Biol.* **73:** 439–444.

Bernstein, H., G. S. Byers, and R. E. Michod. 1981. Evolution of sexual reproduction: Importance of DNA repair, complementation, and variation. *Am. Nat.* **117:** 537–549.

Bertin, R. I. 1982a. Paternity and fruit production in trumpet creeper (*Campsis radicans*). *Am. Nat.* **119:** 694–709.

Bertin, R. I. 1982b. The ecology and maintenance of andromonoecy. *Evol. Theory* **6:** 25–32.

Bertin, R. I. 1982c. The ecology of sex expression red buckeye. *Ecology* **63:** 445–456.

Bierzychudek, P. 1981. The demography of jack-in-the-pulpit, a forest perennial that changes sex. Ph.D. thesis (abstr.), Cornell University, Ithaca, N.Y.

Bonner, J. T. 1958. The relation of spore formation to recombination. *Am. Nat.* **92:** 193–200.

Breese, E. L. 1959. Selection for differing degrees of outbreeding in *Nicotiana rustica. Ann. Bot. N.S.* **23:** 331–334.

Bremermann, H. J. 1980. Sex and polymorphism as strategies in host-pathogen interactions. *J. Theor. Biol.* **87:** 671–702.

Brockman, I. and G. Bocquet. 1978. Ecological influences on the distribution of sexes in *Silene vulgaris* (Moench) Garcke (Caryophyllaceae). English summary. *Ber. Deutch. Botan. Gesell.* **91:**217–230.

Bull, J. J. 1981. Sex ratio evolution when fitness varies. *Heredity* **46:** 9–26.

Bullock, S. H. and K. S. Bawa. 1981. Sexual dimorphism and the annual flowering pattern in *Jacaratia dolichaula* (D. Smith) Woodson (Caricaceae) in a Costa Rican forest. *Ecology* **62:** 1494–1504.

Bulmer, M. G. and P. D. Taylor. 1980. Dispersal and the sex ratio. *Nature* **284:** 448–449.

Burley, N. 1981. Sex ratio manipulation and selection for attractiveness. *Science* **211:** 721–722.

Campbell, J. M. and R. J. Abbott. 1976. Variability of outcrossing frequency in *Senecio vulgaris* L. *Heredity* **36:** 267–274.

Cartier, J. (in press). Heterosis, selfing and outcrossing in the highly self-fertile *Corydalis sempervirens* (L.) Pers. (Fumariaceae). *Can. J. Bot.*

Charlesworth, B. 1980a. The cost of meiosis with alternation of sexual and asexual generations. *J. Theor. Biol.* **87:** 517–528.

Charlesworth, B. 1980b. The cost of sex in relation to mating system. *J. Theor. Biol.* **84:** 655–671.

Charlesworth, B. and D. Charlesworth. 1978a. A model for the evolution of dioecy and gynodioecy. *Am. Nat.* **112:** 975–997.

Charlesworth, B. and D. Charlesworth. 1979b. The maintenance and breakdown of distyly. *Am. Nat.* **114:** 499–513.

Charlesworth, D. 1981. A further study of the problem of the maintenance of females in gynodioecious species. *Heredity* **46:** 27–39.

Charlesworth, D. and B. Charlesworth. 1978b. Population genetics of partial male-sterility and the evolution of monoecy and dioecy. *Heredity* **41:** 137–153.

Charlesworth, D. and B. Charlesworth. 1979a. A model for the evolution of distyly. *Am. Nat.* **114:** 467–498.

Charlesworth, D. and B. Charlesworth. 1979c. The evolution and breakdown of S-allele systems. *Heredity* **43:** 41–55.

Charlesworth, D. and B. Charlesworth. 1979d. The evolutionary genetics of sexual systems in flowering plants. *Proc. R. Soc. Lond.* **B205:** 513–530.

Charlesworth, D. and B. Charlesworth. 1981. Allocation of resources to male and female functions in hermaphrodites. *Biol. J. Linn. Soc.* **15:** 57–74.

Charlesworth, D. and F. R. Ganders. 1979. The population genetics of gynodioecy with cytoplasmic male-sterility. *Heredity* **43:** 213–218.

Charnov, E. L. 1979. Simultaneous hermaphroditism and sexual selection. *PNAS(USA)* **76:** 2480–2484.

Charnov, E. L. and J. J. Bull. 1977. When is sex environmentally determined? *Nature* **266:** 828–830.

Charnov, E. L., J. Maynard Smith, and J. J. Bull. 1976. Why be an hermaphrodite? *Nature* **263:** 125–126.

Charnov, E. L., R. L. Los-den Hartogh, W. T. Jones, and J. van den Assem. 1981. Sex ratio evolution in a variable environment. *Nature* **289:** 27–33.

Clark, R. B. and E. R. Orton. 1967. Sex ratio in *Ilex opaca* Ait. Hort. *Science* **2**(3): 115.

Clay, K. 1982. Environmental and genetic determinants of cleistogamy in a natural population of the grass *Danthonia spicata*. *Evolution* **36:** 734–741.

Clay, K. and N. C. Ellstrand. 1981. Stylar polymorphism in *Epigaea repens*, a dioecious species. *Bull. Torrey Bot. Club* **108:** 305–310.

Colwell, R. K. 1981. Group selection is implicated in the evolution of female-biased sex ratios. *Nature* **290:** 401–404.

Conn, J. S. 1981. Phenological differentiation between the sexes of *Rumex hastatulus*: Niche partitioning or different optimal reproductive strategies? *Bull. Torrey Bot. Club* **108:** 374–378.

Conn, J. S. and U. Blum. 1981a. Sex ratio of *Rumex hastatulus*: The effect of environmental factors and certation. *Evolution* **35:** 1108–1116.

Conn, J. S. and U. Blum. 1981b. Differentiation between the sexes of *Rumex hastatulus* in net energy allocation, flowering and height. *Bull. Torrey Bot. Club* **108:** 446–455.

Conn, J. S., T. R. Wentworth, and U. Blum. 1980. Patterns of dioecism in the flora of the Carolinas. *Am. Midl. Nat.* **103:** 310–315.

Connor, H. E. 1973. Breeding systems in *Cortaderia*. *Evolution* **27:** 663–678.

Connor, H. E. 1979. Breeding systems in the grasses: A survey. *New Zeal. J. Bot.* **17:** 547–574.

Correns, C. 1928. Bestimmung, Vererbung und Verteilung des Geschlechtes bei den höheren Pflanzen. *Handbuch der Vererbungswissenschaften* **2:** 1–138.

Cosmides, L. M. and J. Tooby. 1981. Cytoplasmic inheritance and intragenomic conflict. *J. Theor. Biol.* **89:** 83–129.

Coulter, J. M. 1914. *Evolution of Sex in Plants.* University of Chicago Press, Chicago.

Cox, P. A. 1981. Niche partitioning between sexes of dioecious plants. *Am. Nat.* **117:** 295–307.

Cox, P. A. 1982. Vertebrate pollination and the maintenance of dioecy in *Freycinetia*. *Am. Nat.* **120:** 65–80.

Cruden, R. W. 1977. Pollen-ovule ratios: A conservative indicator of breeding systems in flowering plants. *Evolution* **31:** 32–46.

Cruden, R. W. and S. Miller-Ward. 1981. Pollen-ovule ratio, pollen size, and the ratio of stigmatic area to the pollen-bearing areas of the pollinator: An hypothesis. *Evolution* **35:** 964–974.

Daly, M. 1978. The cost of mating. *Am. Nat.* **112:** 771–774.

Darwin, C. 1871. *The Descent of Man and Selection in Relation to Sex.* Appleton, New York.

Darwin, C. 1877. *The Effects of Cross and Self Fertilization in the Vegetable Kingdom.* Appleton, New York.

Darwin, C. 1884. *The Different Forms of Flowers on Plants of the Same Species.* Reprint of 2nd ed., 1880. Appleton, New York.

deJong, G. 1980. Some numerical aspects of sexuality. *Am. Nat.* **116:** 712–718.

de Nettancourt, D. 1977. *Incompatibility in Angiosperms.* Springer-Verlag, New York.

Dodson, C. H. 1962. Pollination and variation in the subtribe Catasetinae (Orchidaceae). *Ann. Missouri Bot. Gard.* **49:** 35–56.

Dodson, C. H. and G. P. Frymire. 1961. Natural pollination of orchids. *Missouri Bot. Gard. Bull.* **49:** 133–152.

Dommée, G., M. W. Assouad, and G. Valdeyron. 1978. Natural selection and gynodioecy in *Thymus vulgaris* L. *Bot. J. Linn. Soc.* **77:** 17–28.

Drayner, J. K. 1959. Self- and cross-fertility in field beans (*Vicia faba* Linn.). *J. Agric. Sci.* **53:** 387–403.

Dzhaparidze, L. I. 1963. [*Sex in Plants.*] Israel Program for Scientific Translations (1967), Jerusalem.

Ekbohm, G., T. Fagerstrom, and G. I. Agren. 1980. Natural selection for variation in offspring numbers: Comments on a paper by J. H. Gillespie. *Am. Nat.* **115:** 445–447.

Elkington, T. T. and S. R. J. Woodell. 1963. *Potentilla fruticosa* L. *J. Ecol.* **51:** 769–781.

Emlen, J. M. 1968a. A note on natural selection and the sex ratio. *Am. Nat.* **102:** 94–95.

Emlen, J. M. 1968b. Selection for the sex ratio. *Am. Nat.* **102:** 589–591.

Emlen, J. M. 1973. *Ecology: An Evolutionary Approach.* Addison-Wesley, Reading, Mass.

Farmer, R. E. 1964. Sex ratio and sex-related characteristics in eastern cottonwood. *Silv. Genet.* **13:** 116–118.

Fisher, R. A. 1929. *The Genetical Theory of Natural Selection* (2nd ed.). Dover reprint (1958), New York.

Flores, S. and D. W. Schemske. (unpublished). Dioecy and monoecy in the flora of Puerto Rico and the Virgin Islands: Ecological correlates.

Frankel, R. and E. Galun. 1977. *Pollination Mechanisms, Reproduction, and Plant Breeding.* Springer-Verlag, New York.

Freeman, D. C., K. T. Harper, and W. K. Ostler. 1980. Ecology of plant dioecy in the intermountain region of western North America and California. *Oecologia* **44:** 410–417.

Freeman, D. C., L. G. Klikoff, and K. T. Harper. 1976. Differential resource utilization by the sexes of dioecious plants. *Science* **193:** 597–599.

Freeman, D. C., E. D. McArthur, K. T. Harper, and A. C. Blauer. 1981. Influence of environment on the floral sex ratio of monoecious plants. *Evolution* **35:** 194–197.

Ganders, F. R. 1978. The genetics and evolution of gynodioecy in *Nemophila menziesii* (Hydrophyllaceae). *Can. J. Bot.* **56:** 1400–1408.

Ganders, F. R. 1979. The biology of heterostyly. *New Zeal. J. Bot.* **17:** 607–635.

Gerritsen, J. 1980. Sex and parthenogenesis in sparse populations. *Am. Nat.* **115:** 718–742.

Ghiselin, M. T. 1969. The evolution of hermaphroditism among animals. *Q. Rev. Biol.* **44:** 189–208.

Ghiselin, M. T. 1974. *The Economy of Nature and the Evolution of Sex.* University of California Press, Berkeley.

Gilbert, L. E. 1975. Ecological consequences of a coevolved mutualism between butterflies and plants. *In* L. E. Gilbert and P. H. Raven (eds.), *Coevolution of Animals and Plants.* University of Texas Press, Austin.

Gillespie, J. H. 1977. Natural selection for variances in offspring numbers: A new evolutionary principle. *Am. Nat.* **111:** 1010–1014.

Givnish, T. H. 1980. Ecological constraints on the evolution of breeding systems in seed plants: Dioecy and dispersal in gymnosperms. *Evolution* **34:** 959–972.

Givnish, T. H. 1982. Outcrossing versus ecological constraints in the evolution of dioecy. *Am. Nat.* **119:** 849–865.

Glesener, R. R. 1979. Recombination in a simulated predator–prey interaction. *Am. Zool.* **19:** 763–771.

Glesener, R. R. and D. Tilman. 1978. Sexuality and the components of environmental uncertainty: Clues from geographic parthenogenesis in terrestrial animals. *Am. Nat.* **112:** 659–673.

Godley, E. J. 1964. Breeding systems in New Zealand plants, 3: Sex ratios in some natural populations. *New Zeal. J. Bot.* **2:** 205–212.

Godley, E. J. 1976. Sex ratios in *Clematis gentianoides* D.C. *New Zeal. J. Bot.* **14:** 299–306.

Grant, M. C. and J. B. Mitton. 1979. Elevational gradients in adult sex ratios and sexual differentiation in vegetative growth rates of *Populus tremuloides* Michx. *Evolution* **33:** 914–918.

Grant, V. 1975. *Genetics of Flowering Plants.* Columbia University Press, New York.

Greenwood, P. J. 1980. Mating systems, philopatry, and dispersal in birds and mammals. *Anim. Behav.* **28:** 1140–1162.

Gregg, K. B. 1975. The effect of light intensity on sex expression in species of *Cycnoches* and *Catasetum* (Orchidaceae). *Selbyana* **1:** 101–113.

Grewal, M. S. and J. R. Ellis. 1972. Sex determination in *Potentilla fruticosa. Heredity* **29:** 359–362.

Gross, K. L. and J. D. Soule. 1981. Differences in biomass allocation to reproductive and vegetative structures of male and female plants of a dioecious, perennial herb, *Silene alba* (Miller) Krause. *Am. J. Bot.* **68:** 801–807.

Hamilton, W. D. 1967. Extraordinary sex ratios. *Science* **156:** 477–488.

Hamilton, W. D. 1980. Sex versus non-sex versus parasite. *Oikos* **35:** 282–290.

Hamilton, W. D., P. A. Henderson, and N. A. Moran. 1981. Fluctuation of environment and coevolved antagonist polymorphism as factors in the maintenance of sex. *In* R. D. Alexander and D. W. Tinkle (eds.), *Natural Selection and Social Behavior*, Chiron, New York.

Hancock, J. F. and R. S. Bringhurst. 1980. Sexual dimorphism in the strawberry *Fragaria chiloensis. Evolution* **34:** 762–768.

Harding, J. and K. Barnes. 1977. Genetics of *Lupinus*, X: Genetic variability, heterozygosity and outcrossing in colonial populations of *Lupinus succulentus. Evolution* **31:** 247–255.

Harding, J. and C. L. Tucker. 1964. Quantitative studies on mating systems, I: Evidence for the non-randomness of outcrossing in *Phaseolus lunatus. Heredity* **19:** 369–381.

Harding, J. and C. L. Tucker. 1969. Quantitative studies on mating systems, III: Methods for the estimation of male gametophytic selective values and differential outcrossing. *Evolution* **23:** 85–95.

Harpending, H. C. 1979. The population genetics of interactions. *Am. Nat.* **113:** 622–630.

Harris, W. 1968. Environmental effects on the sex ratio of *Rumex acetosella* L. *Proc. New Zeal. Ecol. Soc.* **15:** 51–54.

Hartley, C. W. S. 1970. Some environmental factors affecting flowering and fruiting in the oil palm. *In* L. C. Luckwill and C. V. Cutting (eds.), *Physiology of Tree Crops.* Academic, New York.

Hartung, J. 1981. Genome parliaments and sex with the Red Queen. *In* R. D. Alexander and D. W. Tinkle (eds.), *Natural Selection and Social Behavior.* Chiron, New York.

Heath, D. J. 1977. Simultaneous hermaphroditism; cost and benefit. *J. Theor. Biol.* **64:** 363–373.

Heath, D. J. 1979. Brooding and the evolution of hermaphroditism. *J. Theor. Biol.* **81:** 151–155.

Heslop-Harrison, J. 1957. The experimental modification of sex expression in flowering plants. *Biol. Rev.* (Cambridge) **32:** 38–90.

Heslop-Harrison, J. 1972. Sexuality of angiosperms. *In* F. C. Steward (ed.), *Plant Physiology—A Treatise;* Vol. 6C. Academic, New York.

Horovitz, A. 1978. Is the hermaphroditic flowering plant equisexual? *Am. J. Bot.* **65:** 485–486.

Horovitz, A. and A. Beiles. 1980. Gynodioecy as a possible populational strategy for increasing reproductive output. *Theor. Appl. Genet.* **57:** 11–15.

Horovitz, A. and J. Harding. 1972a. Genetics of *Lupinus*, 5: Intraspecific variability for reproductive traits in *Lupinus nanus*. *Bot. Gaz.* **133:** 155–165.

Horovitz, A. and J. Harding. 1972b. The concept of male outcrossing in hermaphrodite higher plants. *Heredity* **29**(2): 223–236.

Howe, H. F. and D. De Steven. 1979. Fruit production, migrant bird visitation, and seed dispersal of *Gaurea glabra* in Panama. *Oecologia* **38:** 185–196.

Howe, H. F. and G. A. vande Kerckhove. 1979. Fecundity and seed dispersal of a tropical tree. *Ecology* **60:** 180–189.

Jaenike, J. 1978. An hypothesis to account for the maintenance of sex within populations. *Evol. Theory* **3:** 191–194.

Jain, S. K. 1968. Gynodioecy in *Origanum vulgare*: Computer simulation of a model. *Nature* **217:** 764–765.

Janzen, D. H. 1971. Seed predation by animals. *ARES* **2:** 465–492.

Janzen, D. H. 1975. Behavior of *Hymenaea courbaril* when its predispersal seed predator is absent. *Science* **189:** 145–147.

Janzen, D. H. 1977. A note on optimal mate selection by plants. *Am. Nat.* **111:** 365–371.

Janzen, D. H. 1981. Differential visitation of *Catasetum* orchid male and female flowers. *Biotropica* **13** (Suppl.): 77.

Jones, D. F. 1928. *Selective Fertilization.* University of Chicago Press, Chicago.

Kaplan, S. M. 1972. Seed production and sex ratio in anemophilous plants. *Heredity* **28:** 281–285.

Kaul, R. B. 1979. Inflorescence architecture and flower sex ratio in *Sagittaria brevirostra* (Alismataceae). *Am. J. Bot.* **66:** 1062–1066.

Kaur, A., C. O. Ha, K. Jong, V. E. Sands, H. T. Chan, E. Soepadmo, and P. S. Ashton. 1978. Apomixis may be widespread among trees of the climax rain forest. *Nature* **271:** 440–442.

Koller, D. and N. Roth. 1964. Studies on the ecological and physiological significance of amphicarpy in *Gymnarrhena micrantha* (Compositae). *Am. J. Bot.* **51:** 26–35.

Krohne, D. T., I. Baker, and H. G. Baker. 1980. The maintenance of the gynodioecious breeding system in *Plantago lanceolata* L. *Am. Midl. Nat.* **103:** 269–279.

Lamb, R. Y. and R. B. Willey. 1979. Are parthenogenetic and related bisexual insects equal in fertility? *Evolution* **33:** 774–775.

Leigh, E. G., E. L. Charnov, and R. R. Warner. 1976. Sex ratio, sex change, and natural selection. *PNAS(USA)* **73:** 3656–3660.

Lester, D. L. 1963. Variation in sex expression in *Populus tremuloides* Michx. *Silv. Genet.* **12:** 141–151.

Levin, D. A. 1972. Plant density, cleistogamy, and self-fertilization in natural populations of *Lithospermum carolinense. Am. J. Bot.* **59:** 71–77.

Levin, D. A. 1975. Pest pressure and recombination systems in plants. *Am. Nat.* **109:** 437–452.

Levin, D. A. 1976. Consequences of long-term artificial selection, inbreeding and isolation in *Phlox*, I: The evolution of cross-incompatibility. *Evolution* **30:** 335–344.

Levin, D. A. 1981. Dispersal versus gene flow in plants. *Ann. Missouri Bot. Gard.* **68:** 233–253.

Levin, D. A. and H. W. Kerster. 1973. Assortative pollination for stature in *Lythrum salicaria. Evolution* **27:** 144–152.

Levin, D. A. and H. W. Kerster. 1974. Gene flow in seed plants. *Evol. Biol.* **7:** 139–220.

Lewis, D. 1941. Male sterility in natural populations of hermaphroditic plants. *New Phytol.* **40:** 56–63.

Lewis, D. 1942. The evolution of sex in flowering plants. *Biol. Rev.* **17:** 46–67.

Lewis, D. 1949. Incompatibility in flowering plants. *Biol. Rev.* **24:** 472–496.

Lewis, D. 1954. Comparative incompatibility in angiosperms and fungi. *Adv. Genet.* **6:** 235–285.

Lewis, D. 1979. Genetic versatility of incompatibility in plants. *New Zeal. J. Bot.* **17:** 637–644.

Lewis, D. and L. K. Crowe. 1956. The genetics and evolution of gynodioecy. *Evolution* **10:** 115–125.

Lindsey, A. H. 1982. Floral phenology patterns and breeding systems in *Thaspium* and *Zizia* (Apiaceae). *Syst. Bot.* **7:** 1–12.

Lloyd, D. G. 1972. Breeding systems in *Cotula* L. (Compositae, Anthemidae), II: Monoecious populations. *New Phytol.* **71:** 1195–1202.

Lloyd, D. G. 1973. Sex ratios in sexually dimorphic Umbelliferae. *Heredity* **31:** 239–249.

Lloyd, D. G. 1974a. Theoretical sex ratios of dioecious and gynodioecious angiosperms. *Heredity* **32:** 11–34.

Lloyd, D. G. 1974b. Female-predominant sex ratios in angiosperms. *Heredity* **32:** 34–44.

Lloyd, D. G. 1974c. The genetic contributions of individual males and females in dioecious and gynodioecious angiosperms. *Heredity* **32:** 45–51.

Lloyd, D. G. 1975. The maintenance of gynodioecy and androdioecy in angiosperms. *Genetica* **45:** 325–339.

Lloyd, D. G. 1976. The transmission of genes via pollen and ovules in gynodioecious angiosperms. *Theor. Pop. Biol.* **9:** 299–316.

Lloyd, D. G. 1979a. Parental strategies of angiosperms. *New Zeal. J. Bot.* **17:** 595–606.

Lloyd, D. G. 1979b. Evolution toward dioecy in heterostylous populations. *Plant. Syst. Evol.* **131:** 71–80.

Lloyd, D. G. 1979c. Some reproductive factors affecting the selection of self-fertilization in plants. *Am. Nat.* **113:** 67–79.

Lloyd, D. G. 1980a. The distributions of gender in four angiosperm species illustrating two evolutionary pathways to dioecy. *Evolution* **34:** 123–134.

Lloyd, D. G. 1980b. Benefits and handicaps of sexual reproduction. *Evol. Biol.* **13:** 69–111.

Lloyd, D. G. 1980c. Demographic factors and mating patterns in angiosperms. *Bot. Monogr.* **15:** 67–88.

Lloyd, D. G. 1980d. Alternative formulations of the intrinsic cost of sex. *New Zeal. Genet. Soc. Newsl.* No. 6.

Lloyd, D. G. 1981. The distribution of sex in *Myrica gale*. *Plant Syst. Evol.* **138:** 29–45.

Lloyd, D. G. 1982. Selection of combined versus separate sexes in seed plants. *Am. Nat.* **120:** 571–585.

Lloyd, D. G. and C. J. Webb. 1977. Secondary sex characters in seed plants. *Bot. Rev.* **43:** 177–216.

Long, R. W. 1971. Floral polymorphy and amphimictic breeding systems in *Ruellia caroliniensis* (Acanthaceae). *Am. J. Bot.* **58:** 525–531.

Lovett Doust, J. 1980. Floral sex ratios in andromonoecious Umbelliferae. *New Phytol.* **85:** 265–273.

Lynch, M. (unpublished). Genomic incompatibility, general-purpose genotypes, and geographical parthenogenesis.

Manning, J. T. 1975. Gamete dimorphism and the cost of sexual reproduction: Are they separate phenomena? *J. Theor. Biol.* **55:** 393–395.

Manning, J. T. and J. Jenkins. 1980. The "balance" argument and the evolution of sex. *J. Theor. Biol.* **86:** 593–601.

Martin, F. W. 1966. Sex ratios and sex determination in *Dioscorea*. *J. Heredity* **57:** 95–99.

Maynard Smith, J. 1978. *The Evolution of Sex.* Cambridge University Press, Cambridge.

Maynard Smith, J. 1980. A new theory of sexual investment. *Behav. Ecol. Sociobiol.* **7:** 247–251.

McArthur, E. G. 1977. Environmentally induced changes of sex expression in *Atriplex canescens*. *Heredity* **38:** 97–103.

McKenna, M. A. and D. L. Mulcahy. (unpublished). Gametophytic competition in *Dianthus chinensis*: Effect on sporophytic competitive ability.

McNamara, J. and J. A. Quinn. 1977. Resource allocation and reproduction in populations of *Amphicarpum purshii* (Gramineae). *Am. J. Bot.* **64:** 17–23.

Meagher, T. R. 1980. Population biology of *Chamaelirium luteum*, a dioecious lily, I: Spatial distributions of males and females. *Evolution* **34:** 1127–1137.

Meagher, T. R. 1981. Population biology of *Chamaelirium luteum*, a dioecious lily, II: Mechanisms governing sex ratios. *Evolution* **35:** 557–567.

Melampy, M. N. 1981. Sex-linked niche differentiation in two species of *Thalictrum*. *Am. Midl. Nat.* **106:** 325–334.

Melampy, M. N. and H. F. Howe. 1977. Sex ratio in the tropical tree *Triplaris americana* (Polygonaceae). *Evolution* **31:** 867–872.

Moore, L. A. and M. F. Willson. 1982. The effect of microhabitat, spatial distribution, and display size on dispersal of *Lindera* by avian frugivores. *Can. J. Bot.* **60:** 557–560.

Moore, W. W. and W. G. S. Hines. 1981. Sex in a random environment. *J. Theor. Biol.* **92:** 301–316.

Muenchow, G. 1978. A note on the timing of sex in asexual/sexual organisms. *Am. Nat.* **112:** 774–779.

Mulcahy, D. L. 1967. Optimal sex ratio in *Silene alba*. *Heredity* **22:** 411–423.

Mulcahy, D. L. and G. B. Mulcahy. 1975. The influence of gametophytic competition on sporophytic quality in Dianthus chinensis. *Theor. Appl. Genet.* **46:** 277–280.

Myers, J. H. 1978. Sex ratio adjustment under food stress: Maximization of quality or numbers of offspring? *Am. Nat.* **112:** 381–388.

Nyberg, D. 1982. Sex, recombination, and reproductive fitness: An experimental study using *Paramecium*. *Am. Nat.* **120:** 198–217.

Onyekwelu, S. S. and J. L. Harper. 1979. Sex ratio and niche differentiation in spinach (*Spinacia oleracea* L.). *Nature* **282:** 609–611.

Opler, P. A. and K. S. Bawa. 1978. Sex ratios in tropical forest trees. *Evolution* **32:** 812–821.

Ornduff, R. 1980. Heterostyly, population composition, and pollen flow in *Hedyotis caerulea*. *Am. J. Bot.* **67:** 95–103.

Pandey, K. K. 1960. Evolution of gametophytic and sporophytic systems of self-incompatibility in angiosperms. *Evolution* **14:** 98–115.

Parker, G. A. 1978. Selection on non-random fusion of gametes during the evolution of anisogamy. *J. Theor. Biol.* **73:** 1–28.

Parker, G. A., R. R. Baker, and V. G. F. Smith. 1972. The origin and evolution of gamete dimorphism and the male–female phenomenon. *J. Theor. Biol.* **36:** 529–553.

Partridge, L. 1980. Mate choice increases a component of offspring fitness in fruit flies. *Nature* **283:** 290–291.

Philipp, M. 1980. Reproductive biology of *Stellaria longipes* Goldie as revealed by a cultivation experiment. *New Phytol.* **85:** 557–569.

Philipp, M. and O. Schou. 1981. An unusual heteromorphic incompatibility system: Distyly, self-incompatibility, pollen load and fecundity in *Anchusa officinalis* (Boraginaceae). *New Phytol.* **89:** 693–703.

Phillips, J. 1926. Biology of the flowers, fruits and young generation of *Olinia cymosa* Thunb. ("hard peas"). *Ecology* **7:** 338–350.

Pitman, U. and D. A. Levin. (in press). Pollen-pistil relationships in the Polemoniaceae. *Evolution.*

Policansky, D. 1981. Sex choice and the size advantage model in jack-in-the-pulpit. *PNAS (USA)* **78:** 1306–1308.

Price, M. V. and N. M. Waser. 1982. Population structure, frequency, dependent selection, and the maintenance of sexual reproduction. *Evolution* **36:** 35–43.

Primack, R. B. and D. G. Lloyd. 1980a. Andromonoecy in the New Zealand montane shrub manuka, *Leptospermum scoparium* (Myrtaceae). *Am. J. Bot.* **67:** 361–368.

Primack, R. B. and D. G. Lloyd. 1980b. Sexual strategies in plants, IV: The distributions of gender in two monomorphic shrub populations. *New Zeal. J. Bot.* **18:** 109–114.

Purseglove, J. W. 1968. *Tropical Crops: Dicotyledons*, Vol. 2. Longmans, London.

Putwain, P. D. and J. L. Harper. 1972. Studies on the dynamics of plant populations, V: Mechanisms governing the sex ratio in *Rumex acetosa* and *R. acetosella*. *J. Ecol.* **60:** 113–128.

Rathore, J. S. 1969. Distribution patterns of male and female plants of *Diospyros melanoxylon* Roxb. in the forests of Sagar, M. P. *Indian For.* **95:** 701.

Rollins, R. C. 1967. The evolutionary fate of inbreeders and nonsexuals. *Am. Nat.* **101:** 343–351.

Ross, M. D. 1973. Inheritance of self-incompatibility in *Plantago lanceolata*. *Heredity* **30:** 169–176.

Ross, M. D. 1982. Five evolutionary pathways to subdioecy. *Am. Nat.* **119:** 297–318.

Rust, R. W. 1977. Pollination in *Impatiens capensis* and *Impatiens pallida* (Balsaminaceae). *Bull. Torrey Bot. Club* **104:** 361–367.

Rychlewski, J. and K. Zarzycki. 1975. Sex ratio in seeds of *Rumex acetosa* L. as a result of sparse or abundant pollination. *Acta Biol. Cracov. (Bot.)* **18:** 101–114.

Schaaf, H. M. and R. R. Hill. 1979. Cross-fertility differentials in birdsfoot trefoil. *Crop Sci.* **19:** 451–454.

Schemske, D. W. 1978. Evolution of reproductive characteristics in *Impatiens* (Balsaminaceae): The significance of cleistogamy and chasmogamy. *Ecology* **59:** 596–613.

Schemske, D. W. (in press). Limits to specialization and coevolution in plant-animal mutualisms. *In* M. Nitecki (ed.), *Coevolution*. University of Chicago Press, Chicago.

Smith, B. W. 1963. The mechanism of sex determination in *Rumex hastatulus*. *Genetics* **48:** 1265–1288.

Smith, B. W. 1968. Cytogeography and cytotaxonomic relationship of *Rumex paucifolius*. *Am. J. Bot.* **55:** 673–683.

Smith, C. C. 1981. The facultative adjustment of sex ratio in lodgepole pine. *Am. Nat.* **118:** 297–305.

Smith, R. H. 1979. On selection for inbreeding in polygynous animals. *Heredity* **43:** 205–211.

Sobrevila, C. and M. T. K. Arroyo. 1982. Breeding systems in a montane cloud forest in Venezuela. *Plant Syst. Evol.* **140:** 19–37.

Soule, J. D. 1981. Ecological consequences of dioecism in plants: A case study of sex differences, sex ratios and population dynamics of *Valeriana edulis* Nutt. Ph.D. Thesis, Michigan State University.

Sparnaaij, L. D., Y. O. Kho, and J. Baër. 1968. Investigations on seed production in tetraploid freesias. *Euphytica* **17:** 289–297.

Stanley, S. M. 1975. Clades versus clones in evolution: Why we have sex. *Science* **190:** 382–384.

Stanton, D. S. 1981. Soft selection and the evolution of sex. Unpubl. M.S. Thesis, University of Illinois.

Stephenson, A. G. 1979. An evolutionary examination of the floral display of *Catalpa speciosa* (Bignoniaceae). *Evolution* **33:** 1200–1209.

Sterk, A. A. 1969a. Biosystematic studies of *Spergularia media* and *S. marina* in the Netherlands, I: The morphological variability of *S. media*. *Acta Bot. Neerl.* **18:** 325–338.

Sterk, A. A. 1969b. Biosystematic studies of *Spergularia media* and *S. marina* in the Netherlands, II: The morphological variability of *S. marina*. *Acta Bot. Neerl.* **18:** 467–476.

Styles, B. T. 1972. The flower biology of the Meliaceae and its bearing on tree breeding. *Silv. Genet.* **21:** 175–182.

Sutherland, S. and L. Delph. Unpublished manuscript. On the importance of male fitness in plants. I. Patterns in fruit set.

Symon, D. E. 1979. Sex forms in *Solanum* (Solanaceae) and the role of pollen collecting insects. *Linn. Soc. Symp. Ser.* **7:** 385–397.

Taylor, P. D. and M. G. Bulmer. 1980. Local mate competition and the sex ratio. *J. Theor. Biol.* **86:** 409–419.

Taylor, P. D. and A. Sauer. 1980. The selective advantage of sex ratio homeostasis. *Am. Nat.* **116:** 305–310.

Templeton, A. R. 1982. The prophecies of parthenogenesis. *In* H. Dingle and J. P. Hegmann (eds.), *Evolution and Genetics of Life Histories.* Springer-Verlag, New York.

Thomson, J. D. and S. C. M. Barrett. 1981a. Selection for outcrossing, sexual selection, and the evolution of dioecy in plants. *Am. Nat.* **118:** 443–449.

Thomson, J. D. and S. C. M. Barrett. 1981b. Temporal variation of gender in *Aralia hispida* Vent. (Araliaceae). *Evolution* **35:** 1094–1107.

Treisman, M. and R. Dawkins. 1976. The "cost of meiosis": Is there any? *J. Theor. Biol.* **63:** 479–484.

Trivers, R. L. 1972. Parental investment and sexual selection. *In* B. Campbell (ed.), *Sexual Selection and the Descent of Man.* Aldine, Chicago.

Trivers, R. L. and D. E. Willard. 1973. Natural selection of parental ability to vary the sex ratio of offspring. *Science* **179:** 90–91.

Uphof, J. C. T. 1938. Cleistogamic flowers. *Bot. Rev.* **4:** 21–49.

Usberti, J. A. and S. K. Jain. 1978. Variation in *Panicum maximum*: A comparison of sexual and asexual populations. *Bot. Gaz.* **139:** 112–116.

Vaarama, A. and O. Jääskeläinen. 1967. Studies on gynodioecism in the Finnish populations of *Geranium sylvaticum* L. *Ann. Acad. Sci. Fenn., Ser. A. IV Biol.* **108:** 3–39.

Valdeyron, G., B. Dommée, and P. Vernet. 1977. Self-fertilization in male-fertile plants of a gynodioecious species: *Thymus vulgaris* L. *Heredity* **39:** 243–249.

Valentine, F. A. 1975. Genetic control of sex ratio, earliness and frequency of flowering in *Populus tremuloides. NE For. Tree Improv. Conf.* **22:** 111–129.

van den Ende, H. 1976. *Sexual Interactions in Plants.* Academic, New York.

van Leeuwen, B. H. 1981. The role of pollination in the population biology of the monocarpic species *Cirsium palustre* and *Cirsium vulgare. Oecologia* **51:** 28–32.

Van Valen, L. 1973. A new evolutionary law. *Evol. Theory* **1:** 1–30.

Verner, J. 1965. Selection for sex ratio. *Am. Nat.* **99:** 419–421.

Vuilleumier, B. S. 1967. The origin and evolutionary development of heterostyly in the angiosperms. *Evolution* **21:** 210–226.

Wade, M. J. 1979. Sexual selection and variance in reproductive success. *Am. Nat.* **114:** 747–764.

Wallace, C. S. and P. W. Rundel. 1979. Sexual dimorphism and resource allocation in male and female shrubs of *Simmondsia chinensis. Oecologia* **44:** 34–39.

Waller, D. M. 1979. The relative costs of self- and cross-fertilized seeds in *Impatiens capensis* (Balsaminaceae). *Am. J. Bot.* **66:** 313–320.

Waller, D. M. 1980. Environmental determinants of outcrossing in *Impatiens capensis* (Balsaminaceae). *Evolution* **34:** 747–761.

Warner, R. R. 1975. The adaptive significances of sequential hermaphroditism in animals. *Am. Nat.* **109:** 61–82.

Warner, R. R., D. R. Robertson, and E. G. Leigh. 1975. Sex change and sexual selection. *Science* **190:** 633–638.

Webb, C. J. 1979a. Breeding system and seed set in *Euonymus europaeus* (Celastraceae). *Plant Syst. Evol.* **132:** 299–303.

Webb, C. J. 1979b. Breeding systems and the evolution of dioecy in New Zealand apioid Umbelliferae. *Evolution* **33:** 662–672.

Webb, C. J. 1981. Test of a model predicting equilibrium frequencies of females in populations of gynodioecious angiosperms. *Heredity* **46:** 397–405.

Webb, C. J. and D. G. Lloyd. 1980. Sex ratios in New Zealand apioid Umbelliferae. *New Zeal. J. Bot.* **18:** 121–126.

Weiss, P. W. 1980. Germination, reproduction and interference in the amphicarpic annual *Emex spinosa* (L.) Campd. *Oecologia* **45:** 244–251.

Weller, S. G. and R. Ornduff. 1977. Cryptic self-incompatibility in *Amsinckia grandiflora. Evolution* **31:** 47–51.

Wells, H. 1979. Self-fertilization: Advantageous or deleterious? *Evolution* **33:** 252–255.

Werren, J. H. 1980. Sex ratio adaptations to local mate competition in a parasitic wasp. *Science* **208:** 1157–1159.

Werren, J. H. and E. L. Charnov. 1978. Facultative sex ratios and population dynamics. *Nature* **272:** 349–350.

Westergaard, M. 1958. The mechanism of sex determination in dioecious flowering plants. *Adv. Genet.* **9:** 217–281.

Whitehouse, H. L. K. 1959. Cross- and self-fertilization in plants. *In* P. R. Bell (ed.), *Darwin's Biological Work.* Cambridge University Press, Cambridge.

Whitham, T. G. and C. N. Slobodchikoff. 1981. Evolution by individuals, plant-herbivore interactions, and mosaics of genetic variability: The adaptive significance of somatic mutations in plants. *Oecologia* **49:** 287–292.

Wiens, D. and B. A. Barlow. 1975. Permanent translocation heterozygosity and sex determination in East African mistletoes. *Science* **187:** 1208–1209.

Williams, G. C. 1975. *Sex and Evolution.* Princeton University Press, Princeton, N.J.

Williams, G. C. 1979. The question of adaptive sex ratio in outcrossed vertebrates. *Proc. R. Soc. Lond.* **B205:** 567–580.

Williams, G. C. 1980. Kin selection and the paradox of sexuality. *In* G. W. Barlow and J. Silverberg (eds.), *Sociobiology: Beyond Nature/Nurture?* AAAS Symposium, Westview Press, Boulder, Colo.

Willson, M. F. 1979. Sexual selection in plants. *Am. Nat.* **113:** 777–790.

Willson, M. F. 1981a. Complex life cycles in plants: A review and an evolutionary perspective. *Ann. Missouri Bot. Gard.* **68:** 275–300.

Willson, M. F. 1981b. Sex expression in fern gametophytes: Some evolutionary possibilities. *J. Theor. Biol.* **93:** 403–409.

Willson, M. F. 1982. Sexual selection and dicliny in angiosperms. *Am. Nat.* **119:** 579–583.

Willson, M. F. (in press). *Vertebrate Natural History.* Saunders, Philadelphia.

Willson, M. F. (unpublished). Mating patterns in seed plants.

Willson, M. F. and N. Burley. (in press). Mate choice in plants: Tactics, mechanisms and consequences. Princeton University Press, Princeton, N.J.

Willson, M. F. and P. W. Price. 1977. The evolution of inflorescence size in *Asclepias* (Asclepiadaceae). *Evolution* **31:** 495–511.

Willson, M. F. and B. J. Rathcke. 1974. Adaptive design of floral display in *Asclepias syriaca. Am. Midl. Nat.* **92:** 47–57.

Wilson, D. S. and R. K. Colwell. 1981. Evolution of sex ratio in structured demes. *Evolution* **35:** 882–897.

Wright, S. 1977. *Evolution and the Genetics of Populations*, Vol. 3: *Experimental Results and Evolutionary Deductions.* University of Chicago Press, Chicago.

Yampolsky, C. and H. Yampolsky. 1922. Distribution of sex forms in the phanerogamic flora. *Bibl. Genet., Lpz.* **3:** 1–62.

Young, D. A. 1972. The reproductive biology of *Rhus integrifolia* and *Rhus ovata* (Anacardiaceae). *Evolution* **26:** 406–414.

Zapata, T. R. and M. T. K. Arroyo. 1978. Plant reproductive ecology of a secondary deciduous tropical forest in Venezuela. *Biotropica* **10:** 221–230.

Zarzycki, K. and J. Rychlewski. 1972. Sex ratios in Polish natural populations and in seedling samples of *Rumex acetosa* L. and *R. thyrsiflorus* Fing. *Acta Biol. Cracov.* **15:** 135–151.

Zeide, B. 1978. Reproductive behavior of plants in time. *Am. Nat.* **112:** 636–639.

Zimmerman, J. K. and M. J. Lechowicz. 1982. Responses to moisture stress in male and female plants of *Rumex acetosella* L. (Polygonaceae). *Oecologia* **53:** 305–309.

Chapter 3

MATING

Mating by the lower plants is a relatively simple matter of the release and union of gametes. However, in the seed plants another step intervenes—pollination precedes fertilization. Phylogenetically, the interpolation of this event in the sexual cycle represents a drastic reduction of the male gametophyte from an independent or semi-independent entity to a tiny being, parasitic upon its sporophytic parent until it is released. This change in the life cycle was undoubtedly one central element in the adaptive radiation of terrestrial seed plants, because it freed the process of fertilization from dependence on water. The sperm of lower plants are generally actively motile, propelling themselves through water to find an egg. For terrestrial algae, mosses, and ferns and their allies, this necessitates an accessible film of water at the time of sperm release. As a result, gametophytes may be required to remain small, ground-hugging organisms in order to have access to the occasional water-film. But the seed plants have packaged up their sperm in a passively transported gametophyte that now has the additional function of delivering the sperm to their destination. The packaging temporarily protects the sperm from desiccation and, in addition, often provides a supply of cellular organelles and material that may participate in zygote development. The transformation of male gametophytes into pollen grains opened the airways as a medium for sperm transfer and, in effect, created new adaptive zones of growth form and mating mechanisms.

In the seed plants, both pollination and fertilization are part of the mating process, although they are sometimes widely separated in time. Furthermore, if females of any species selectively abort the zygotes of certain males, the process of mating can be extended through the period of selective abortion, because only then have the fathers of the zygotes been determined. One useful definition of mating includes both the bringing together of male and female and the conjugal events of parenthood. Males that pollinate but do not fertilize, or that fertilize but fail to father surviving embryos are mates only in a limited sense.

SEXUAL PERSPECTIVES

Plants exhibit remarkable variation in the degree of differentiation of male and female gametes. A number of algae are morphologically isogamous, although the gametes may be chemically different. Most plants, however, are anisogamous, with male gametes smaller than female; botanists often distinguish extreme anisogamy as "oogamy." Interestingly, a great range of gamete dimorphism sometimes occurs within a single taxon, even within a genus such as *Chlamydomonas,* a unicellular green alga (Bold et al. 1980). Ecological and evolutionary explanations for this variation are poorly developed. A general theoretical model is provided by Parker et al. (1972), reinforced by Bell (1978), Maynard Smith (1978), and Parker (1978). If the fitness of zygotes is related to their size, there will be selection for large gametes that provision the zygote well. However, for a given allocation of resources to gamete production, obviously fewer large gametes can be produced than small ones. Thus there is likely to be counterselection of some sort, favoring decreases of gamete size and production of large numbers of them. Rather than arriving at some uniform intermediate size and number of gametes, Parker et al. suggest that disruptive selection favored the production of large, nutrient-loaded gametes by some genotypes and small ones (but numerous and motile) by others. Maynard Smith imagined a hypothetical primitive population of isogamous microgamete producers being invaded by genotypes that produced larger gametes and thus enhanced their zygote survival. Then it will profit the producers of microgametes to produce many of them to increase the probability of meeting and fusing with a macrogamete. In short, once large gametes exist and confer advantages on zygotes, there can immediately be competition among microgametes to mate with the macrogametes. Fusions of two macrogametes to make a giant zygote might be countered by diminishing advantages of greater and greater zygote size or by rapid evolution of countermeasures of microgametes, made possible by their higher numbers and potentially greater variation.

Plants differ greatly in the amount of "care" lavished on zygotes before they are released into the environment. At one extreme are the isogamous unicellular forms, in which "parent" and "gamete" are represented by the same entity, and the zygote is free-living from its inception. Some algae retain the zygote in protective walls (e.g., *Oedogonium*) and some apparently may nourish it as well; in some of the red algae, for instance, the diploid zygote is retained on the parent and grows, perhaps using parental resources (references in Willson 1981). The gametophyte of the mosses and ferns commonly provides at least some resources to the developing sporophyte. And seed plants, of course, typically endow the zygote with stored nutrition, the amount varying from virtually none in orchids to many grams. The quality of nutrition stored in seeds may also vary; some seeds contain a great deal of protein, for instance, whereas others are very oily. Protective structures and dispersal devices entail still more expenditure. The cost of parental care (entirely a female function, at least after fertilization) is a major factor creating differences in sexual perspectives.

Reproductive success of an hermaphroditic plant must be measured through both its female and its male functions. When sexes are separate, the necessity of separate measures of RS is obvious. The same principles apply in the case of hermaphroditism, but male RS is ignored too frequently in favor of measures of offspring numbers (seeds, zygotes, etc.) or female RS. Male and female RS of an individual are unlikely to be equal, except in obligate selfers. Historically, the emphasis on offspring production may have emerged in part because of researchers' concern with population recruitment and growth, although passage of genes by males can have important effects on the genetic structure of the population and even on the RS of other species, if foreign sperm or pollen block access by sperm or pollen of the appropriate species (Crosby 1966, Levin and Anderson 1970). Furthermore, male success in fertilizing eggs and, ultimately, in fathering viable juveniles is notoriously difficult to measure. It may be tedious, but not impossible, to count or estimate the total number of offspring as an index of female RS (neglecting post dispersal survival here); to ascertain sperm arrival and fecundation rates, not to mention ensuing zygote survival, for individual males is very difficult and possibly has never been done. Meeting the challenge of measuring male RS will permit major advances in our understanding of plant reproductive ecology.

At least in principle, however, it is easily seen not only that male and female RS must be measured separately, but that the evolutionary pressures on male and female functions are sometimes different. From a male point of view, it is important to father as many viable embryos as possible. This may entail producing and delivering copious sperm, preventing other males from successfully fecundating, and preventing abortion of zygotes by the receiving female. Once sperm has arrived at a receptive site, selection should favor an all-out effort to secure eggs, because at that point the sperm has no other options—it cannot get up and move over to another female, to try again (barring a fortuitous retransfer by a vector, K, Garbutt unpublished manuscript). From a female point of view, on the other hand, it is important to receive not only enough sperm to fertilize all the eggs she has enough resources to mature, but also enough sperm of sufficient quality. Because the female expends more resources in maturing offspring than does the male, it is in the female's interest that these resources be devoted to the best possible offspring. Therefore, females should be expected to evolve mechanisms of attracting and sorting males.

Although male and female interests may conflict in some cases, by no means should they always do so, and male and female functions can then be served by the same evolved feature. In the next section, we begin to explore some means of accomplishing these varied reproductive ends.

Seed plants are discussed to the virtual exclusion of other plants, simply because these questions have been addressed so seldom in the "lower" plants.

Pollen Quantity and Quality

There may be selection to increase receipt of pollen as a means of increasing female RS if pollen supply limits seed production and resources are available to produce a larger seed crop, or if receipt of more pollen increases the array of males from which the female may "choose" fathers for her offspring. At the same time, selection may favor enhanced pollen delivery to increase male RS. Some hermaphroditic plants have ways of accomplishing both ends by the same means, although it is not always apparent whether the prevailing selection pressures were strongest for male function, female function, or some combination of the two.

To enhance pollen donation or deposition, animal-pollinated species may improve pollinator attraction and increase pollinator rewards: producing more or richer nectar, presenting more pollen per flower, or by adding flowers to the floral display. Such tactics can only be successful if there is an available population of unemployed, underemployed, or divertible pollen vectors. It is highly unlikely that any of these means can be called into play just when a plant finds itself to be pollen-limited, although some plants (e.g., *Sanguinaria canadensis*, Schemske 1978) fertilize their seeds autogamously in the absence of incoming pollen from other individuals. It is possible that nectar production patterns might shift in response to differences in physiology of plants growing in different conditions, but whether such hypothetical differences are actually related to differences in levels of pollination remains to be seen. The number of stamens per flower is commonly, though not always, fixed for a species; therefore increased pollen rewards might be most likely in terms of number of grains per anther. Whether this feature varies among individuals and among ecological circumstances within a species is apparently unknown.

We do know, however, that increasing the size of the floral display, which presumes an appropriate availability of resources, can sometimes (but not always, e.g., Roubik et al. 1982) lead to greater seed production. Large inflorescences of *Phlox divaricata* produced more seeds than small ones (Figure 3.1; Willson et al. 1979), in keeping the common expectation; the same was true for *Yucca whipplei* (Udovic 1981), and many other species will be found to do the same. In contrast, the correlation of umbel size with seed and pod production was far poorer in *Asclepias syriaca*, in which large umbels seldom produced more pods than smaller

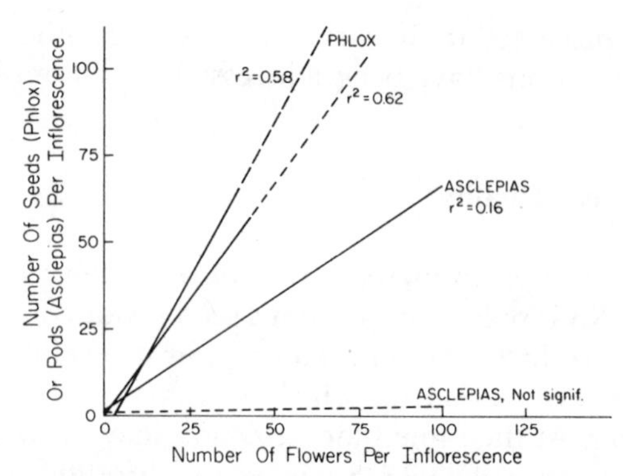

Figure 3.1 The rates of increase of offspring production (an index of female fitness) as a function of number of flowers in *Phlox divaricata* and *Asclepias syriaca.*

ones (Figure 3.1; Willson and Price 1977). More pollinators were, indeed, attracted and more pollinia inserted into the stigmatic chamber (Willson and Bertin 1979), but the number of pods produced did not increase correspondingly. If the "extra" flowers on larger inflorescences served a female function, it was not mainly the production of more pods; the possibility that pollen came from a greater variety of pollen sources was not tested. Similar results were obtained for *A. quadrifolia* (Chaplin and Walker in press). Sutherland and Delph (unpublished manuscript) suggest that, because plants with hermaphroditic flowers may commonly produce fruits from a lower percentage of their flowers than do plants with unisexual flowers, selection on the size of floral display frequently may be based on increasing male RS more than female (unless larger floral displays also increase offspring quality and diversity). However, if female RS of an hermaphrodite is pollen-limited, it seems likely that the relative importance of female function to the evolution of the floral display would be much greater.

Male and female functions may last different amounts of time. Flowers of *Asimina triloba* are hermaphroditic but protogynous; the female phase lasts at least twice as long as the male phase (Willson and Schemske 1980). Dioecious orchids often differ greatly in length of floral life, with females lasting far longer than males (Chapter 2). It may be relevant to observe that, probably in both cases, pollen availability limits seed production and female RS. The extent to which pollinator activity and pollen removal (and deposition) limit male RS is not known, but male RS is presumably less limited than female (see Janzen 1981 for tantalizing but regrettably sketchy observations). Whether such limitations apply to other cases of differing lengths of male and female function remains to be seen. A potentially interesting contrast is seen in *Impatiens capensis,* in which seed-set apparently is not pollen-limited, the male phase of the flower is much larger than the female, and a chief function of chasmogamy might be the male role (Chapter 2).

An evolutionary convergence in floral morphology may allow one species to capitalize on the pollinators of another. The success of such mimicry obviously depends on the adequacy of pollinator activity: the mimic must be good enough to draw some of the model's pollinators without diminishing pollination success of the model, or else selection will favor a divergence of the model away from resemblance to the mimic. A plausible example of this kind of floral mimicry is provided by two unrelated desert annuals, *Mohavea confertiflora* (Scrophulariaceae) and *Mentzelia involucrata* (Loasaceae; Little 1980; Figure 3.2). *Mohavea confertiflora* is considered to be the mimic, producing no pollinator rewards and rarely visited. But its resemblance to *M. involucrata* is apparently sufficient that four species of bee will visit the putative mimic occasionally, perhaps effecting pollination; other visitors to the model seem not to be fooled. The system is open to experimental manipulation to test the mimicry hypothesis; such tests are not necessarily easy (Bierzychudek 1981).

Alternatively, several species may converge toward a single morphology to their mutual advantage; together they provide a resource sufficient to support certain pollinators that otherwise might be rare or visit other flowers. Two bee-pollinated herbs of the forest understory in Panama, *Costus allenii* and *C. laevis,* share habitats, flowering season, floral morphology and color, nectar secretion patterns, and a single pollinator (*Euglossa imperialis*). They are each self-compatible but seldom hydridize, even though they frequently receive each other's pollen. Both species occur at low densities, exacerbated by destruction of flowers by weevils—these factors may have selected for floral convergence and sharing of the pollinator (Schemske 1981). Eight species of hummingbird-pollinated flowers in the Arizona mountains have apparently converged in floral morphology: all are red, tubular, and at least some of them offer similar nectar rewards. A ninth species is similar in morphology, but nectarless, and may mimic some of the eight (Brown and Kodric-Brown 1979). It is reasonable to guess that floral mimicry (both mutual and unilateral convergence) is one evolutionary solution to problems of pollinator limitation, in this case affecting both female and male RS.

On the other side of the coin, however, competition for pollinators could produce selection for divergence of floral morphology and/or flowering times. This suggestion opens up a real Pandora's box, because so much has been written on the subject and so little has been success-

Figure 3.2 A putative floral model and mimic system: *Mentzelia involucrata,* on the right, is considered to be the model; *Mohavea confertiflora,* on the left, the mimic. Photos by R. J. Little.

fully demonstrated (Waser in press). Nevertheless, in principle, if pollinator activity is so low that seed production and pollen donation are limited, a shift to a new pollinator, with attendant changes in signals and rewards, could provide an escape from that limitation. The conditions favoring convergence on the same pollinator over divergence toward use of different pollinators is still moot (Brown and Kodric-Brown 1979; see also Brown and Kodric-Brown 1981, Williamson and Black 1981).

Interestingly, the efficiency of pollen tranfer can sometimes affect male function more than female: seed-set in a North Carolina population of *Claytonia virginica* was seldom limited by pollinators; a very specialized insect pollinator and a very generalized one were equally effective in causing seeds to be set (Motten et al. 1981). In this case, any quantitative differences in transfer efficiency would affect only male function.

Mate quality is important to both males and females. Male functions should be subject to selection against pollen wastage—selection for mechanisms that improve the probability of pollen delivery to a suitable female. Female functions should be selected to enhance the arrival of appropriate pollen and, perhaps, subsequently to discriminate among the arrivals. Thus both male and female functions are involved with the evolution of the pollen delivery system, but females, more than males, assess quality after pollination has occurred.

Note, however, that selection for accuracy of pollen delivery affects males and females in somewhat different ways. The problem for males is the wastage of pollen, that is, the allocation of resources to pollen that never arrives in the right place. The greater the accuracy of delivery, the more efficient can be the operation of male function. Efficiency of pollen delivery might mean that males can afford to reduce pollen production (Cruden 1976a,b, 1977; Cruden and Miller-Ward 1981) in some circumstances, because vast quantities of pollen are then unnecessary to assure the successful arrival of some of it, and those resources can then be allocated to other functions. However, if there is intermale competition for mates and selection for increased pollen donation, an increase of pollen production would be much more effective than when pollen delivery is inaccurate. The possible effects of efficiency and of selection on the male role have yet to be disentangled. For females, the problem is to prevent the cluttering of receptive surfaces by useless pollen and potential reallocation of conserved resources is less likely to be involved.

An important aspect of pollen quality is, of course, the appropriate species. Except insofar as the delivery of pollen to a foreign species might be viewed as a means of interspecific competition, by depressing a competitor's seed production, conspecificity is essential. Sometimes pollen transfer between taxonomic entities called species is, in fact, successful in yielding fertile hybrids; this fact tells us more about the arbitrariness of taxonomic delineation than it does about the basic biological problem at hand. The question is: To what extent is pollen deposited on individuals of the appropriate sort and how do plants achieve the observed degree of accuracy?

The flowering of both wind- and animal-pollinated species must be synchronous among at least some of the members of each population, for obvious reasons. For wind-pollinated species, strong synchrony may be the only way to improve the probability of pollen delivery. However, for animal-pollinated species there are a number of additional potential means of improving the accuracy of pollen delivery. Thus we should expect that at least some animal-pollinated species achieve far higher degrees of accuracy than is possible for wind-pollinated species. Indeed, animal-pollination ranges from highly accurate to very inaccurate (Grant 1950), perhaps little more accurate than wind-pollination. Inaccuracy can result from the flowers being visited by animals that may carry and transfer relatively small amounts of pollen (e.g., Primack and Silander 1975), or that visit and deposit pollen on other flower species, or from the loss of pollen to animals that consume the pollen themselves, take it home to their kin, or groom it off their bodies and drop it. Some plant species, such as apples and roses, have rewards that appeal to a spectrum of visitors and a floral morphology that excludes none. But others, including *Dicentra* and *Pedicularis,* have hidden the sexual organs away inside a complex flower that can be pried open and visited "legitimately" only by certain bees (e.g., Macior 1978, see above); and some (e.g., *Linanthus parviflorus, Cantera candelilla*) conceal the working parts deep in a long corolla tube accessible only to vectors with long probes (e.g., long-tongued flies or bees, hawkmoths, hummingbirds; Grant and Grant 1965); other plants are pollinated by vectors that exhibit high degrees of constancy to a particular species for reasons of foraging efficiency (Heinrich 1976, Real 1981). Some plants are visited by extremely specific pollinators (Beattie 1972, Free 1970), and we can suppose that the accuracy of pollen transfer in these instances is quite high. Examples include the western American *Yuccas,* some pollinated only by one species of *Tegeticula* moth, the Mediterranean orchid *Ophrys speculum,* apparently pollinated solely by male wasps of a single species (see Proctor and Yeo 1973) and many other orchids as well (Dressler 1968), and the famous case of the figs and figwasps discussed below. Differences in floral signals facilitate such specificity (e.g., Jones 1978).

Why some species have such specialized flowers (and visitors) and others do not, we can only guess. Although the benefits presumably include efficiency of pollen transfer, the costs may include ecological restriction of habitat and flowering season to those also suitable for the vector and complete failure of sexual reproduction whenever conditions become unsuitable for the pollinator. I do not believe we know what conditions make it possible for just certain species to pay these costs or why certain species have found it possible to evolve such specialization. At this point, we can say that, clearly, high pollinator specificity is not restricted to a few geographic regions, a few habitats, or a few taxa.

Moreover, as noted in Chapter 2, both forms of SI bring with them the inevitable consequence that certain other individuals (bearing the same compatibility genes) will also be rejected (partially or completely) as fathers for the offspring of the female in question. Therefore, SI systems reduce the range of mates available to females, no doubt excluding some pollen donors that would be highly suitable except for the incompatibility gene(s). Perhaps pollen availability, mate quality, and/or variety are not limiting in these cases. SI systems also reduce the number and variety of mates accessible to males and increase the risk of wasting pollen on unsuitable stigmas.

Architectural modifications may encourage the arrival of nonself pollen, to the advantage of both male and female functions. *Cyanella alba* and *C. lutea* in South Africa are pollinated by large bees and have the curious feature that the style is deflected to one side or the other and the reproductively functional stamens to the opposite side, right-handed and left-handed flowers often occurring on the same individual, especially in *C. lutea.* Pollen accordingly is deposited on and received from one side of the pollinator by each floral morph. *Cyanella alba* usually only bears one open flower at a time, but on the rare occasions when there are two, they are usually of the same morph, and geitonogamous pollen transfers are thus reduced. *Cyanella lutea* bears more open flowers, of both types, but is strongly SI. Thus in both species the possibility and/or the success of within-plant pollination is low; the different floral morphs may provide a means of reducing the amount of pollen wasted on geitonogamous transfers. In line with this suggestion, the amount of pollen produced per flower is much lower in these two species than in two congeners lacking the floral dimorphism (Dulberger and Ornduff 1980). Such an interpretation may be too simplistic, however, in view of the existence of both floral morphs in single individuals of some self-compatible *Solanum* and *Cassia* (Ornduff and Dulberger 1978).

Sequential maturation of flowers on an elongate inflorescence may induce pollinator visits to follow a sequence that improves the likelihood of cross-pollination. Four species of Rocky Mountain plants (*Aconitum columbianium, Delphinium nelsonii, D. barbeyi,* and *Epilobium angustifolium*) have flowers arranged in spikes, in which the older, lower flowers are functionally female and the upper ones male. Pollinating bees work the inflorescence from bottom to top, thus tending to deposit pollen from other inflorescences on the lower flowers and remove pollen from the upper ones to be deposited elsewhere (Pyke 1978). However, another, rarer, species, *Penstemon strictus,* in the same region is visited by bees in a similar way, but the flowers of this species are not arrayed with functional females lower than males, so some other factors must be involved (Pyke 1978). The proximate factor guiding the bees' path sometimes might be the greater nectar reward available in the lower flowers (Best and Bierzychudek 1982, Haddock and Chaplin in press), although this is not the case in *Scrophularia aquatica* or *Linaria vulgaris* (Corbet et al. 1981). The orientation of the inflorescence is the guiding factor in some cases (Heinrich 1979), and the characteristic foraging habits of the pollinating insects may constrain their movement patterns (Corbet et al. 1981).

Male Competition

Some mechanisms of (potentially) increasing RS are operative chiefly in terms of male function. Male plants, or the male function of hermaphrodites may compete with each other for ovules. Male–male competition can occur whenever pollen grains from more than one male land on a stigma in excess of the number of available ovules. Competition among males may occur by several means. Pollen tubes might interfere with each other, as shown for self- and cross-pollen in *Brassica oleracea* (Ockendon and Currah 1977). When sibling pollen grains travel in groups, which is very common in some plants (see below), the efficiency of pollination may be improved by ensuring the arrival of multiple pollen grains on a stigma and permitting the fertilization of multiple ovules for each successful deposition of pollen (Cruden and Jensen 1979, Schlising et al. 1980). By the same token, however, pollen might be *lost* in batches, so it seems necessary to invoke the presence of relatively efficient pollinators, which reliably visit the flowers, at the same time (Willson 1979). Of equal possible importance is that arrival of clumped pollen might also occupy enough of the stigmatic surface to reduce or even prevent the arrival of other pollen (Willson 1979). This could be advantageous even if there are not enough ovules for each of the sibling pollen grains to fertilize. (The fertilization of sibling ovules in an ovary by sibling pollen grains may also have important consequences for interactions among developing seeds because the competitive patterns among full sibs may be different than among half-sibs; Kress 1981, Willson and Burley in press). A possible countertactic in female function might be a reduction of the number of ovules per ovary and an increase in the number of ovaries. By increasing the number of pollinating events required, the genetic diversity of the offspring is potentially enhanced (Wilbur 1977).

The well-known pollen population effect might function in pollen competition, particularly if sibling pollen grains evince a greater effect than unrelated grains. The pollen population effect occurs, at least in vitro, in a wide variety of angiosperm species (Brewbaker and Kwack 1963, Brewbaker and Majumder 1961): as the number of pollen grains increases, the germination success and pollen tube lengths also increase. The numerical response varies among species and, sometimes, conspecific populations; sometimes the response is a linear function of numbers, sometimes not. Certain species have a minimum number of grains required before any germination occurs. A pollen population effect is indicated for wild *Passiflora vitifolia* in Costa Rica, in which 25–50 grains are required for fruit set and over 500 are needed for full seed-set—a ratio of about 1.6 grains per seed (Snow 1982). How common is the so-called population effect in other populations in nature is still unknown. If sibling pollen grains can, indeed, "cooperate" when they land on a stigma, this could be a potent means of male–male competition.

A curious situation in certain gymnosperms suggests some possible parallels with clumped pollen in angiosperms. Several species of *Cupressus* and *Microcycas calocoma* have numerous functional sperm within each pollen grain (references in Willson and Burley in press). The egg-bearing sites in *Cupressus* and its relatives are aggregated into archegonial complexes in the female gametophyte so that it is possible for each of the 16 or 20 sperm from a single pollen grain to fertilize an egg (Figure 3.3). Similar multiple fertilizations might occur also in *Microcycas*. If this actually happens, the number of *other* pollen grains that can claim eggs is obviously reduced—but usually not to zero, inasmuch as the archegonial complexes sometimes contain many more than 16 or 20 eggs. A most interesting feature of these gymnosperms is that no matter how many eggs are fertilized in each female gametophyte, only one zygote usually matures. All the rest are aborted and may be absorbed by the survivor. We (Willson and Burley in press) have suggested that multiple sperm may be a mechanism by which one male (even if he delivers only one pollen grain to each female gametophyte) may be able to partially block access by other males. Furthermore, since all of the sperm in one pollen grain are descendants of a single haploid microspore, they are virtually identical in genetic constitution, and so are all the eggs on the female side. Therefore, it is an ideal system for kin-selected sacrifice of some zygotes for the benefit of their identical sibs. Because the genes that are passed on are the same, it makes no evolutionary difference which of the sperm from a pollen grain is successful. Please note that this argument is entirely hypothetical; we have no concrete data demonstrating that blockage is effective or that one zygote can make use of its doomed siblings. Unfortunately (for now) or perhaps fortunately (in the long run), the testing of these possibilities requires an intersection of ecology and embryology, an interaction almost entirely lacking at present.

In the grand phylogenetic scheme, males may have had still another

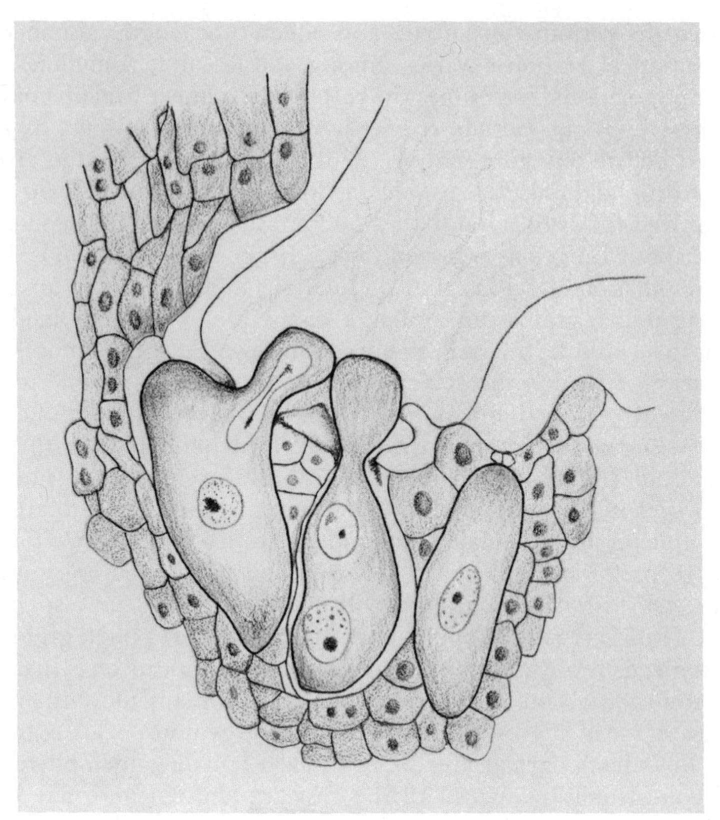

Figure 3.3 An archegonial complex of *Sequoia sempervirens,* showing three archegonia; the one on the left is being entered by a sperm nucleus (and cytoplasm from the pollen grain); the one in the middle contains a sperm nucleus and male cytoplasm. Redrawn from Buchholz (1939).

tactic to improve their success. Angiosperms are characterized by "double fertilization," in which pollen grains each carry two sperm, one of which fertilizes the egg, whereas the other joins with one to several female polar nuclei to form the endosperm that nourishes the developing embryo. Several authors have suggested that, in various ways, this feature may have evolved as a means of improving the survival of zygotes (Charnov 1979, Willson and Burley in press) and may be nutritionally analogous to the multiple fertilizations in the gymnosperms *Cupressus* and *Microcycas.* However, I have found no evidence that the male side of double fertilization varies among the angiosperms, although the female side does—different numbers of polar nuclei are involved. A possible interpretation is that females can counter the male influence by adding nuclei to the endosperm but that males do not exhibit the reciprocal countermeasure (Willson and Burley, in press).

Pollen might also compete through interference of pollen tubes, chemically or physically, and in rapidity of pollen tube growth. To some extent rapid growth might depend on interactions with female tissue (and thus partly reflect female choice), but the resources (minerals, oil, starch) contained in the grain itself might contribute to growth as well. If pollen resources affect the outcome of pollen competition, then there may be selection for increased pollen grain size (to store those resources), although there would be limits set by the ability to disperse (especially for wind-borne pollens), the cost of the stored resources themselves, and selection on pollen numbers. Minerals seem less likely to require much storage space than oil or starch, so the potential effect on pollen size should be less.

Female Choice

Female plants (or the female function of hermaphrodites) may be able to evaluate potential mates by several potentially interacting criteria: genetic quality, genetic complementarity, and material resources contributed to the zygote by the pollen grain (Willson and Burley in press). There is, at present, little concrete evidence that female choice of mates actually occurs (except in the cases of cross- and self-incompatibility already discussed), but there is good reason to suppose that mate choice is feasible. The occurrence of apparent mating "preferences" between particular individuals supports the idea (Willson and Burley in press). Particularly if female RS is resource-limited, females have the opportunity to accept sperm differentially, although even pollen-limited females may exercise some choice (on the presumably rare occasions when pollen grains from different males exceed the number of ovules available).

The array of mates from which to choose may vary enormously in response to many factors. Whole groups of potential mates can be excluded by an SI system. Behavior of animal pollen vectors also affects the diversity of possible mates. The number of flowers and individual plants visited per foraging trip of each individual pollinator and the degree of pollen carryover (see below) are involved and can be influenced by conspecific plant density and the degree of flowering synchrony among members of the population. Plants growing relatively near to conspecifics sometimes receive more pollen or sperm than those at greater distances (although the absolute scale may be small indeed); this can result in higher seed-set (*Arisaema triphyllum,* Rust 1980) and could also, in some instances, provide a wider array of possible mates. Note, however, that the distance effect on levels of fertilization is not always apparent (Duckett and Duckett 1980, Melampy 1979, Willson et al. 1979). Foraging animals sometimes have particular behavior patterns that create a degree of assortative mating; for instance, honeybees visiting *Lythrum salicaria* in Indiana showed a tendency, on any one flight, to forage at a relatively consistent height above the ground, so that tall plants were most likely to mate with other tall plants, and short with short (Levin and Kerster 1973). Red wattlebirds (*Anthochaera carunculata,* Meliphagidae) in Australia use different methods of locomotion while foraging on two *Anigozanthos* species of different stature, which contributes to assortative mating of these species (Hopper and Burbidge 1978).

Genetic quality of a male may be exhibited by the growth rate of the pollen tube, or other indices of vigor, by stamina (the ability to maintain fertility), and by competitive interactions with other pollen grains. Females might erect barriers that make pollen-tube growth more difficult or delay acceptance to test for stamina, and thus select the males of higher quality. Several authors have previously noted that the development of a style between stigma and ovary in angiosperm flowers provides a site for testing of incoming males. The length of a style need not be an adequate measure of the filtration ability of the female, however, inasmuch as long styles may evolve in response to several different selection pressures (including, for example, those involving the kind of pollen vectors employed). In some cases pollen quality, in terms of rates of pollen tube growth, is correlated with quality of ensuing progeny (e.g., Bertin 1982a, Mulcahy and Mulcahy 1975, Ottaviano et al. 1980), and some genes are expressed in both gametophyte and sporophyte (e.g., Tanksley et al. 1981). Genetic complementarity is probably best assessed after fertilization, when maternal and paternal chromosome sets begin to work together. Complementarity may extend well beyond matters of genetic similarity of relatives to the functional working of any two haploid sets of chromosomes in the embryo. Reports of optimal outcrossing in *Delphinium nelsonii* and *Ipomopsis aggregata* indicate that seed-set is greatest when mates are growing neither very close together nor very far apart (Waser and Price in press), perhaps indicating that inbreeding with near neighbors and outbreeding with individuals from other locally differentiated populations both lead to reduced RS.

In addition, males may be judged by the investment they make in the zygote or by other means of speeding up zygote development after fertilization (Willson and Burley in press). It becomes unprofitable for a female to abort well-developed embryos because these require less additional investment from her than do more retarded embryos. Males of otherwise inferior genetic quality might even improve their acceptability by investing in zygotes.

Males could, perhaps, influence developmental rates of their offspring by genetic or paragenetic means, either nuclear or cytoplasmic. Paternal genes in zygotes might be activated earlier than maternal alleles; males might provide duplicated genes or chromosomes that increase availability of template for RNA synthesis, or transmit "B" or accessory chromosomes that can affect cell volume or mitotic rates but carry fewer active genes than the typical "A" chromosomes. Paternal transmission of cytoplasmic genes is known for several kinds of plants (algae, angiosperms, gymnosperms) and fungi. There is not a shred of evidence that males actually use any of these mechanisms, but the potential is there; gene amplification and B-chromosome transmission are often unusually frequent during gamete formation. These possibilities remain to be explored.

Furthermore, male cytoplasm enters the egg along with the fertilizing sperm far more commonly than is usually believed. It is not impossible, but is still undocumented, that male cytoplasmic material could block the action of female RNA. Mechanisms for RNA blockage are known, and there exist cases of zygotes (not necessarily the paternal aspects of them) tying up maternal ribosomes. More probably, male cytoplasmic materials contribute to synthesis of metabolic products in the zygote. Zygote cytoplasm is largely of male origin in some gymnosperms and consists of a mix of male and zygote-synthesized products in some others. Even when maternally derived cytoplasm predominates, male microtubules, mitochondria, protochloroplasts, starch grains, or fat droplets often enter the egg and could contribute significantly to zygote growth and development. Again, the necessary observations have not been made, because the question has not been asked.

Ovules sometimes contain more than one embryo, a condition known as polyembryony. Multiple embryos may arise from multiple eggs in each female gametophyte, from cleavage of a single egg after fertilization, or even from maternal tissue. Normally only one of these embryos survives. When plural embryos result from the fertilization of several eggs by genetically different sperm, then clearly the female has a basis for choice among them. In certain gymnosperms, however, pollen grains transmit up to 20 identical sperm, and here the basis for choice is obviously reduced. If female choice influences female RS, selection might then favor higher degrees of polyembryony (or some other means of maintaining a range of choices) to counter the multiple fertilization from a single pollen grain. Cleavage polyembryony may be viewed as a means of intermale competition and/or female choice among the embryonic units. If multiple cleavage embryos can synthesize proteins and other materials more rapidly because of the greater surface/volume ratio and the greater amount of nucleic acid template, development of the complex embryonic unit should proceed rapidly. Females should then allocate more resources to rapidly growing embryonic units and be less likely to abort them. The surviving cleavage embryo can absorb its genetically identical siblings and metabolize their substance, thus profiting again from their earlier synthetic activity. When the sibling embryos are identical, the sacrifice of one for another is easily seen to involve no loss of fitness.

Of necessity, these suggestions are speculative. The question of pollen quality at the level of the individual female has been almost untouched. Is a *variety* of fathers for a seed crop important to female RS (Janzen 1977)? Could the evolution of smaller fruits containing fewer seeds occur in response to selection for increasing the variety of fathers for the brood of a given individual (Wilbur 1977)? How fussy are females about male quality, and does the choosiness vary with circumstances, including the array of males and condition of the female?

POLLEN VECTORS

When pollen is released, it may be borne by water, by wind, or by a variety of animals. Some aquatic plants have emergent flowers and pollen transfer through the air. Others release pollen either on the water surface or underwater. The ecology of water pollination is little studied and I will not discuss it further; Proctor and Yeo (1973) present some specific examples.

Wind-pollination prevails in the gymnosperms, with the possible exception of some cycads and some species in the Gnetales (s.l.; Faegri and van der Pijl 1979). This mode of pollen transfer is also common among angiosperms, predominating in the orders Amentiferae and Urticales, in the families Gramineae (Poaceae), Cyperaceae, Juncaceae, and in many Chenopodiaceae and Polygonaceae; it occurs sporadically in many other taxa as well. Wind-pollination in angiosperms is generally thought to be derived phylogenetically from a basic condition of pollination by animals, specifically insects.

Wind-borne pollen is characteristically dry and smooth on the exterior (Figure 3.4), so that grains are usually dispersed as single units or, as in the Typhaceae and some others, in tetrads (a unit of four pollen grains). Buoyancy is enhanced by small size, low density, or the addition of air sacs. Because buoyancy is of obvious importance, any tendency to

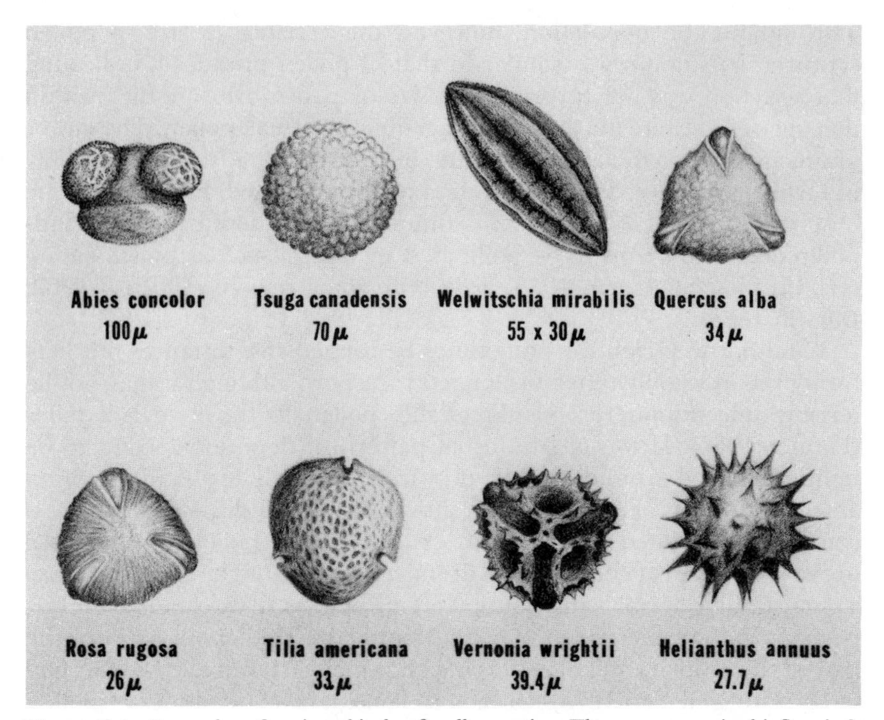

Figure 3.4 Examples of various kinds of pollen grains. The upper row is chiefly wind-borne, the lower row mostly carried by insect vectors. Redrawn from Wodehouse (1935).

endow pollen grains with additional nutrients or metabolic machinery would probably necessitate a concomitant compensatory change in density of the storage products or in density of the whole grain (e.g., air sacs). Presumably as a result of these size constraints coupled with selection for vast numbers, wind-borne pollens are typically smaller (10–60 μm, excluding the inflated air sacs of *Pinus*) than those borne by animals, which range up to about 300 μm. On the other hand, small pollen grains are more easily whisked past the receptive female surfaces than are large ones, and this factor presumably sets some lower size limit. A detailed consideration of the physics of pollen dispersal and deposition is found in Gregory (1973).

Primack (1978a) found that the pollen grains of outcrossing *Plantago* species varied less in average diameter than did pollen grains of cleistogamous species. He interpreted this as evidence of selection constraining the size of wind-borne pollen. Within a single species (*P. lanceolata*), small pollen grains traveled farther than large ones.

Throughout the population, however, the variance in size of pollen captured by stigmas was similar to that of pollen produced, indicating that selection was not favoring one size of pollen over another within that species, despite the greater dispersibility of small pollen. The larger grains probably had a compensating, but unknown, advantage, possibly in fertilizing ability. The interpretation is complicated, however, by the observation that *P. lanceolata,* commonly considered to be wind-pollinated, also seems to be pollinated by insects such as pollen-eating syrphid flies (Stelleman 1978) at least in some regions (Clifford 1962, Darwin 1884).

Wind-borne pollen can sometimes be carried vast distances but may rarely retain viability after prolonged transport, although it may not be terribly uncommon for possibly viable pollen to travel several miles (Lanner 1966). However, the usual pattern of deposition seems to be tightly centered around the parent, with a relatively few grains ranging afar. A number of studies have shown that wind-dispersed pollen is commonly deposited in a leptokurtic pattern: more pollen is found close to the source *and* at considerable distances than would be expected from a statistically normal distribution (Figure 3.5). For some conifers, perhaps 90% of the pollen may be deposited within about 36 m of the parent; for poplars and elms this distance may be far greater—over 600 m (Proctor and Yeo 1973)—although considerable quantities of pollen may travel much farther than this (Wright 1978). A woodland herb, *Carex platyphylla,* has extremely restricted pollen flow, almost all of it landing within 1 m of the source (Handel 1976). As a result of this deposition pattern, most matings are believed to be among neighbors, a relatively small (but not unimportant) number occurring over great distances.

Wind-pollinated plants frequently release their pollen only in favorable conditions, namely, rainless and breezy. Nevertheless, most of the pollen released never reaches its target and quickly dies. Only if huge quantities of pollen are produced will wind pollination be successful. Because stigmas are rather small targets (although some have expanded surfaces), the density of air-borne pollen must be very high to ensure that pollination occurs. Thousands upon thousands of pollen grains per square meter must be available; thus a single birch catkin may contain over 5 million pollen grains and a birch tree will normally bear many catkins (Proctor and Yeo 1973).

For wind-pollinated plants, the primary ways to improve individual success in pollination involve the production of greater amounts of pollen (for the male viewpoint), the production of better catching surfaces (e.g., Niklas and Paw U 1982; for the female viewpoint), and perhaps the

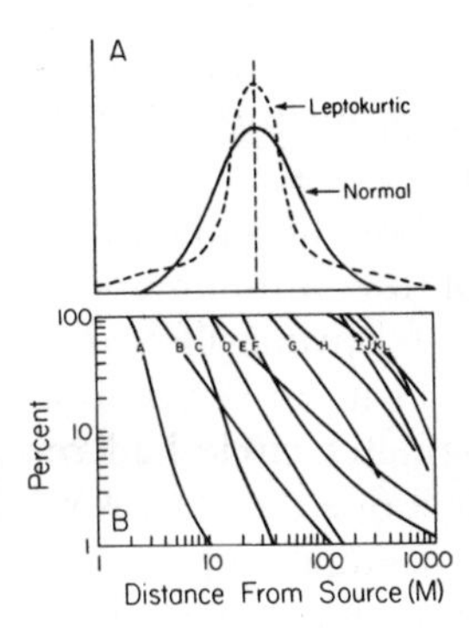

Figure 3.5 (*a*) Normal distribution pattern compared to a leptokurtic one, on an arithmetic plot. (*b*) Examples of some leptokurtic pollen-distribution patterns of various species, on a logarithmic plot. Redrawn from Levin and Kerster (1974) by permission of Plenum Publishing Corporation and D. A. Levin.

placement of flowers or cones on the plant. The "reach" of male gametes could be enhanced by placing them high on the plant, but a countering pressure may derive from an increased probability of self-pollination; as discussed in Chapter 2, the question of spatial arrangement of male and female portions of monoecious plants is unanswered.

Although it is not true that most plants with unisexual flowers are wind-pollinated, the converse appears to be valid: most wind-pollinated flowers (or cones) are unisexual. That is, the plants tend to be monoecious or dioecious. This is usually interpreted as a mechanism for the increase of outcrossing. Clearly, wind-pollination of hermaphroditic flowers in which male and female functioned simultaneously would result in very high degrees of self-pollination given the usual dispersion patterns of wind-borne pollen. But it seems that differences in timing of male and female functions within an individual would avoid selfing equally well in many cases. Therefore, we need to invoke additional factors favoring unisexual wind-pollinated flowers or cones. Perhaps, for some reason, temporal separation is not always developmentally feasible or ecologically favorable. Or perhaps unisexual flowers bring other advantages, such as manipulation of the floral sex ratio or the differential placement of male and female flowers.

The most common form of animal-pollination involves the insects, but some angiosperms are pollinated by vertebrates, specifically birds and mammals. Animal-borne pollen is sometimes larger, is often sticky and/or highly ornamented with spines and bumps, and sometimes adheres in clumps of several grains. The usual pattern of pollen deposition appears to be leptokurtic (e.g., Hopper and Burbidge 1978), as for wind-pollinated species. However, the distances involved can sometimes be greater, relative to the stature of the plant (see Levin 1981, Levin and Kerster 1974).

Clearly, both plant density and the habits of the animal pollinators can affect the scale of the distribution of pollen (e.g., Beattie 1976). If a plant population grows at rather low density and has some pollinators that preferentially visit its flowers, the "reach" of the pollen may be considerable. For instance, marked pollen of the hummingbird-pollinated herb *Delphinium cardinale* was concentrated primarily within about 4 m of the source, but some traveled at least 11 m; in fact, in one trial, over 60% of all the plants within a 22 m-diameter circle had received marked pollen from the central source in just one day (Schlising and Turpin 1971). If Janzen (1971) is correct that euglossine bees in tropical forests travel a regular route measured in kilometers, then clearly the length of the "pollen-shadow," or even of the first 90% of it, can be very long, the shadow may be highly asymmetrical, and pollen flow will tend to be unidirectional, so that certain individuals regularly father offspring with particular maternal parents. Such "trap-lining" is recorded for temperate-zone bumblebees foraging on several kinds of flowers (e.g., R. A. Johnson pers. comm., Thomson et al. 1982) and suggested for other pollinators (e.g., Feinsinger and Colwell 1978, Gilbert 1975, Stiles 1978). Furthermore, a seldom-measured but important factor influencing the "reach" of pollen is termed "pollen carryover." Not all pollen from one plant is deposited on the next one visited; much of it may be carried over to subsequently visited flowers (and some may even be recollected from one site of deposition and transferred to another, Garbutt unpublished manuscript). Pollen carryover by large bees can be quite impressive—flowers ranked seventh or eighth in a visitation sequence may receive good loads of pollen derived from the first flower, and in one remarkable case pollen was carried to the fifty-fourth flower in the sequence (Thomson and Plowright 1980, see also Levin 1981, Price and Waser 1982).

On the other hand, the foraging habits of the pollinator may severely restrict the pollen shadow. Patches of *Heliconia* blossoms defended by territorial hummingbirds exhibited nonleptokurtic but still constricted patterns of pollen deposition (Linhart 1973). Territorial hummingbirds on large, many-flowered individuals of *Combretum farinosum* can restrict pollen flow completely to the source plant itself, except when nonterritorial hummingbirds evade the residents' defense and steal visits to the flowers (Schemske 1980). Ecologists often claim that very large floral displays or very rich nectar sources may restrict pollinator movements, reducing cross-pollination and constricting the pollen shadow. This may indeed occur, as seen in the territorial hummingbirds, and some other nectar-feeding birds (e.g., Carpenter and MacMillen 1976), and as reported for small opportunistic bees drawn to many-flowered trees (Frankie et al. 1976) and bumblebees foraging on *Delphinium virescens* (Waddington 1981) and *Trifolium repens* (Heinrich 1979), but the effect is not universal; individual *Melipona* bees visiting *Hybanthus prunifolius,* a tropical shrub related to violets, did not seem to stay long at any one shrub nor did they visit a large proportion of flowers during any one visit (Augspurger 1980). Clearly, we need to know a great deal more about the relationship between reward and visitation rate, as well as the pollen shadows of animal-pollinated plants, before sound generalizations are possible.

Animals can carry remarkably large loads of pollen in various places on their bodies, with the result that several to many pollen grains may land on a receptive surface. Thus one stigma (in angiosperms) can capture pollen for a number of ovules, in contrast to wind-pollinated species in which the number of ovules per stigma is commonly low and is often reduced to just one (Faegri and van der Pijl 1979). The ability of animal pollinators to carry loads has permitted the evolution, in some plants, of pollen grains that are packaged in groups. Most Asclepiadaceae and Orchidaceae bundle many pollen grains into sacs called pollinia, which are transported singly or in groups. Other plants stick pollen grains together in clumps, sometimes containing a standard number of grains (e.g., 4, 8, or 16, Kress 1981; Willson 1979), sometimes an indeterminate number (e.g., some composites, Stebbins 1974; and Onagraceae, Cruden and Jensen 1979).

The geographic distribution of wind- and animal-pollinated plants is decidedly nonrandom. Wind-pollination is likely to be most effective where conspecifics grow relatively near each other and, hence, communities of low species richness may permit the reliance on wind as a vector. There appears to be a general correlation of increasing frequency of wind-pollination at high latitudes and high elevations, both of which support communities of relatively few species in many parts of the world (Regal 1982). Sometimes this trend is associated with the predominance of conifers, all of which seem to be wind-pollinated, but the trend is supported also by the distributions of wind-pollinated angiosperms such as oaks (*Quercus*) and the southern beeches (*Nothofagus*). Low species diversity is a permissive rather than a driving factor (Regal 1982): just because conspecifics are often close neighbors in certain areas does not force them to become wind-pollinated.

Changes in the relative frequencies of the two major pollination modes also occur on a small scale. Regal (1982) noted that a transect, only a few meters long in some instances, out from a "tropical hammock" in southern Florida exhibits a trend from no wind pollination (trees with West Indian tropical affinities) to predominant wind-pollination (oaks and pines). There are no good explanations of such trends based simply on the availability of winds or the absence of high rainfall—many rainforests have lots of wind-pollinated species, and even tropical rainforests have some wind-pollinated members (Bawa and Crisp 1980). Regal advanced the hypothesis that high levels of environmental uncertainty may select for high levels (not necessarily obligate) outbreeding and that this, in turn, selects for the use of animals as pollen vectors. In general, animal vectors increase the likelihood that some pollen will travel far (especially when the plants are far apart) and increase outcrossing with more distant individuals, and this should increase the genetic diversity of the ensuing offspring. When some wind-pollinated species persist in an area dominated by animal-pollinated ones, one then might predict that these species have some special means of reducing the effective environmental variability (e.g., long seed dormancies and/or restricted conditions for germination (Regal 1982). However, this example seems to concern variability chiefly in the physical environment; biological unpredictabilities could easily produce countering pressures that obscure any effects of physical environments. Furthermore, at least some wind-pollinated trees have long pollen shadows (Wright 1978), so the differences in outcrossing rates of wind- and animal-pollinated plants may be smaller than they are presumed to be. And, of course, seed dispersal also affects the potential for outcrossing, and the length of the seed shadow may be very different from the reach of pollen. Global changes in plant distribution and historical factors may have some influence on present-day distributors of pollination modes, but their relative importance is difficult to evaluate. In short, a thorough-going explanation for the distributions of pollination modes is still lacking.

There are broad geographic patterns in the frequency of different kinds of animal vectors of pollen. Pollination by bats is largely a tropical and subtropical phenomenon, edging as far north as southern Arizona in desert habitats. Pollination by nonvolant mammals is known or strongly suspected in marsupials (the Australian honeypossum *Tarsipes spencerae,* neotropical opossums, and perhaps others), primates (lemurs on Madagascar, galagos in Africa, marmosets and monkeys in Central and South America), rodents in Australia and South Africa, and perhaps procyonids in South America (Armstrong 1979, Carpenter 1978, Janson et al. 1981, Roarke and Wiens 1977, Sussman and Raven 1978, Wiens and Roarke 1978). Bird-pollination is likewise concentrated at lower latitudes and in the southern hemisphere, the major exception being found in the geologically recent invasion of North America by hummingbirds. Although many different kinds of birds include nectar as part of their diet, a few different families have concentrated on this diet (see also Stiles 1981): hummingbirds (Trochilidae) and some orioles (Icteridae) in the Americas, sunbirds (Nectariniidae) in Africa (and a few in southern Asia), honeycreepers ("Coerebidae") in the tropical Americas, and Hawaiian honeycreepers (Drepanididae) in the Hawaiian archipelago, honey-eaters (Meliphagidae) and a few lorikeets (Psittacidae) in Australia (Armstrong 1979, Christiansen 1971, Ford et al. 1979, Paton and Ford 1977), and the sugarbirds of southern Africa, whose taxonomy is as uncertain as the "coerebids." Nectar-feeding birds are noticeably uncommon in the Palearctic (e.g., Lien 1972, Stiles 1981) and, if interspecific competition in nectar-feeding animals is a significant constraint on adaptive radiation (Brown et al. 1978), their absence may permit diversification of the bee (and other pollinating insects) fauna (Inouye 1977), or perhaps in some cases the radiation of insects prevents a corresponding radiation in nectar-feeding vertebrates. In still other instances, such as the movement of hummingbirds into North America, the presence of new kinds of pollinators may permit new waves of adaptive radiation by the pollinated plants (see e.g., Grant and Grant 1965), although, on the other hand, the small size and competitive interactions of hummingbirds (e.g., Carpenter 1978a) may restrict the value of hummingbirds as pollinators for small plants because a single flowering canopy might be occupied by territorial hummingbirds that reduce pollen-flow (Stiles 1978, 1981).

The explanation(s) for known geographic patterns are little explored. Biogeographic and historical factors are clearly implicated in the case of North American hummingbirds, and perhaps of the Palearctic paucity of nectar-feeding vertebrates, and in the exploitation of flowers by lemurs in Madagascar and marsupials in Australia. Vertebrate pollination is costly, requiring large flowers or inflorescences and copious nectar (Recher 1981, Stiles 1978). These costs must be outweighed by some advantage (Brown et al. 1978). Ecological explanations have only begun to be advanced: selection for outcrossing may favor the case of volant far-traveling vertebrates as pollinators (Cruden et al. 1976, Ford et al. 1979); birds—and sometimes mammals—often consume insects in and around the flowers they are visiting, and may have an additional antiherbivore function, perhaps especially in nutrient-poor habitats where the protein cost of new leaves is high (Ford et al. 1979); competition for pollinators could provide an additional evolutionary thrust (Recher 1981). These are, no doubt, geographic patterns in variety of insect pollination (e.g., an absence of hawkmoths in Arctic and sub-Arctic regions and at high elevations), but I have found little discussion of such patterns (but see Cruden et al. 1976, Primack 1978b).

Unlike the wind, animal pollen vectors can evolve, and so the interactions between pollinator and pollinated can be far more intricate when animals are used as vectors. Animal behavior is so varied that a diversity of pollination adaptations may exploit this variety and animal behavior may be "manipulated" to advantage by morphological features of the flowers. As a result, there are a number of ways to increase (evolutionarily) both male and female reproductive success through the pollination system.

SIGNALS AND REWARDS

Animal-pollinated plants attract their pollen vectors in a variety of ways. A reward is provided in most cases, but sometimes only a fake reward, or none at all, is presented and the vector deceived.

A common reward is nectar, secreted by the flower and consumed by the animal visitor; in one case, extrafloral nectaries may reward the pollinator (Faegri and van der Pijl 1979). Nectar volume differs with species, pollinator foraging behavior, microsite, floral age, time of day, sex, and no doubt other factors (e.g., Bawa 1980, Feinsinger 1978, Haddock and Chaplin in press, Pleasants and Chaplin unpublished manuscript). *Asclepias quadrifolia,* a milkweed of wooded habitats, exhibits great variation in nectar volumes among individuals, among the flowers on an umbel, and with umbel size (Pleasants and Chaplin unpublished manuscript). Plants with larger roots produced more nectar per flower; above the average umbel size, floral nectar volumes decreased. Interestingly, pollinia removal rates per plant were correlated with nectar production rates, so male RS is likely to benefit directly from levels of available rewards for pollinators (mainly an anthophorid bee, also three species of skipper and a butterfly; Chaplin and Walker in press). However, neither pod initiation nor eventual female RS were associated with nectar levels (Pleasants and Chaplin unpublished manuscript). This is the only study I know that relates intraspecific, between-individual, differences in pollinator rewards to potential reproductive success of each sex.

Concerning both nectar volume and nectar composition (below), it is possible, for plants that customarily grow in groups, that pollinator rewards are assayed by the pollinator at the level of the group of plants (a foraging patch for the pollinator, rather than the individual plant). Although this may open the door to "cheating" among conspecifics, whereby individuals with low levels of reward are visited as successfully as their neighbors that expend more resources on pollinator rewards, this possibility and its consequences have not been studied. There is evidence, however, that bees may prefer continuous to intermittent rewards (e.g., Waddington et al. 1981), whereas some hummingbirds may favor variable rewards (Feinsinger 1978).

The chemical composition of nectar varies enormously among species, and in some cases also varies with floral age, sex, season, and location (Baker 1978, Percival 1961, Shuel 1975). The biological significance of intraspecific variation in composition is unexplored, but some interspecific patterns are beginning to emerge.

Major constituents of apparently all floral nectars are sugars, mainly sucrose, glucose, and fructose (Percival 1961). Amino acids are also extremely common (Baker and Baker 1975). Proteins, lipids, anti-oxidants (such as vitamin C), other vitamins, and even toxins occur (Baker 1978, Baker and Baker 1975, Rhoads and Bergdahl 1981, Stephenson 1981). Flowers with deeply concealed nectar, generally at the base of a long-tubed corolla, especially those pollinated by lepidopterans and hummingbirds (in Costa Rica, Baker 1978), show some tendency to produce nectars with sucrose predominant among the sugars (Percival 1961) and lepidopteran-pollinated flowers often have higher amino acid contents than open flowers with unconcealed nectar (Baker and Baker 1975). These trends are interpreted to reflect a higher degree of specialization of concealed-nectar flowers to their pollinators.

Within these broad trends, however, considerable variation is found. The nectars of flowers pollinated by specialized carrion flies and by beetles commonly have very high amino acid contents (although these flower corollas are not especially long), whereas nectars from flowers pollinated by bats and birds (except the North American hummingbirds) have low average amino acid contents (and often long corollas). Hummingbird flowers in California are thought to be exceptional perhaps because they have only recently evolved from being pollinated by bees; tropical hummingbird flowers have long-standing relationships with their pollen vectors and low amino acid scores (Baker 1978, Baker and Baker 1975). Amino acid rewards in flowers pollinated by long-tongued bees, butterflies and some moths, and wasps are also relatively large; interestingly, certain low-scoring butterfly flowers happen to be those in which the floral morphology causes the butterflies to push pollen down into the nectar as the proboscis is extended, and amino acids are quickly leached into the nectar and consumed by the visiting butterfly (Baker and Baker 1975). Nectar rewards for hummingbirds are not especially rich in sugars; bee-pollinated flowers seem to produce the most concentrated nectars (Baker 1975, Pyke and Waser 1981). However, the adaptive basis for differences in concentration is still in dispute (Baker 1975, Pyke and Waser 1981).

Nectar is an attractive food resource (and *Trigona* bees even use it in nest-building, Ramirez and Gomez 1978) and some animals have found ways of getting it without pollinating the flowers. In some cases, the flowers are visited by insects (thrips, ants, etc.) whose bodies are so small that they do not touch the anthers. More interesting are the habitual mutilators of flowers, who chew holes in the corolla and obtain nectar without entering the flower. These holes may be subsequently used by other nectar-stealing animals. Many species of large bees (*Bombus, Xylocopa*) are regular nectar thieves on the flowers of such things as *Aquilegia* and *Impatiens*. Even birds (the neotropical flower-piercers, *Diglossa* spp.) regularly forage by puncturing floral corollas and others do so at least sometimes. These larcenous habits may make it advantageous to protect the nectar supply. Thieves could be excluded by such features as thick enveloping sepals or other morphological devices (Kerner 1878), perhaps the use of insect guardians (e.g., biting ants) attracted by special rewards outside the flower, and toxins in the nectar. Various nectar chemicals exert a toxic effect; alkaloids are known to occur in the nectars of several bee-pollinated flowers and to have narcotic, toxic effects on lepidopterans (Baker and Baker 1975, Stephenson 1981). Bees themselves may be poisoned; in the case of nectar from *Sophor microphyllus*, the toxicity varied among individual trees (Clinch et al. 1972, in Godley 1979).

The possibility that nectar depletion by thieves forces legitimate pollinators to visit more flowers and hence to be more effective has been discussed but not established in many cases: the system appears *not* to work in *Asclepias curassavica* robbed by ants (Wyatt 1980) or in *Impatiens capensis* robbed by bees (Rust 1979). Bumblebees utilized fewer flowers per inflorescence of *Trifolium repens* and by-passed more neighboring inflorescences when florets contained little nectar (Heinrich 1979). Pollen-flow patterns must then have been more wide-ranging, but the effects on seed-set and pollen donation were not recorded. McDade and Kinsman (1980) provide a review of literature on nectar and pollen thievery and raise the more likely alternative possibility that thievery may turn away the legitimate pollinators altogether, thus having a negative effect on RS of the plant (see also Rust 1979). Ants on the umbels of *Asclepias syriaca* were associated with reduced pod production (Fritz and Morse 1981). Robbery by *Trigona* bees and bee aggression against the pollinating hummingbirds resulted in reduced seed production in *Pavonia dasypetala* in Panama (Roubik 1982). Thieves might even affect the numbers and distributions of flowers on a plant: one existing hypothesis is that nectar thievery by flower-piercers (*Diglossa plumbea*) selects for *Centropogon valerii* plants that produce enough flowers in certain positions that the remaining flowers still contain enough nectar to attract hummingbirds, which are the principal pollinator (Colwell et al. 1974). And Free (1970) noted that corolla-piercing and nectar-robbing by bumblebees foraging on *Phaseolus multiflorus* might attract honeybees, which begin foraging by using the holes pierced by the bumblebees but eventually find the proper floral entrance and pollinate the flowers. There seems to be no paucity of available hypotheses regarding the effects of nectar larceny on patterns of pollen flow, but general patterns of these are lacking.

Typically, we expect daily patterns of nectar availability to match the activity of the effective pollinators (e.g., Schaffer and Schaffer 1977). However, there are exceptions (*Tilia,* Anderson 1976; *Asclepias,* Bertin and Willson 1980, Willson et al. 1979) that may indicate systems in transition or disruption, although it is possible that conflicting selection pressures might prevent nectar presentation from closely tracking pollinator activity. If the costs of having nectar available at times of low pollinator activity are small, there may be little selection to restrict the daily pattern (Miller 1981). There has been little analytical study of the costs and benefits of circadian patterns of nectar production and availability.

Pollen is another very common reward for pollinators; it provides protein, lipids, sugars, and starches (Baker and Baker 1979, Faegri and van der Pijl 1979). Pollen contents may be subject to a variety of selection pressures, including the reward for pollen vectors. A survey of almost 1000 California species demonstrated that starch was a major constituent of most wind-borne pollens but was much less frequent in animal-borne pollens (Baker and Baker 1979). Where pollen was the reward presented to bee and fly pollinators, starch was far less frequently present, and oil was a major component. These pollen grains were relatively small, perhaps because selection might favor production of large numbers of grains so that some escape predation (Baker and Baker 1979). On the other hand, the large pollen of hummingbird-pollinated flowers likewise contained starch only rarely. Hummingbirds are not known to use pollen as food; in this case the Bakers suggest that the oily, energy-rich pollen storage products provide resources needed for pollen-tube growth through the long style. Flowers pollinated by lepidopterans also tend to have long styles, but their pollens are not predominantly oil-rich; instead the pollen grains are unusually large. The basis for the evolutionary choice between quantity and quality of the pollen reserves is not clear (Baker and Baker 1979).

Some insects consume pollen on the spot, by chewing (beetles) and by leaching out the contents in nectar (*Heliconius* butterflies, Gilbert 1972).

Pollen consumption enhances fecundity of female *Heliconius* (Gilbert 1972), and different pollens affect female fecundity of ichneumon wasps to different degrees (Leius 1963). Bats also ingest pollen along with nectar and then extract the nutrients internally (Howell 1974). When pollen is a reward for foraging bats, the protein content is high (Howell 1972, in Schaffer and Schaffer 1977). Small Australian parrots called lorikeets feed on nectar and especially pollen of eucalyptus trees; at least some species even feed their young on this diet (Christiansen 1971). Some bees (*Bombus, Apis*) collect large quantities of pollen and take it back to their nests for larval food. Use of pollen as a pollinator reward often necessitates its copious production; even wind-pollinated species (e.g., *Zizania aquatica*) are subject to pollen collection by bees and probably are selected to produce compensating amounts. Some plants bear anthers offering pollen modified especially for consumption by vectors, a different substance altogether, or nothing at all (Baker 1978, Faegri and van der Pijl 1979, Mori et al. 1980, Simpson and Neff 1981, Vogel 1978). The offering of modified pollen and the fraudulent presentation of anthers may protect the functional pollen from the floral visitors. In some instances the pollen itself is repellent to would-be collectors (e.g., *Gossypium, Kallstroemia grandiflora*; Faegri and van der Pijl 1979).

The other major food reward known is oil, commonly secreted by the corolla or other floral parts (Simpson and Neff 1981). Members of at least eight families (reported especially from South American representatives) offer oils to solitary bees (anthophorids in the neotropics). Male bees may eat the oil; females apparently carry it back to the nest, probably as larval food (Buchmann and Buchmann 1981, Faegri and van der Pijl 1979, Simpson and Neff 1981). Resin is offered by the flowers of *Dalechampia magnistipulata* in Mexico; this is collected by female euglossine bees for use in nest-building (Armbruster and Webster 1979). In some cases the pollinators consume major floral components: the petals and sepals of male flowers of the palm *Bactris major* are eaten by beetles that then carry pollen to female flowers (Baker 1978).

One of the more intriguing pollinator rewards is found in a number of orchids and members of several other families (Armbruster and Webster 1979, Simpson and Neff 1981). This is an oil-based perfume collected by male euglossine bees (Dodson 1975, Dodson et al. 1969) that may use the odor to attract females (Kimsey 1980, Simpson and Neff 1981).

Some flowers are used as shelters by insects that then pollinate them. Heliotropic flowers track the sun during all or part of the day. The temperature inside the open corollas is higher than ambient when the sun is shining and the wind is slow. Insects bask in these flowers and often pollinate them in the process (Kevan 1975, Smith 1975). Heliotropism seems to be most often reported for arctic and alpine flowers, but many kinds of dish-shaped flowers, even if not heliotropic, can be used for insect basking. Bees of several species sleep in the flowers of a terrestrial orchid, *Serapias vomeracea*, in Israel, receiving shelter and morning warmth, and providing pollination (Dafni et al. 1981). Sun-trapping flowers may also increase the rate of seed maturation (e.g., Knutson 1981).

Heat generated by the plant itself may attract insects. Skunk cabbage (*Symplocarpus foetidus*) blooms in early spring; its hooded inflorescence is thermogenic and is sought by early-flying insects, which then often pollinate the flowers (Knutson 1979). Many other members of this chiefly tropical family (Araceae) also use heat as a pollinator attractant.

Brood sites for insect breeding are provided (with no reciprocal benefit in pollination) by some flowers, for example, nitidulid beetles in cactus flowers (Grant and Connell 1979). In contrast, amazing complex and reciprocal relationships of sheltering are known to occur between *Tegeticula* moths and *Yucca* flowers and between figs (*Ficus*) and fig wasps (Agaonidae). In both cases, the plant "pays a price" for specialized pollination, because both moth and wasp larvae eat developing seeds of the host plant. Predation rates on seeds are sometimes rather high (Janzen 1979a,b; Wallen and Ludwig 1978), and the cost of pollination may be significant.

Yucca is distributed in dry habitats across the southern United States and south through Mexico into Central America. *Tegeticula* pollinates all of them, but only *Y. whipplei* and perhaps a few others are known to have a *Tegeticula* specific to them. Female moths cross-pollinate the flowers and, during the same visit, oviposit in the ovary (Aker and Udovic 1981). The larvae—usually one or two per ovary but sometimes as many as six—consume a fraction of the seeds as they mature (Powell and Mackie 1966).

Ficus is a tropical genus of many species; typically, each fig species is pollinated by one species of fig wasp, but there are exceptions. The inflorescence is globular, and the flowers are internal. Female wasps gain entry by way of a small pore and lay eggs, through the styles, in the ovules. They bear pollen from other figs and pollinate flowers in the inflorescence used for egg-laying. When their female offspring mature, they are fertilized by new, wingless males (often their brothers). The new females collect a load of pollen and leave home for another fig. where they, in turn, lay eggs. *Ficus* inflorescences are strongly protogynous, the stamens maturing weeks or months (depending on the species) after the stigmas. In general, the figs on a single tree mature synchronously, but individual trees in any region are asynchronous, so that pollen movement is commonly between trees (see also Milton et al. 1982).

Most figs have three kinds of flowers in one inflorescence: male flowers, female flowers with long styles, which are pollinated by the wasps but only sometimes receive wasp eggs because the style is too long for oviposition, and female flowers with short styles, which may be pollinated and could produce mature seeds, but which commonly are parasitized by the wasp. Less commonly, and reportedly only in the Old World, long-styled flowers are borne on some, female-functioning, individuals, whereas male and short-styled flowers are borne on others, which function largely as males. Wasps apparently confuse the two types of figs and enter both, but they cannot oviposit very successfully in the long-styled "seed fig." These types of figs thus tend to be functionally dioecious. It is tempting to speculate that functional separation of the sexes may have evolved as a means of increasing the number of pollen-carrying female wasps while lowering seed predation (Janzen 1979a).

I have summarized only part of a remarkable and still-emerging story regarding figs and fig wasps. An entrée into the literature is provided by Janzen (1979a) and Valdeyron and Lloyd (1979).

The offering of bona-fide rewards by most animal-pollinated flowers makes possible the evolution of floral types that offer no food reward but deceive the pollinators at least long enough to achieve the transfer of pollen (see also Wiens 1978). For such a system to work, the rewardless flowers should be relatively rare, so that the animals do not learn to recognize and avoid them; pollination may rely on seasonal flushes of naive pollinators (e.g., Ackerman 1981), or success in pollination may be achieved principally by means of compensating features, such as long floral life (e.g., Melampy and Hayworth 1980). A false signal may be presented to the pollinators, deceptively indicating the presence of a reward. Because the pollinator then obtains no reward, at best its visits represent a waste of time or effort, perhaps small.

Orchids include some past masters of deception. Several species have flowers that mimic female bees, wasps, or flies. The males of these insects are attracted (visually and olfactorily, Bergström 1978) to the flowers, try to copulate with them, and may go so far as to prefer the flowers to real females (van der Pijl and Dodson 1966). An Australian ichneumon wasp (*Lissopimpla*) pollinates *Cryptostylis* this way; male tachinid flies service South American *Trichoceros*; male scoliid wasps (*Campsoscolia ciliata*) in the Mediterranean area pollinate *Ophrys speculum,* which is fringed with red hairs like the abdomen of the female. Many other examples are known (Dressler 1981, Proctor and Yeo 1973, van der Pijl 1966, van der Pijl and Dodson 1966). Other species (e.g., *Oncidium*) resemble male bees and, when the flowers sway in the wind, are attacked (and pollinated) by aggressive, perhaps territorial, male *Centris* bees. Still others (*Brassia, Calochilius*) appear to resemble prey and are attacked by female *Camp-*

Figure 3.6 A *Brassia* orchid, thought to resemble the arachnid prey of large wasps that pollinate it.

someris wasps foraging to provision their young (Figure 3.6, van der Pijl and Dodson 1966).

More serious losses (from the pollinator's perspective) occur when response to a false signal entails the loss of an entire brood of young. Fungus gnats (Mycetophilidae) normally oviposit on fungi, but some angiosperms seem to have spongy, white, odoriferous, fungoid tissues within their flowers. The gnats mate and oviposit on this fake fungus just as if it were the real thing, but the larvae are doomed to an early death. In the process of mating, however, the gnats apparently pollinate the flowers of *Arisarum proboscideum* and *Asarum caudatum* (Vogel 1973).

The signals presented by flowers for the attraction of pollinators may be visual or olfactory. Olfactory signals have not been studied in much detail although they clearly vary greatly, from the common sweet aroma of most odor-bearing flowers (and congeners may vary perceptibly even to the relatively poor olfactory perceptions of humans; *Asclepias* is a case in point) to the carrion stench of *Stapelia*. Insects can discern many odors not perceptible to humans at all. Odors can have a long-range function, to elicit searching behavior and bring in the pollinators, and a short-range function, to orient them properly in the flower and elicit feeding behavior.

Visual signals can be shapes (and sizes) or colors. Floral colors are complicated to study, because insects (at least bees), and perhaps other pollinating animals as well, perceive a different color spectrum than we do. Bees perceive wavelengths in the ultraviolet range (which humans do not) and fail to perceive the differences among the shades of red; thus their spectrum of color sensitivity is shifted toward the shorter wavelengths compared to the human spectrum. At least some butterflies can see red, however (Bernard 1979). Recently, we have learned that hummingbirds also can see ultraviolet light (Goldsmith 1980). Furthermore, insects can distinguish smaller differences in wavelengths than can humans, so that what we might see as one color might be four or five or even more discriminable colors to insects (Kevan 1978). The "color" or hue of a flower is determined by the position of the reflected wavelength in the visual spectrum. Two other measures that may enhance the ability to discriminate one flower from another or a flower from its background are saturation (the degree to which particular wavelengths predominate) and luminance (the amount of light energy reflected; Kevan 1978).

The frequency distribution of flower colors—in terms of the human visual spectrum—are known to vary geographically and seasonally. For instance, white and yellow flowers predominate in Japan (Utech and Kawano 1975) and Canada (Kevan 1972), and white flowers dominate the flora of New Zealand, especially in the mountains, but are less common in the floras of Tierra del Fuego and still less common in the British Isles where yellow is better represented (Godley 1979). White flowers prevail in the early spring flora of the understory in eastern deciduous forest in Illinois (Schemske et al. 1978), whereas later-blooming flowers are more often yellow, blue, or reddish. The frequency distribution of floral colors differs in coniferous forests on either side of the Cascade Range, a higher diversity of colors being correlated with greater diversity of pollinators and of plants (del Moral and Standley 1979). Even if we could describe such patterns as perceived by the pollinators (as Kevan does), there is no explanation for them yet. Although relative attractiveness to available pollinators may be one factor, other factors including historical and biogeographical areas may be also involved. The issue is almost unstudied scientifically.

In a very general way, floral colors can be correlated with classes of pollinators: bird flowers are often predominantly red; nocturnal moth or bat flowers are commonly white, and some kinds of bees are apparently particularly attracted to blue. But such correlations are by no means tight; bees visit flowers of many colors in addition to blue, birds do not fancy red to the exclusion of all else, and butterflies also use red flowers. The extent to which pollinators choose flowers on the basis of color varies with the diversity and reward levels of available flowers, the availability of other perceptible signals, and the foraging habits of pollinators (e.g., Free 1966, Heinrich 1976, Kay 1978, Stiles 1976).

The use of color signals by flower visitors varies enormously, as do the consequences. *Raphanus raphanistrum* in Britain has a conspicuous floral color polymorphism; the flowers are white or yellow (to humans, degrees of purple to bees). In mixed populations, the butterfly *Pieris rapae* and syrphid flies, *Eristalis* spp., strongly preferred yellow flowers, whereas some *Bombus* spp. in some areas favored white ones (Kay 1978). Such behavioral differences may maintain the polymorphism in these populations—both color morphs are pollinated, although relative seed production was apparently not measured. Another example of an apparently stable polymorphism is found in the Hawaiian tree, *Metrosideros collina*. Red-flowered individuals are pollinated primarily by Hawaiian honeycreepers, yellow-flowered ones (found mostly in lower altitudes) may be pollinated by insects or, in part, by autogamy (Carpenter 1976). White-flowered morphs of *Phlox pilosa* set as many or even more seeds that the normal pink-flowered morph in mixed populations (Levin and Kerster 1970). In contrast, rare albino flowers of *Delphinium nelsonii* in the Colorado mountains were discriminated against by their hummingbird and bumblebee pollinators, which preferred and foraged more efficiently on the normal blue flowers; the blue flowers set more seed as a result (Waser and Price 1981). Whether the pale morphs are maintained in the population at a low frequency simply by mutation or perhaps by some compensating advantages in particular situations is not known. Lepidopteran pollinators discriminated among corolla-color variants of *Phlox drummondii*; very rare phenotypes were at a disadvantage (Levin 1972a,b). Pink-flowered variants of *Lupinus nanus* were rare in five California populations and suffered significantly lower fecundity than the normal blue floral types (Harding 1970). Again in contrast, insect pollinators (bees, sphingid moths, flies) failed to discriminate among color morphs (light blue to purple) of *Polemonium viscosum* in Colorado (although they did respond to a dimorphism in scent; Galen and Kevan 1980). The self-incompatible *Platystemon californicus* has five color morphs distributed largely in different regions, but no consistent differences in pollinator visitation or in seed-set were observed (the species is also wind-pollinated; Hannon 1981). Kay (1978) discusses several other examples of floral polymorphism that likewise provide a diversity of results. Obviously, it is impossible to generalize meaningfully at this point.

At close range, colors may serve as signals focusing the attention of the pollinator in a direction appropriate for pollination. A blue–white

Figure 3.7 Two UV-color morphs of *Rudbeckia hirta*. UV wavelengths are reflected by the light parts of the petals and absorbed by the dark portions. The photographs of course cannot show the yellow color of the petals in the "visible" spectrum, which can also be seen by insects. Photos by K. D. McCrea.

contrast in normal flowers of *Delphinium nelsonii* marks the opening to the nectaries and seems to be important in orienting the foraging pollinators (Waser and Price 1981). Nectar guides, in the form of contrasting colors (including UV) in zones, spots, or lines (and sometimes olfactory cues as well) are well known (e.g., Jones and Buchmann 1974, Procter and Yeo 1973). At least in some cases, the development of these guide marks varies among individuals of a population; examples include *Rapistrum rugosum*, Horovitz and Cohen 1972; *Rudbeckia hirta*, an obligate outbreeder (Figure 3.7, Abrahamson and McCrea 1977), and birdfoot violet (*Viola pedata*). However, the extent and importance of such variation in different species seems to be unstudied. Color is also used to signal the end of floral receptivity: for example, a yellow ring around the throat of the lavender corolla of *Lantana trifoliata* fades away after the flower is pollinated (Schemske 1976). These changes presumably serve to focus the pollinators' attention on the remaining receptive flowers.

Lupinus argenteus flowers have on the banner petal a yellow spot that turns purple as the flower ages (Gori unpublished manuscript). Although the older flowers function in attraction of pollinators, they are not actually visited by the pollen-foraging bees that land on the inflorescence. Although this ensures that incoming pollen is deposited on younger flowers, at least in *L. argenteus* seed-set is not pollen-limited, so restriction of the bees' activities to the younger flowers has little impact on female RS. Instead, Gori suggests that pollen removal and male RS are probably enhanced.

As a general rule, and with other factors controlled (see Pyke 1981), large flowers are more attractive to pollinators than small ones (e.g., Faegri and van der Pijl 1979). This is no doubt why outcrossing types generally have larger flowers than their inbreeding relatives (e.g., Rick et al. 1979, Schoen 1977, Spira 1980, and many others—but Schoen 1982 reports that the levels of outcrossing actually achieved in *Gilia achilleifolia* was not correlated with flower size). Clusters of flowers also are more attractive than single ones (e.g., Gori unpublished manuscript). *Trillium erectum* individuals with large rootstocks produced more and larger flowers than those with small roots, and their flowers were visited more often by pollinating insects (Davis 1981). Furthermore, large inflorescences of *Asclepias syriaca* attracted more visitors and more potential pollinators than small ones (Willson and Bertin 1979). Bumblebees prefer large umbels of *Aralia hispida* (Thomson et al. 1982). However, flowers may be clustered not merely for pollinator attraction (to increase the amount of pollen deposited and removed, and the variety of pollen sources), but possibly for architectural reasons or to protect the innermost flowers from predators or nectar thieves (Barrows 1976).

Floral shapes may serve both as features for identification and as determinants of which visitors can be effective pollinators. In some cases, the correspondence between floral shape and some feature of the pollinators is reasonably good. *Heliconia* flowers with relatively short corollas, often straight ones, are commonly pollinated by hummingbirds with relatively short, straight bills, in Costa Rica (Stiles 1975). But species with long, often curved, corollas are pollinated mainly by hermit hummingbirds, with large curved bills (Stiles 1975). (The hermits are quite different in ecology from other hummingbirds, being nonterritorial, not

Figure 3.8 Variation in floral size and shape among races of *Gilia splendens*. (*a*) Cyrtid fly and the associated floral design. (*b*) Bombyliid fly and the common floral form. (*c*) Hummingbird and another corolla variant. (*d*) Autogamous form. Redrawn from Grant and Grant (1965) by permission of Columbia University Press.

sexually dimorphic, courting on communal display areas called leks, and foraging on traplines rather than in patches, Stiles and Wolf 1979.)

A number of examples of apparently adaptive floral shapes are provided in Grant and Grant (1965), who emphasize that although no correspondences are perfect, clear patterns of floral shapes associated with primary pollinators emerge, not only among species and genera of the Polemoniaceae but even within species. Perhaps a classic example is found in *Gilia splendens* in California (Figure 3.8). The widespread race has pink funnel-shaped flowers with moderately long corollas and is pollinated mainly by bee-flies (*Bombylius*). A high mountain race is pollinated mainly by a cyrtid fly with a long, needlelike proboscis and has, correspondingly, a long, slender corolla. Another race in another mountain range has larger, broader, brilliant pink flowers and is pollinated primarily by hummingbirds, and yet a fourth race, in the desert, has small pale flowers and is autogamous. On a larger taxonomic scale, the Grants (1965) describe one section of the genus *Ipomopsis* with different pollinators and floral characteristics among the species. *Ipomopsis rubra, I. arizonica,* and *I. aggregata* have red (or pink) trumpet-shaped flowers usually pollinated by hummingbirds, although some are visited also by moths. However, *I. tenuituba, I. candida, I. thurberi,* and *I. longiflora* have white or violet flowers with very long, narrow, tubular corollas and are pollinated by hawkmoths. A last species, *I. macombii,* bears smaller versions of the moth-flowers and is pollinated mostly by skippers.

On the other side of the coin, we can find examples of flowers with very different morphologies being pollinated by similar or even the same pollinators. Most *Pedicularis* species in North America are pollinated almost exclusively by bumblebees (*Bombus*; Macior 1970, 1973a,b). Yet the *Bombus*-pollinated species exhibit a dramatic array of flower shapes, all variations on a theme, and the bees obtain pollen, and nectar when present, with a variety of methods: the flower may be entered right side up or upside down; pollen may be obtained by passive rubbing onto the body, by scraping with the legs, or by vibrating the wings and shaking it out. These species of *Pedicularis* do not show strong sympatry and it is hard in most instances to make a case for present-day interspecific competition driving the divergence. Yet diverge they have—and have kept their pollinators. Even very differently shaped flowers are sometimes visited and pollinated by the same individual bees: *Pedicularis groenlandica* and *Dodecatheon pulchellum* in Colorado both have pink flowers, grow sympatrically, flower synchronously, and in some instances bees will collect pollen (by wing-beat vibration) from both kinds of flowers on the same foraging trip (Figure 3.9, Macior 1968, 1971).

Clearly, although some coevolution may occur between pollinator and flower, such pressures are probably not alone in shaping floral characters. What we need is an integrated picture of floral evolution, which is likely to be a long way down the road.

SEASONALITY

Seasonal patterns of flowering can be regulated by many factors, some of which have been documented and others not. Seed-set by *Claytonia vir-*

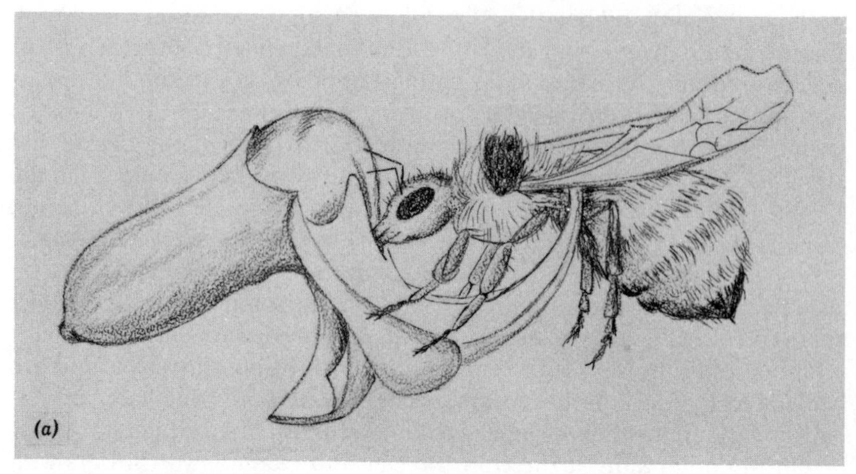

Figure 3.9 Bees foraging on and pollinating flowers of (*a*) *Pedicularis groenlandica* and (*b*) *Dodecatheon pulchellum,* showing the very different floral morphology of flowers often visited by the same individual bee. Redrawn from Macior (1971) by permission of International Bureau for Plant Taxonomy and Nomenclature.

ginica in Illinois appears to be limited early in the season by pollinator activity and late by resource limitation when the canopy leafs out (Schemske 1977). But another species (*Allium tricoccum*) in the same woods produces its ephemeral leaves early and its flowers long after the canopy has closed, and so meets its problem in a very different way, as yet unstudied.

The early flowering of many wind-pollinated trees in the deciduous forests of eastern North America takes advantage of the period before the leaves are well-developed when the breezes have full access to the flowers; whether early flowering was encouraged by competition for insect pollinators later in the season is a moot issue (e.g., Regal 1982). Zimmerman (1980) presented evidence that pollinator activity was limiting to seed-set in *Polemonium foliosissimum* during the later portion of the flowering season and inferred that intraspecific competition was occurring. However, he did not imply that the flowering season was restricted by this factor. More commonly, interspecific competition for pollinators is suggested to shape the flowering season but, in general, the importance of competition for pollinators in determining flowering season is unestablished (Agren and Fagerstrom 1980; Cole 1981; Gleeson 1981; Pleasants 1980; Rabinowitz et al. 1981; Tepedino and Stanton 1981; Thomson 1981; Waser 1978, in press). Although it is likely to be important, in some cases, I have not been able to find any really well-documented instances (see also Waser in press).

Figure 3.9 *(Continued)*

Waser (1978) showed that interspecific pollen transfer (chiefly by hummingbirds) between *Delphinium nelsonii* and *Ipomopsis aggregata* is associated with lowered seed-set of both species. To complete the story, one needs to know also whether seed-set is reduced because of the lower supply of conspecific pollen in the stigma or because of interference by the foreign pollen. Waser suggested that the reduction of seed-set was sufficient to provide selection against simultaneous flowering and to maintain the differences in average flowering times of the two species. To confirm the latter suggestion, however, ideally one would eliminate other possible factors controlling temporal patterns of seed production. However, the story is more complex than this: *D. nelsonii* flowers mainly before *I. aggregata* and, when *D. nelsonii* is flowering abundantly, many hummingbirds are attracted to this resource. Under these conditions, *I. aggregata* achieves good seed-set and the situation has been called an "effective mutualism" (Waser and Real 1979). Perhaps in some years the supply of pollinators would be sufficient to allow simultaneous flowering of both species; only long-term studies could determine how frequently that occurs.

We commonly assume that the flowering season occurs at the time when opportunities for pollination are best, but this view is probably too narrow. The costs of fruit maturation and/or vegetative growth may select for flowering that occurs outside the best season for these activities, if resources for them are not simply stored up (Janzen 1967). Species that flower in early spring in the temperate zone, or shortly after an unfavorable season in general, commonly develop flower primordia at the end of the previous active season. This enables them to reproduce early, before accumulating photosynthate from the current season, but it means that the flowering effort is adjusted to conditions of the previous season, not the present one. Many other species, however, grow before flowering, and flower buds develop during the course of each active season. Such scheduling differences could be responses both to selection for the timing of flowering and for the timing of vegetative growth; their relative importance is unknown. Moreover, temperate trees exhibit two patterns of xylem production (Zimmerman and Brown 1971): some (in which old xylem is inactive) produce new xylem vessels each spring before producing leaves, whereas others, whose old xylem remains functional, leaf out immediately in spring, perhaps because they do not need to take time to lay down new xylem. It seems possible that delayed leaf-out might be as much associated with the requirements of transport anatomy as with those of the pollination system; the association of pollination phenology with seasonal aspects of woody growth in such species might repay closer scrutiny.

The availability of alternative food resources for animal pollinators or the activities of seed predators may also be important (Feinsinger et al. 1982, Janzen 1967). The consequences of asynchrony of flowering may be severe: individuals of the neotropical shrub *Hybanthus prunifolius* suffered both lower seed-set and higher seed predation when experimentally induced to flower out of synchrony with the rest of the population; smaller individuals suffered more than large ones (Augspurger 1981).

Note that we are concerned here with the ultimate factors that give selective value to a given reproductive tactic, not with the proximate factors that provide the appropriate cues. For every species that is constrained by some factor, there is another that has eluded that constraint, and the pressures favoring a certain response in one species but not in another are seldom understood.

LITERATURE CITED

Abrahamson, W. G. and K. D. McCrea. 1977. Ultraviolet light reflection and absorption patterns in populations of *Rudbeckia* (Compositae). *Rhodora* **79:** 269–277.

Ackerman, J. D. 1981. Pollination biology of *Calypso bulbosa* var. *occidentalis* (Orchidaceae): A food-deception system. *Madroño* **28:** 101–110.

Agren, G. I. and T. Fagerström. 1980. Increased or decreased separation of flowering times? The joint effect of competition for space and pollination in plants. *Oikos* **35:** 161–164.

Aker, C. L. and D. Udovic. 1981. Oviposition and pollination behavior of the yucca moth, *Tegeticula maculata* (Lepidoptera: Prodoxidae), and its relation to the reproductive biology of *Yucca whipplei* (Agavaceae). *Oecologia* **49:** 96–101.

Anderson, G. J. 1976. The pollination biology of *Tilia. Am. J. Bot.* **63:** 1203–1212.

Armbruster, W. S. and G. L. Webster. 1979. Pollination of two species of *Dalechampia* (Euphorbiaceae) in Mexico by euglossine bees. *Biotropica* **II:** 278–283.

Armstrong, J. A. 1979. Biotic pollination mechanisms in the Australian flora—a review. *New Zeal. J. Bot.* **17:** 467–508.

Augspurger, C. K. 1980. Mass-flowering of a tropical shrub (*Hybanthus prunifolius*): Influence on pollinator attraction and movement. *Evolution* **34:** 475–488.

Augspurger, C. K. 1981. Reproductive synchrony of a tropical shrub: Experimental studies on effects of pollinators and seed predators on *Hybanthus prunifolius* (Violaceae). *Ecology* **62:** 775–788.

Baker, H. G. 1975. Sugar concentrations in nectars from hummingbird flowers. *Biotropica* **7:** 37–41.

Baker, H. G. 1978. Chemical aspects of the pollination biology of woody plants in the tropics. *In* P. B. Tomlinson and M. H. Zimmerman (eds.), *Tropical Trees as Living Systems.* Cambridge University Press, Cambridge.

Baker, H. G. and I. Baker. 1975. Studies of nectar-constitution and pollinator-plant coevolution. *In* L. E. Gilbert and P. H. Raven (eds.), *Coevolution of Animals and Plants.* University of Texas Press, Austin.

Baker, H. G. and I. Baker. 1979. Starch in angiosperm pollen grains and its evolutionary significance. *Am. J. Bot.* **66:** 591–600.

Barrows, E. M. 1976. Nectar robbing and pollination of *Lantana camara* (Verbenaceae). *Biotropica* **8:** 132–135.

Bawa, K. S. 1980. Mimicry of male by female flowers and intrasexual competition for pollinators in *Jacaratia dolichaula* (D. Smith) Woodson (Caricaceae). *Evolution* **34:** 467–474.

Bawa, K. S. and J. E. Crisp. 1980. Wind-pollination in the understory of a rainforest in Costa Rica. *J. Ecol.* **68:** 871–876.

Beattie, A. J. 1972. The pollination ecology of *Viola*, 2: Pollen loads of insect visitors. *Watsonia* **9:** 13–25.

Beattie, A. J. 1976. Plant dispersion, pollination and gene flow in *Viola*. *Oecologia* **25:** 291–300.

Bell, G. 1978. The evolution of anisogamy. *J. Theor. Biol.* **73:** 247–270.

Bergström, G. 1978. Role of volatile chemicals in *Ophrys*-pollinator interactions. *In* J. B. Harborne (ed.), *Biochemical Aspects of Plant and Animal Coevolution*. Academic, New York.

Bernard, G. D. 1979. Red-absorbing visual pigments in butterflies. *Science* **203:** 1125–1127.

Bertin, R. I. 1982. Paternity and fruit production in trumpet creeper (*Campsis radicans*). *Am. Nat.* **119:** 694–709.

Bertin, R. I. and M. F. Willson. 1980. Effectiveness of diurnal and nocturnal pollination of two milkweeds. *Can. J. Bot.* **58:** 1744–1746.

Best, L. S. and P. Bierzychudek. 1982. Pollinator foraging on foxglove (*Digitalis purpurea*): A test of a new model. *Evolution* **36:** 70–79.

Bierzychudek, P. 1981. *Asclepias, Lantana,* and *Epidendrum*: A floral mimicry complex? *Biotropica* **13** (Suppl.): 54–58.

Bold, H. C., C. J. Alexopoulos, and T. Delevoryas. 1980. *Morphology of Plants and Fungi*. Harper & Row, New York.

Brewbaker, J. L. and B. H. Kwack. 1963. The essential role of calcium ion in pollen germination and pollen tube growth. *Am. J. Bot.* **50:** 859–865.

Brewbaker, J. L. and S. K. Majumder. 1961. Cultural studies of the pollen population effect and the self-incompatibility inhibition. *Am. J. Bot.* **48:** 457–464.

Brown, J. H. and A. Kodric-Brown. 1979. Convergence, competition and mimicry in a temperate community of hummingbird-pollinated flowers. *Ecology* **60:** 1022–1035.

Brown, J. H. and A. Kodric-Brown. 1981. Reply to Williamson and Black's comment. *Ecology* **62:** 497–498.

Brown, J. H., W. A. Calder, and A. Kodric-Brown. 1978. Correlates and consequences of body size in nectar-feeding birds. *Am. Zool.* **18:** 687–700.

Buchholz, J. T. 1939. The embryogeny of *Sequoia sempervirens* with a comparison of the sequoias. *Am. J. Bot.* **26:** 248–257.

Buchmann, S. L. and M. D. Buchmann. 1981. Anthecology of *Moariri myrtilloides* (Melastomaceae: Memecyleae), an oil flower in Panama. *Biotropica* **13** (Suppl.): 7–24.

Carpenter, F. L. 1976. Plant-pollinator interactions in Hawaii: pollination energetics of *Metrosideros collina* (Myrtaceae). *Ecology* **57:** 1125–1144.

Carpenter, F. L. 1978a. A spectrum of nectar-eater communities. *Am. Zool.* **18:** 805–819.

Carpenter, F. L. 1978b. Hooks for mammal pollination? *Oecologia* **35:** 123–132.

Carpenter, F. L. and R. E. MacMillen. 1976. Threshold model of feeding territoriality and test with a Hawaiian honeycreeper. *Science* **194:** 639–642.

Chaplin, S. J. and J. L. Walker. (in press). Energetic constraints and adaptive significance of the floral display of a forest milkweed. *Ecology*.

Charnov, E. L. 1979. Simultaneous hermaphroditism and sexual selection. *Proc. Natl. Acad. Sci. USA* **76:** 2480–2484.

Christiansen, P. 1971. The purple-crowned lorikeet and eucalypt pollination. *Aust. For.* **35:** 263–270.

Clifford, H. T. 1962. Insect pollinators of *Plantago lanceolata. Nature* **193:** 196.

Cole, B. J. 1981. Overlap, regularity, and flowering phenologies. *Am. Nat.* **117:** 993–997.

Colwell, R. K., B. J. Betts, P. Bunnell, F. L. Carpenter, and P. Feinsinger. 1974. Competition for the nectar of *Centropogon valerii* by the hummingbird *Colibri thalassinus* and the flower-piercer *Diglossa plumbea*, and its evolutionary implications. *Condor* **76:** 447–452.

Corbet, S. A., I. Cuthill, M. Fallows, T. Harrison, and G. Hartley. 1981. Why do nectar-foraging bees and wasps work upward? *Oecologia* **51:** 79–83.

Crosby, J. L. 1966. Reproductive capacity in the study of evolutionary processes. *In* J. G. Hawkes (ed.). *Reproductive Biology and Taxonomy of Vascular Plants* (Bot. Soc. Brit. Isles Conf. Rep. 9). Pergamon, Oxford.

Cruden, R. W. 1976a. Intraspecific variation in pollen-ovule ratios and nectar secretion—preliminary evidence of ecotypic adaptation. *Ann. Missouri Bot. Gard.* **63:** 277–289.

Cruden, R. W. 1976b. Fecundity as a function of nectar production and pollen-ovule ratios. *Linn. Soc. Symp. Ser.* **2:** 171–178.

Cruden, R. W. and K. G. Jensen. 1979. Viscin threads, pollination efficiency and low pollen-ovule ratios. *Am. J. Bot.* **66:** 875–879.

Cruden, R. W. and S. Miller-Ward. 1981. Pollen-ovule ratio, pollen size, and the ratio of stigmatic area to the pollen-bearing areas of the pollinator: An hypothesis. *Evolution* **35:** 964–974.

Cruden, R. W., S. Kinsman, R. E. Stockhouse, and Y. B. Linhart. 1976. Pollination, fecundity, and the distribution of moth-flowered plants. *Biotropica* **8:** 204–210.

Dafni, A., Y. Iuri, and N. B. M. Brantjes. 1981. Pollination of *Serapias vomeracea* Briq. (Orchidaceae) by imitation of holes for sleeping solitary male bees (Hymenoptera). *Acta Bot. Neerl.* **30:** 69–73.

Darwin, C. 1884. The Different Forms of Flowers on Plants of the Same Species (Reprint of 2nd, 1880, ed.). Appleton, New York.

Davis, M. A. 1981. The effect of pollinators, predators, and energy constraints on the floral ecology and evolution of *Trillium erectum. Oecologia* **48:** 400–406.

del Moral, R. and L. A. Standley. 1979. Pollination of angiosperms in contrasting coniferous forests. *Am. J. Bot.* **66:** 26–35.

Dodson, C. H. 1975. Coevolution of orchids and bees. *In* L. E. Gilbert and P. H. Raven (eds.), *Coevolution of Animals and Plants.* University of Texas Press, Austin.

Dodson, C. H., R. L. Dressler, H. G. Hills, R. M. Adams, and N. H. Williams. 1969. Biologically active compounds in orchid fragrances. *Science* **164:** 1243–1249.

Dressler, R. L. 1968. Pollination by euglossine bees. *Evolution* **22:** 202–210.

Dressler, R. L. 1981. *The Orchids.* Harvard University Press, Cambridge, Mass.

Duckett, J. G. and A. R. Duckett. 1980. Reproductive biology and population dynamics of wild gametophytes of *Equisetum. Bot. J. Linn. Soc.* **80:** 1–40.

Dulberger, R. and R. Ornduff. 1980. Floral morphology and reproductive biology of four species of *Cyanella* (Tecophilaeaceae). *New Phytol.* **86:** 45–56.

Faegri, K. and L. van der Pijl. 1979. *The Principles of Pollination Ecology* (3rd rev. ed.). Pergamon, Oxford.

Feinsinger, P. 1978. Ecological interactions between plants and hummingbirds in a successful tropical community. *Ecol. Monogr.* **48:** 269–287.

Feinsinger, P. and R. K. Colwell. 1978. Community organization among neotropical nectar-feeding birds. *Am. Zool.* **18:** 779–795.

Feinsinger, P., J. A. Wolfe, and L. A. Swarm. 1982. Island ecology: Reduced hummingbird diversity and the pollination biology of plants, Trinidad and Tobago, West Indies. *Ecology* **63:** 494–506.

Ford, H. D., D. C. Paton, and N. Forde. 1979. Birds as pollinators of Australian plants. *New Zeal. J. Bot.* **17:** 509–519.

Frankie, G. W., P. A. Opler, and K. S. Bawa. 1976. Foraging behavior of solitary bees: Implications for outcrossing of neotropical tree species. *J. Ecol.* **64:** 1049–1057.

Free, J. B. 1966. The foraging behavior of bees and its effect on the isolation and speciation of plants. *In* J. G. Hawkes (ed.), *Reproductive Biology and Taxonomy of Vascular Plants* (Bot. Soc. Brit. Isles Conf. Rep. 9). Pergamon, Oxford.

Free, J. B. 1970. *Insect Pollination of Crops.* Academic, New York.

Fritz, R. S. and D. H. Morse. 1981. Nectar parasitism of *Asclepias syriaca* by ants: Effect on nectar levels, pollinia insertion, pollinaria removal and pod production. *Oecologia* **50:** 316–319.

Galen, C. and P. G. Kevan. 1980. Scent and color, floral polymorphisms and pollination biology in *Polemonium viscosum* Nutt. *Am. Midl. Nat.* **104:** 281–289.

Garbutt, K. (unpublished). Secondary pollen transfer: Two vector mediated pollen transfer demonstrated in *Sinapsis arvensis* L.

Gilbert, L. E. 1972. Pollen feeding and reproductive biology of *Heliconius* butterflies. *Proc. Natl. Acad. Sci. USA* **69:** 1403–1407.

Gilbert, L. E. 1975. Ecological consequences of a coevolved mutualism between butterflies and plants. *In* L. I. Gilbert and P. H. Raven (eds.), *Coevolution of Animals and Plants.* University of Texas Press, Austin.

Gleeson, S. K. 1981. Character displacement in flowering phenologies. *Oecologia* **51:** 294–295.

Godley, E. J. 1979. Flower biology in New Zealand. *New Zeal. J. Bot.* **17:** 441–466.

Goldsmith, T. H. 1980. Hummingbirds see near ultraviolet light. *Science* **207:** 786–788.

Gori, D. F. (unpublished manuscript). The evolution of floral display in *Lupinus argenteus* (Leguminosae). I. Floral retention and color change.

Grant, V. 1950. The flower constancy of bees. *Bot. Rev.* **16:** 379–398.

Grant, V. and W. A. Connell. 1979. The association between *Carpophilus* beetles and cactus flowers. *Plant Syst. Evol.* **133:** 99–102.

Grant, V. and K. Grant. 1965. *Flower Pollination in the Phlox Family.* Columbia University Press, New York.

Gregory, P. H. 1973. *The Microbiology of the Atmosphere* (2nd ed.). Wiley, New York.

Haddock, R. C. and S. J. Chaplin. (in press). Pollination and seed production in two phenologically divergent prairie legumes (*Baptisia leucophaea* and *B. leucantha*). *Am. Midl. Nat.*

Handel, S. N. 1976. Restricted pollen flow of two woodland herbs determined by neutron-activation analysis. *Nature* **260:** 422–423.

Hannon, G. L. 1981. Flower color polymorphism and pollination biology of *Platystemon californicus* Benth (Papaveraceae). *Am. J. Bot.* **68:** 233–243.

Harding, J. 1970. Genetics of *Lupinus*, II: The selective disadvantage of the pink flower color mutant in *Lupinus nanus*. *Evolution* **24:** 120–127.

Heinrich, B. 1976. The foraging specializations of individual bumblebees. *Ecol. Monogr.* **46:** 105–128.

Heinrich, B. 1979. Resource heterogeneity and patterns of movement in foraging bees. *Oecologia* **40:** 235–245.

Hopper, S. D. and A. H. Burbidge. 1978. Assortative pollination by red wattlebirds in a hybrid population of *Anigozanthus* Labill. (Haemodoraceae). *Aust. J. Bot.* **26:** 335–350.

Horovitz, A. and Y. Cohen. 1972. Ultraviolet reflectance characteristics in flowers of crucifers. *Am. J. Bot.* **59:** 706–713.

Howell, D. J. 1974. Bats and pollen: Physiological aspects of the syndrome of chiropterophily. *Comp. Biochem. Physiol.* **48A:** 263–276.

Inouye, D. W. 1977. Species structure of bumblebee communities in North America and Europe. *In* W. J. Mattson (ed.), *The Role of Arthropods in Forest Ecosystems.* Springer-Verlag, Berlin.

Janson, C. H., J. Terborgh, and L. H. Emmons. 1981. Non-flying mammals as pollinating agents in the Amazonian forest. *Biotropica* **13** (Suppl.): 1–6.

Janzen, D. H. 1967. Synchronization of sexual reproduction of trees within the dry season in Central America. *Evolution* **21:** 620–637.

Janzen, D. H. 1971. Euglossine bees as long-distance pollinators of tropical plants. *Science* **171:** 203–205.

Janzen, D. H. 1977. A note on optimal mate selection by plants. *Am. Nat.* **111:** 365–371.

Janzen, D. H. 1979a. How to be a fig. *ARES* **10:** 13–51.

Janzen, D. H. 1979b. How many babies do figs pay for babies? *Biotropica* **11:** 48–50.

Janzen, D. H. 1981. Differential visitation of *Catasetum* orchid male and female flowers. *Biotropica* **13** (Suppl.): 77.

Jones, C. E. 1978. Pollinator constancy as a pre-pollination isolating mechanism between sympatric species of *Cercidium. Evolution* **32:** 189–198.

Jones, C. E. and S. L. Buchmann. 1974. Ultraviolet floral patterns as functional orientation cues in hymenopterous pollination systems. *Anim. Behav.* **22:** 481–485.

Kay, O. O. N. 1978. The role of preferential and assortative pollination in the maintenance of flower color polymorphisms. *Linn. Soc. Symp. Ser.* **6:** 175–190.

Kerner, A. 1878. *Flowers and Their Unbidden Guests* (transl. and ed. by W. Ogle). Paul, London.

Kevan, P. G. 1972. Floral colors in the high arctic with reference to insect–flower relations and pollination. *Can. J. Bot.* **50:** 2289–2316.

Kevan, P. G. 1975. Sun-tracking solar furnaces in high Arctic flowers: Significance for pollination and insects. *Science* **189:** 723–726.

Kevan, P. G. 1978. Floral coloration. Its colorimetric analysis and significance in anthecology. *Linn. Soc. Symp. Ser.* **6:** 51–78.

Kimsey, L. S. 1980. The behavior of male orchid bees (Apidae, Hymenoptera, Insecta) and the question of leks. *Anim. Behav.* **28:** 996–1004.

Knutson, R. M. 1979. Plants in heat. *Nat. Hist.* **88**(3): 42–47.

Knutson, R. M. 1981. Flowers that make heat while the sun shines. *Nat. Hist.* **90**(10): 75–81.

Kress, W. J. 1981. Sibling interaction and the evolution of pollen unit, ovule number and pollen vector in angiosperms. *Syst. Bot.* **6:** 101–112.

Lanner, R. M. 1966. Needed: A new approach to the study of pollen dispersion. *Silv. Genet.* **15:** 50–52.

Leius, K. 1963. Effects of pollens on fecundity and longevity of adult *Scambus buolianae* (Htg.) (Hymenoptera: Ichneumonidae). *Canad. Entom.* **95:** 202–207.

Levin, D. A. 1972a. The adaptedness of corolla-color variants in experimental and natural populations of *Phlox drummondii. Am. Nat.* **106:** 57–70.

Levin, D. A. 1972b. Low frequency disadvantage in the exploitation of pollinators by corolla variants in *Phlox. Am. Nat.* **106:** 453–460.

Levin, D. A. 1981. Dispersal versus gene flow in plants. *Ann. Missouri Bot. Gard.* **68:** 233–253.

Levin, D. A. and W. W. Anderson. 1970. Competition for pollinators between simultaneously flowering species. *Am. Nat.* **104:** 455–467.

Levin, D. A. and H. W. Kerster. 1970. Phenotypic dimorphism and populational fitness in *Phlox. Evolution* **24:** 128–134.

Levin, D. A. and H. W. Kerster. 1973. Assortative mating for stature in *Lythrum salicaria. Evolution* 27: 144–152.

Levin, D. A. and H. W. Kerster. 1974. Gene flow in seed plants. *Evol. Biol.* **7:** 139–220.

Lien, M. R. 1972. A trophic comparison of avifaunas. *Syst. Zool.* **21:** 135–150.

Linhart, Y. B. 1973. Ecological and behavioral determinants of pollen dispersal in hummingbird-pollinated *Heliconia. Am. Nat.* **107:** 511–523.

Little, R. J. 1980. Floral mimicry between two desert annuals, *Mohavea confertifolia* (Scrophulariaceae) and *Mentzelia involucrata* (Loasaceae). Ph.D. Thesis, Claremont Graduate School.

Macior, L. W. 1968. Pollination adaptation in *Pedicularis groenlandica. Am. J. Bot.* **55:** 927–932.

Macior, L. W. 1970. The pollination ecology of *Pedicularis* in Colorado. *Am. J. Bot.* **57:** 716–728.

Macior, L. W. 1971. Co-evolution of plants and animals—systematic insights from plant–insect interactions. *Taxon* **20:** 17–28.

Macior, L. W. 1973a. The pollination ecology of *Pedicularis* on Mount Rainier. *Am. J. Bot.* **60:** 863–871.

Macior, L. W. 1973b. Pollination ecology—the study of cooperative interactions in evolution. *In* N. B. M. Brantjes and H. F. Linskens (eds.), *Pollination and Dispersal.* Publ. Dept. Bot., University of Nijmegen.

Macior, L. W. 1978. Pollination interactions in sympatric *Dicentra* species. *Am. J. Bot.* **65:** 57–62.

Maynard Smith, J. 1978. *The Evolution of Sex.* Cambridge University Press, Cambridge.

McDade, L. A. and S. Kinsman. 1980. The impact of floral parasitism in two neotropical hummingbird-pollinated plant species. *Evolution* **34:** 944–958.

Melampy, M. N. 1979. An evolutionary interpretation of diverse modes of reproduction among certain species in the Ranunculaceae. Ph.D. Thesis, University of Illinois.

Melampy, M. N. and A. M. Hayworth. 1980. Seed production and pollen vectors in several nectarless plants. *Evolution* **34:** 1144–1154.

Miller, R. B. 1981. Hawkmoths and the geographic patterns of floral variation in *Aquilegia caerulea. Evolution* **35:** 763–774.

Milton, K., D. M. Windsor, D. W. Morrison, and M. A. Estribi. 1982. Fruiting phenologies of two neotropical *Ficus* species. *Ecology* **63:** 752–762.

Mori, S. A., J. E. Orchard, and G. T. Prance. 1980. Intrafloral pollen differentiation in the New World Lecythidaceae, subfamily Lecythidoideae. *Science* **209:** 400–402.

Motten, A. F., D. R. Campbell, D. E. Alexander, and H. L. Miller. 1981. Pollination effectiveness of specialist and generalist visitors to a North Carolina population of *Claytonia virginica*. *Ecology* **62:** 1278–1287.

Mulcahy, D. L. and G. B. Mulcahy. 1975. The influence of gametophytic competition on sporophytic quality in *Dianthus chinensis*. *Theor. Appl. Genet.* **46:** 277–280.

Niklas, K. J. and K. T. Paw U. 1982. Pollination and airflow patterns around conifer ovulate cones. *Science* **217:** 442–444.

Ockendon, D. J. and L. Currah. 1977. Self-pollen reduces the number of cross-pollen tubes in the styles of *Brassica oleracea*. *New Phytol.* **78:** 675–680.

Ornduff, R. and R. Dulberger. 1978. Floral enantiomorphy and reproductive system of *Wachendorfia paniculata* (Haemodoraceae). *New Phytol.* **80:** 427–434.

Ottaviano, E., M. Sari-Gorla, and D. L. Mulcahy. 1980. Pollen tube growth rates in *Zea mays*: Implications for genetic improvement of crops. *Science* **210:** 437–438.

Parker, G. A. 1978. Selection on non-random fusion of gametes during the evolution of anisogamy. *J. Theor. Biol.* **73:** 1–28.

Parker, G. A., R. R. Baker, and V. G. F. Smith. 1972. The origin and evolution of gamete dimorphism and the male–female phenoneuron. *J. Theor. Biol.* **36:** 529–553.

Paton, D. C. and H. A. Ford. 1977. Pollination by birds of native plants in South Australia. *Emu* **77:** 73–85.

Percival, M. S. 1961. Types of nectar in angiosperms. *New Phytol.* **60:** 235–281.

Pleasants, J. M. 1980. Competition for bumblebee pollinators in Rocky Mountain plant communities. *Ecology* **61:** 1440–1459.

Pleasants, J. M. and S. J. Chaplin. (unpublished). Nectar production rates in *Asclepias quadrifolia*: The adaptive significance of individual variation.

Plitmann, U. and D. A. Levin. (in press). Pollen-pistil relationships in the Polemoniaceae. *Evolution.*

Powell, J. A. and R. A. Mackie. 1966. Biological interrelationships of moths and *Yucca whipplei* (Lepidoptera: Gelechiidae, Blastopasidae, Prodoxidae). *Univ. Calif. Publ. Entomol.* **42:** 1–46.

Price, M. V. and N. M. Waser. 1982. Experimental studies of pollen carryover: Hummingbirds and *Ipomopsis aggregata*. *Decologia* **54:** 353–358.

Primack, R. B. 1978a. Evolutionary aspects of wind pollination in the genus *Plantago* (Plantaginaceae). *New Phytol.* **81:** 449–458.

Primack, R. B. 1978b. Variability in New Zealand montane and alpine pollination assemblages. *New Zeal. J. Ecol.* **1:** 66–73.

Primack, R. B. and J. A. Silander. 1975. Measuring the relative importance of different pollinators to plants. *Nature* **255:** 143–144.

Proctor, M. and P. Yeo. 1973. *The Pollination of Flowers.* Collins, London.

Pyke, G. H. 1978. Optimal foraging in bumblebees and coevolution with their plants. *Oecologia* **36:** 281–291.

Pyke, G. H. 1981. Effects of inflorescence height and number of flowers per inflorescence on fruit set in waratahs (*Telopea speciosissima*). *Aust. J. Bot.* **29:** 419–424.

Pyke, G. H. and N. M. Waser. 1981. The production of dilute nectars by hummingbird and honeyeater flowers. *Biotropica* **13:** 260–270.

Rabinowitz, D., J. K. Rapp, V. L. Sork, B. J. Rathcke, G. A. Reese, and J. C. Weaver. 1981. Phenological properties of wind- and insect-pollinated prairie plants. *Ecology* **62:** 49–56.

Ramirez, B. W. and P. L. D. Gomez. 1978. Production of nectar and gums by flowers of *Monstera deliciosa* (Araceae), and of some species of *Clusia* (Guttiferae) collected by New World *Trigona* bees. *Brenesia* **14–15:** 407–412.

Real, L. A. 1981. Uncertainty and pollinator-plant interactions, the foraging behavior of bees and wasps on artificial flowers. *Ecology* **62:** 20–26.

Recher, H. 1981. Nectar-feeding and its evolution among Australian vertebrates. *In* A. J. Keast, *Ecological Biogeography of Australia.* Junk, The Hague.

Regal, P. J. 1982. Pollination by winds and by animals: Ecology of geographic patterns. *ARES* **13:** 497–524.

Rhoades, D. F. and J. C. Bergdahl. 1981. Adaptive significance of toxic nectar. *Am. Nat.* **117:** 798–803.

Rick, C. M., M. Holle, and R. W. Thorp. 1979. Rates of cross-pollination in *Lycopersicon pimpinellifolium*—impact of genetic variation in floral characters. *Plant Syst. Evol.* **129:** 31–44.

Roarke, J. and D. Wiens. 1977. Convergent floral evolution in South African and Australian Proteaceae and its possible bearing on pollination by nonflying mammals. *Ann. Missouri Bot. Gard.* **64:** 1–17.

Roubik, D. W. 1982. The ecological impact of nectar-robbing bees and pollinating hummingbirds on a tropical shrub. *Ecology* **63:** 354–360.

Roubik, D. W., J. D. Ackerman, C. Copenhaver, and B. H. Smith. 1982. Stratum, tree, and flower selection by tropical bees: implications for the reproductive biology of outcrossing *Cochlospermum vitifolium* in Panama. *Ecology* **63:** 712–720.

Rust, R. W. 1979. Pollination of *Impatiens capensis*: Pollinators and nectar robbers. *J. Kansas Entomol. Soc.* **52:** 297–308.

Rust, R. W. 1980. Pollen movement and reproduction in *Arisaema triphyllum. Bull. Torrey Bot. Club* **107:** 539–542.

Schaffer, W. M. and M. V. Schaffer. 1977. The reproductive biology of Agavaceae, I: Pollen and nectar production in four Arizona agaves. *Southwest. Nat.* **22:** 157–168.

Schemske, D. W. 1976. Pollinator specificity in *Lantana camara* and *L. trifolia* (Verbenaceae). *Biotropica* **8:** 260–264.

Schemske, D. W. 1977. Flowering phenology and seed set in *Claytonia virginica* (Portulacaceae). *Bull. Torrey Bot. Club* **104:** 254–263.

Schemske, D. W. 1978. Sexual reproduction in an Illinois population of *Sanguinaria canadensis* L. *Am. Midl. Nat.* **100:** 261–268.

Schemske, D. W. 1980. Floral ecology and hummingbird pollination of *Combretum farinosum* in Costa Rica. *Biotropica* **12:** 169–181.

Schemske, D. W. 1981. Floral convergence and pollinator sharing in two bee-pollinated tropical herbs. *Ecology* **62:** 946–954.

Schemske, D. W., M. F. Willson, M. N. Melampy, L. H. Miller, L. Verner, K. M. Schemske, and L. B. Best. 1978. Flowering ecology of some spring woodland herbs. *Ecology* **59:** 351–366.

Schlising, R. A. and R. A. Turpin. 1971. Hummingbird dispersal of *Delphinium cardinale* pollen treated with radioactive iodine. *Am. J. Bot.* **58:** 401–406.

Schlising, R. A., D. H. Ikeda, and S. C. Morey. 1980. Reproduction in a Great Basin evening primrose, *Camissonia tanacetifolia* (Onograceae). *Bot. Gaz.* **141:** 290–293.

Schoen, D. J. 1977. Morphological, phenological, and pollen-distribution evidence of autogamy and xenogamy in *Gilia achilleafolia* (Polemoniaceae). *Syst. Bot.* **2:** 280–286.

Schoen, D. J. 1982. The breeding system of *Gilia achilleifolia:* Variation in floral characteristics and outcrossing rate. *Evolution* **36:** 352–360.

Shuel, R. W. 1975. The production of nectar. *In* Dadant and Sons (eds.), *The Hive and the Honeybee.* Dadant and Sons, Hamilton, Ill.

Simpson, B. B. and J. L. Neff. 1981. Floral rewards: Alternatives to pollen and nectar. *Ann. Missouri Bot. Gard.* **68:** 301–322.

Smith, A. P. 1975. Insect pollination and heliotropism in *Oritrophium limnophilum* (Compositae) of the Andean paramo. *Biotropica* **7:** 284–286.

Snow, A. A. 1982. Pollination intensity and seed set in *Passiflora vitifolia. Oecologia* **55:** 231–237.

Spira, T. P. 1980. Floral parameters, breeding system and pollinator type in *Trichostema* (Labiatae). *Am. J. Bot.* **67:** 278–284.

Stebbins, G. L. 1974. *Flowering Plants/Evolution Above the Species Level.* Belknap (Harvard), Cambridge, Mass.

Stelleman, P. 1978. The possible role of insect visits in pollination of reputedly anemophilous plants, exemplified by *Plantago lanceolata,* and syrphid flies. *Linn. Soc. Symp. Ser.* **6:** 41–46.

Stephenson, A. G. 1981. Toxic nectar deters nectar thieves of *Catalpa speciosa. Am. Midl. Nat.* **105:** 381–383.

Stiles, F. G. 1975. Ecology, flowering phenology, and hummingbird pollination of some Costa Rican *Heliconia* species. *Ecology* **56:** 285–301.

Stiles, F. G. 1976. Taste preferences, color preferences, and flower choice in hummingbirds. *Condor* **78:** 10–26.

Stiles, F. G. 1978. Ecological and evolutionary implications of bird pollination. *Am. Zool.* **18:** 715–727.

Stiles, F. G. 1981. Geographical aspects of bird-flower coevolution, with particular reference to Central America. *Ann. Missouri Bot. Gard.* **68:** 323–351.

Stiles, F. G. and L. L. Wolf. 1979. Ecology and evolution of lek mating behavior in long-tailed hermit hummingbirds. *Ornith. Monog.* **27:** 1–77.

Sussman, R. W. and P. H. Raven. 1978. Pollination by lemurs and marsupials: An archaic coevolutionary system. *Science* **200:** 731–736.

Sutherland and Delph. Unpublished manuscript. On the importance of male fitness in plants. I. Patterns in fruit set.

Tanksley, S. D., D. Zamir, C. M. Rick. 1981. Evidence for extensive cover of sporophytic and gametophytic gene expression in *Lycopersicon esculentum. Science* **213:** 453–455.

Tepedino, V. J. and N. L. Stanton. 1981. Diversity and competition in bee-plant communities on short-grass prairie. *Oikos* **36:** 35–44.

Thomson, J. D. 1980. Implications of different sorts of evidence for competition. *Am. Nat.* **116:** 719–726.

Thomson, J. D. 1981. Spatial and temporal components of resource assessment by flower-feeding insects. *J. Anim. Ecol.* **50:** 49–59.

Thomson, J. D. and R. C. Plowright. 1980. Pollen carryover, nectar rewards, and pollinator behavior with special reference to *Diervilla lonicera. Oecologia* **46:** 68–74.

Thomson, J. D., W. P. Maddison, and R. C. Plowright. 1982. Behavior of bumble bee pollinators of *Aralia hispida* Vent. (Araliaceae). *Oecologia* **54:** 326–336.

Udovic, D. 1981. Determinants of fruit set in *Yucca whipplei*: Reproductive expenditure vs. pollinator availability. *Oecologia* **48:** 389–399.

Utech, F. H. and S. Kawano. 1975. Spectral polymorphisms in angiosperm flowers determined by differential ultraviolet reflectance. *Bot. Mag. Tokyo* **88:** 9–30.

Valdeyron, G. and D. G. Lloyd. 1979. Sex differences and flowering phenology in the common fig. *Ficus carica* L. *Evolution* **33:** 673–675.

van der Pijl, L. 1966. Pollination mechanisms in orchids. *In* J. G. Hawkes (ed.), *Reproductive Biology and Taxonomy of Vascular Plants* (Bot. Soc. Brit. Isles Conf. Rep. 9). Pergamon, Oxford.

van der Pijl, L. and C. H. Dodson. 1966. *Orchid Flowers/Their Pollination and Evolution.* University of Miami Press, Coral Gables, Fla.

Vogel, S. 1973. Fungus gnat flowers and fungus mimesis. *In* N. B. M. Brantjes and H. F. Linskens (eds.), *Pollination and Dispersal.* Publ. Dept. Bot., University of Nijmegen.

Vogel, S. 1978. Evolutionary shifts from reward to deception in pollen flowers. *Linn. Soc. Symp. Ser.* **6:** 89–96.

Waddington, K. D. 1981. Factors influencing pollen flow in bumblebee-pollinated *Delphinium virescens. Oikos* **37:** 153–159.

Waddington, K. D., T. Allen, and B. Heinrich. 1981. Floral preferences of bumblebees (*Bombus edwardsii*) in relation to intermittent versus continuous rewards. *Anim. Behav.* **29:** 779–784.

Wallen, D. R. and J. A. Ludwig. 1978. Energy dynamics of vegetative and reproductive growth in Spanish bayonet (*Yucca baccata* Torr). *Southwest Nat.* **23:** 409–422.

Waser, N. M. 1978. Interspecific pollen transfer and competition between co-occurring plant species. *Oecologia* **36:** 223–236.

Waser, N. M. (in press). Competition for pollination and floral character differences among sympatric plant species: a review of evidence. *In* C. E. Jones and R. J. Little (eds.), *Handbook of Experimental Pollination Ecology.* Von Nostrand Reinhold, New York.

Waser, N. M. and M. V. Price. 1981. Pollinator choice and stabilizing selection for flower color in *Delphinium nelsonii. Evolution* **35:** 376–390.

Waser, N. M. and M. V. Price. (in press). Optimal and actual outcrossing in plants and the nature of plant-pollinator interaction. *In* C. E. Jones and R. J. Little (eds.), *Handbook of Experimental Pollination Ecology.* Van Nostrand Reinhold, New York.

Waser, N. M. and L. A. Real. 1979. Effective mutualism between sequentially flowering plant species. *Nature* **281:** 670–672.

Wiens, D. 1978. Mimicry in plants. *Evol. Biol.* **11:** 365–403.

Wiens, D. and J. P. Rourke. 1978. Rodent pollination in southern African *Protea* spp. *Nature* **276:** 71–73.

Wilbur, H. M. 1977. Propagule size, number, and dispersion pattern in *Ambystoma* and *Asclepias. Am. Nat.* **11:** 43–68.

Williamson, G. B. and E. M. Black. 1981. Mimicry in hummingbird-pollinated plants? *Ecology* **62:** 494–496.

Willson, M. F. 1979. Sexual selection in plants. *Am. Nat.* **113:** 777–790.

Willson, M. F. 1981. On the evolution of complex life cycles in plants: A review and an ecological perspective. *Ann. Missouri Bot. Gard.* **68:** 275–300.

Willson, M. F. and R. I. Bertin. 1979. Flower-visitors, nectar production, and inflorescence size of *Asclepias syriaca. Can. J. Bot.* **57:** 1380–1388.

Willson, M. F. and N. Burley. (in press). Mate choice in plants: Tactics, mechanisms, and consequences. Princeton Biological Monographs, Princeton, N.J.

Willson, M. F. and P. W. Price. 1977. The evolution of inflorescence size in *Asclepias* (Asclepiadaceae). *Evolution* **31:** 495–511.

Willson, M. F. and D. W. Schemske. 1980. Pollinator limitation, fruit production, and floral display in pawpaw (*Asimina triloba*). *Bull. Torrey Bot. Club* **107:** 401–408.

Willson, M. F., R. I. Bertin, and P. W. Price. 1979. Nectar production and flower visitors of *Asclepias verticillata. Am. Midl. Nat.* **102:** 23–35.

Willson, M. F., L. J. Miller, and B. J. Rathcke. 1979. Floral display in *Phlox* and *Geranium:* Adaptive aspects. *Evolution* **33:** 41–51.

Wodehouse, R. P. 1935. *Pollen Grains.* McGraw-Hill, New York.

Wright, S. 1978. *Evolution and the Genetics of Populations,* Vol. 4: *Variability Within and Among Natural Populations.* University of Chicago Press, Chicago.

Wyatt, R. 1980. The impact of nectar-robbing ants in the pollination system of *Asclepias curassavica. Bull. Torrey Bot. Club* **107:** 24–28.

Zimmerman, M. 1980. Reproduction in *Polemonium*: Competition for pollinators. *Ecology* **61:** 497–501.

Zimmerman, M. H. and C. L. Brown. 1971. *Trees: Structure and Function.* Springer-Verlag, New York.

Chapter 4

OFFSPRING

The "lower" plants, many of which have a functional alternation of generations, often produce "offspring" during both a sporophytic and a gametophytic phase of the cycle. Haploid meiotically produced spores are the offspring of the sporophyte; they disperse and claim a site for establishment. Spores are numerous, as befits their role in dispersal, and small (often tiny), having received relatively little endowment from the sporophytic parent. In contrast, gametophytes (functioning as females) commonly produce a relatively small number of eggs and, after fertilization of those eggs, may support the growing sporophyte both physically and nutritionally. The cost of maintaining the young sporophyte, even briefly, may be considerable and thus preclude the production of very large numbers. And the greater the chance of offspring survival—as is the case when parental care is involved—the larger the initial investment in the offspring is likely to be (Shine 1978). Such a cycle is common among mosses, ferns and other lower vascular plants, and some sessile algae.

The seed plants, in contrast, have no independent gametophytic phase; the female gametophyte is retained on the sporophytic parent, the male (in the form of a pollen grain) is sent forth to fertilize a female. Sporophytic parents generally endow their offspring with some amount of nourishment. In gymnosperms the allocation to offspring is committed before fertilization, but in angiosperms, the nutritional material develops through double fertilization; in both cases the endowment may be retrieved if the embryo is aborted. The embryo and its nutritional store are enclosed in maternal tissues. Maternal investment in tissues around the embryo varies enormously among species, as does the amount of nutrition provided to the embryo. Although double fertilization in angiosperms means that the father of a zygote has contributed to its nutrition by providing a nucleus that with the corresponding maternal nuclei forms the endosperm, the materials for endosperm growth are probably derived from the maternal sporophyte. And, although pollen grains of both angiosperms and gymnosperms often provide not only a sperm nucleus, but also an array of stored nutrients and metabolic machinery to the egg at fertilization, the amount of material provided by the male (however important it may be to initial zygote growth) is probably small in comparison to the amount of investment by the mother.

The seeds of angiosperms are unusual entities in that they comprise tissues of three distinct genetic compositions: the embryo, usually formed by the union of paternal and maternal gametes; the endosperm, also derived from male and female components but usually in proportions different from those in the embryo; the surrounding tissues, of solely maternal origin. There is the possibility, then, of conflicts of interest among these components of the seed. Westoby and Rice (1982) have argued that conflict exists between embryo and mother—a parent–offspring conflict—and that maternal tissues around the embryo and endosperm limit the amount of maternal endowment the embryonic unit can capture. In addition, Willson and Burley (in press) have argued that a conflict may lie between father and mother over the control of the embryo's fate. Whether one argues for parent–offspring conflict, as have Westoby and Rice (1982) and Queller (in press), or for male–female conflict, as have Willson and Burley, a fundamental point is that a seed probably is not merely a passive recipient of whatever its mother chooses to give it. This question has only begun to be explored theoretically and should yield fascinating developments as research continues. Note that the seeds of gymnosperms comprise tissues of two distinct compositions, zygotic and maternal, and that some conflict may also occur here. Among the lower plants in which the gametophyte supports the young sporophyte, conflict is also possible after fertilization, but no conflict of this sort should occur during the time of spore production.

This chapter deals primarily with the offspring of seed plants because there is much more ecological information available for these than for the lower plants. Of course, many of the same principles can be applied to any plant species. Until a more general treatment becomes feasible, a discussion of the ecology of seed-plant offspring sets forth a number of interesting problems. A seed does many things once it is mature; it travels, it lands somewhere, it may persist for some time, it germinates, and it supports a seedling. A seed's several roles necessitate that a variety of functions resulting from divers and sometimes conflicting selection pressures be bundled together in a deceptively simple-looking and often tiny package.

ENDOWMENT OF THE YOUNG

Two aspects of parental endowment of offspring can be examined here. The quantity is reflected in seed size, the quality in composition.

Seed Size

The size of a seed is an index of the material and energetic investment in each offspring. Clearly, it is an imperfect index because these investments will change with differing compositions (the amount of lipid, of protein, of specific compounds). Nevertheless, as a rule of thumb, a first approximation, we can assume that large seeds are costly to produce.

They may often require relatively long development times (Stebbins 1971, 1974), their cost may necessitate smaller crop sizes or longer intervals between seed crops, and they may be subject to predation by a host of consumers. These potential difficulties all may be surmounted by a variety of adaptive devices. To the extent that seed size represents variation in stored nutrition, it indicates the entire parental endowment for the new individual. Once released from the parent, the seed and seedling are on their own.

Seed size obviously refers directly to the volume of a seed. However, because seed shapes are so varied, volume is often difficult to measure, and weight is commonly used as an index of size. For some purposes, this is fine; but when the external dimensions of a seed influence its dispersibility, landing orientation, or the availability of safe microsites for germination, then clearly weight may be a poor index of size.

The range of seed weights among the seed plants is enormous (Harper et al. 1970). At one extreme, orchids produce minuscule seeds (e.g., *Goodyera repens,* 2×10^{-6} g). At the other are the coconuts; the double coconut (*Lodoicea maldivica*) of the Seychelle Islands is especially impressive; one fruit weighs about 2×10^4 g and contains a huge, bilobed seed in a dispersal unit almost half a meter long (Corner 1966, McCurrach 1960). I have not been able to find the weight of the seed itself.

Although seed weight within a population may vary considerably, the *mean* seed weight is often relatively invariant compared to other features of the parent plant (such as plant size, number of inflorescences, number of seeds; Harper et al. 1970). In some cases, however, mean seed weights differ markedly among populations of the same species (Cook 1972, Crouch and Vander Kloet 1980, Schimpf 1977, P. Thompson 1981, Winstead et al. 1977, Wyatt 1981), or even among conspecific coexisting individuals (e.g., Zohary and Imber 1963). Closely related species sometimes exhibit quite different patterns of regulation of seed sizes and numbers. For example, *Sesbania macrocarpa* varies seed weight more than seed number in response to low levels of nutrition, whereas *S. drummondii* seed numbers and weights are equally variable (Marshall 1982). *Sesbania vesicaria,* in contrast, conforms to the Harperian generalization and maintains a relatively constant seed weight.

If and when mean seed size is relatively constant as stressed by Harper et al. (1970), we cannot argue that, therefore, seed size must be more *important* to the plant's fitness than more variable traits. We can only infer that mean seed size is more constrained than other traits and the degree of constraint appears to be similar among populations. When mean seed size differs among conspecific populations, we cannot know (without further examination) whether the most adaptive seed size itself is different, seed size varies in correlation with some other trait that is directly subject to selection, or the achieved seed size merely reflects constraints from other factors.

Even among the seeds of a single parent, two (or more) distinctly different kinds of seeds sometimes are produced. Seed polymorphisms involving differences in dispersal devices are well known (Hannan 1980; Harper et al. 1970; Sorenson 1978; Stebbins 1971, 1974), as are those with different shapes and germination requirements (Harper et al. 1970, Weiss 1980). Susceptibility to predation may differ (Cook et al. 1971). Polymorphism in seed size is also known in many genera (e.g., *Chenopodium, Atriplex*); in *Xanthium* a large and a small seed travel in a single dispersal unit (Harper et al. 1970). Some plants that reproduce both cleistogamously and chasmogamously produce seeds of two different sizes. Cleistogamously produced seeds of *Gymnarrhena micrantha* and *Emex spinosa* are large and subterranean, whereas the aerial seeds are tiny and air-borne (Weiss 1980). In *Impatiens capensis,* however, the chasmogamously produced seeds are the heavier; in *I. pallida* there is no difference in seed size (Schemske 1978). Although size morphs commonly differ in many aspects of their ecology, the actual adaptive value of seed size polymorphism has been little assessed in detail.

A degree of genetic control of seed size and responsiveness to selection is known for crop and weed plants of agricultural importance. An example of adaptation in seed size (and shape) is provided by *Camelina sativa,* a cruciferous weed that grows with flax (*Linum usitatissimum*) crops. When flax is cultivated for fiber in the northern USSR, the seeds are thin and small, but in the southern part of the country, flax is grown for oil and the seeds (from which the oil is extracted) are plump. When the crop is harvested, the seeds are sorted from the chaff by winnowing. Northern *Camelina* seeds are of a size and shape that allow them to be blown the same distance as the northern flax seeds; southern *Camelina* seeds have a different size and shape, which allows them to match the winnowing behavior of southern flax seeds. Interestingly, southern *Camelina* seeds are much smaller than those of the flax plants with which they coexist and it is their shape more than their size that governs their behavior when harvested (Stebbins 1950). Thus in each region the weed seed has become adapted to the seed harvesting process in such a way that it will be resown with its "host" crop.

There are several important factors that select for particular seed sizes in natural populations. Seed size is a function both of the amount of nutrition stored for the young plant and of the enveloping tissues, which can have several functions, including protection, dispersal, control of orientation when the seed lands, and uptake of water for germination.

Sometimes total seed size may be constrained (e.g., by mode of dispersal) and any changes in nutritional content must be compensated for by corresponding changes in the maternal envelope. But if an increase of nutritional content increases the vulnerability of the seed to predators, there may be selection for an accompanying increase in the thickness of the protective coverings; if total size is constrained, then selection might favor alternate forms of protection (e.g., chemical). In most instances, studies on seed size have emphasized total seed weight to the neglect of possible alterations of internal proportions, thus this discussion centers on total seed size; in rare cases the problem of internal allocations can be addressed.

Germination and Establishment. After a seed disperses and lands, it must eventually germinate. To do so, many seeds must take up water from the soil. Therefore, the amount of contact between seed and soil may be important. The shape of the seed and smoothness of its exterior coat can influence the amount of contact. In addition, seed size is probably critical. Large seeds have a smaller surface/volume ratio than small ones, yet they require more water; furthermore, large seeds may find fewer available crevices in the soil where they can find a safe site for germination (Harper et al. 1970). Harper and colleagues suggest that these problems may be the reasons that large seeds apparently contain more water with them as they disperse; some of the largest known seeds (coconuts) contain a fluid endosperm, perhaps as a solution to the problem of obtaining water for germination. The germination and establishment of seeds of different sizes are affected differentially by substrates of various particle sizes; large seeds of *Plantago* species did better on substrates of large particles (Blom 1978). The thickness, hardness, and composition of the investing envelopes around the embryo/endosperm unit can also influence water uptake (Harper et al. 1970). However, the evolutionary contribution of these factors to seed size have apparently not been studied.

Successful germination and establishment of the young seedling also depend on the nutritional reserves within the seed. An example is provided by Thomas' (1966) study on *Lolium perenne,* a perennial grass. Six different varieties were tested in a greenhouse experiment in which seeds of known size were sown at a standard density. For all six varieties there was a significant positive correlation of seed weight with the length, width, and area of the first few leaves to appear on the main tiller; for most varieties there was also a good correlation of seed weight with tiller number. For all varieties, the effect of seed weight on leaf size diminished as later leaves appeared but, interestingly, the rate of attenu-

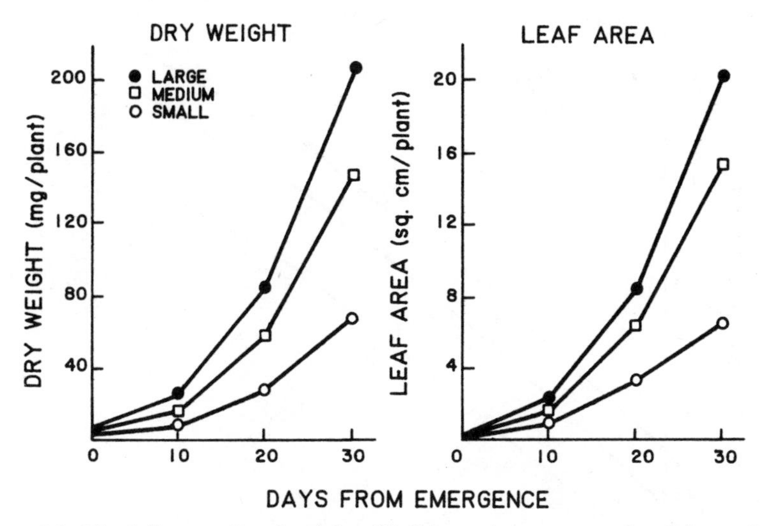

Figure 4.1 The influence of seed weight of *Trifolium subterraneum* on size of the seedling, indexed both as seedling weight and leaf area. Redrawn from Black (1956) by permission of CSIRO.

ation differed among the varieties, suggesting genetic control of the effect. It is also worth noting that the described effects did not appear under all experimental conditions; when environmental conditions were such that growth was slow, seed size seldom produced significant effects on leaf size.

The classic, and perhaps most often cited, study is that of Black (1956, 1957a, 1957b, 1958) on subterranean clover (*Trifolium subterraneum*) in greenhouse culture. The first two papers demonstrated for several varieties of clover that large seeds can produce seedlings from greater depths and large seeds produce heavier seedlings with greater cotyledon area than is true for small seeds (Black 1956, 1957a; Figure 4.1). Greater cotyledon area was associated with greater leaf area for seedlings up to 30 days old. The effect was due to the initial size of the emergent seedling, not to differences in daily relative growth rate, which was similar for both large and small seeds (Black 1956; Figure 4.2). Black (1958) then sowed large ($\bar{x}$ = *10 mg*) and small ($\bar{x}$ = 4 mg) seeds at a density such that the density of seedlings was about 2.3 per square inch. Large and small seeds were sown both separately, in pure stands, and together, in mixed stands. The stands were first sampled almost 40 days after seedling emergence, by which time the initial advantage of seedlings from large seeds was no longer evident—in pure stands. However, in mixed plantings, weight gain and leaf area for seedlings emerged from

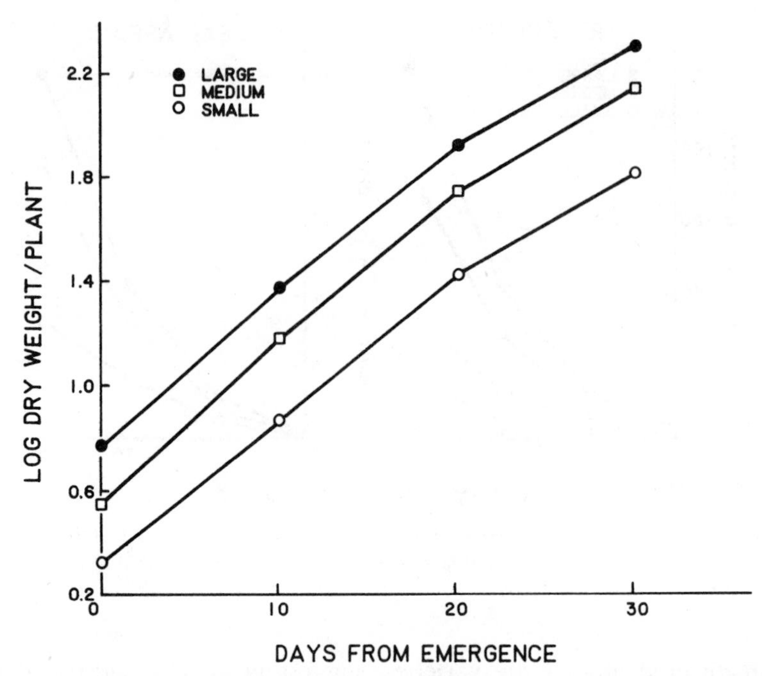

Figure 4.2 Rates of seedling growth from large, medium, and small seeds of *Trifolium subterraneum*. The slope of the lines is the same, only the starting points are different. Redrawn from Black (1956) by permission of CSIRO.

large seeds was markedly greater than for small seeds and mortality of small-seed seedlings was greater (Figure 4.3). Another potentially important observation was that, although the dry weights of plants grown in dense pure stands from seeds of different sizes were proportional to seed weight over the first part of the growing season, this proportionality did not continue indefinitely (Black 1957b). After about two months, competition for light among the large plants grown from large seeds resulted in markedly slower growth rates, but this point was not reached for small plants (from small seeds) until after about three months of growth. After about four months, there were again no differences in relative growth rates; all were very slow. Thus the small plants could continue growing longer than the large ones, in these circumstances.

Experimental reductions of the amount of endosperm per seed have variable results. Removal of up to 10% of the stored reserves of *Mucuna andreana* seeds resulted in seedlings with lowered capacity to recover from simulated herbivory (Janzen 1976b). Removal of various proportions of the cotyledons of *Cucumis melo* markedly reduced subsequent

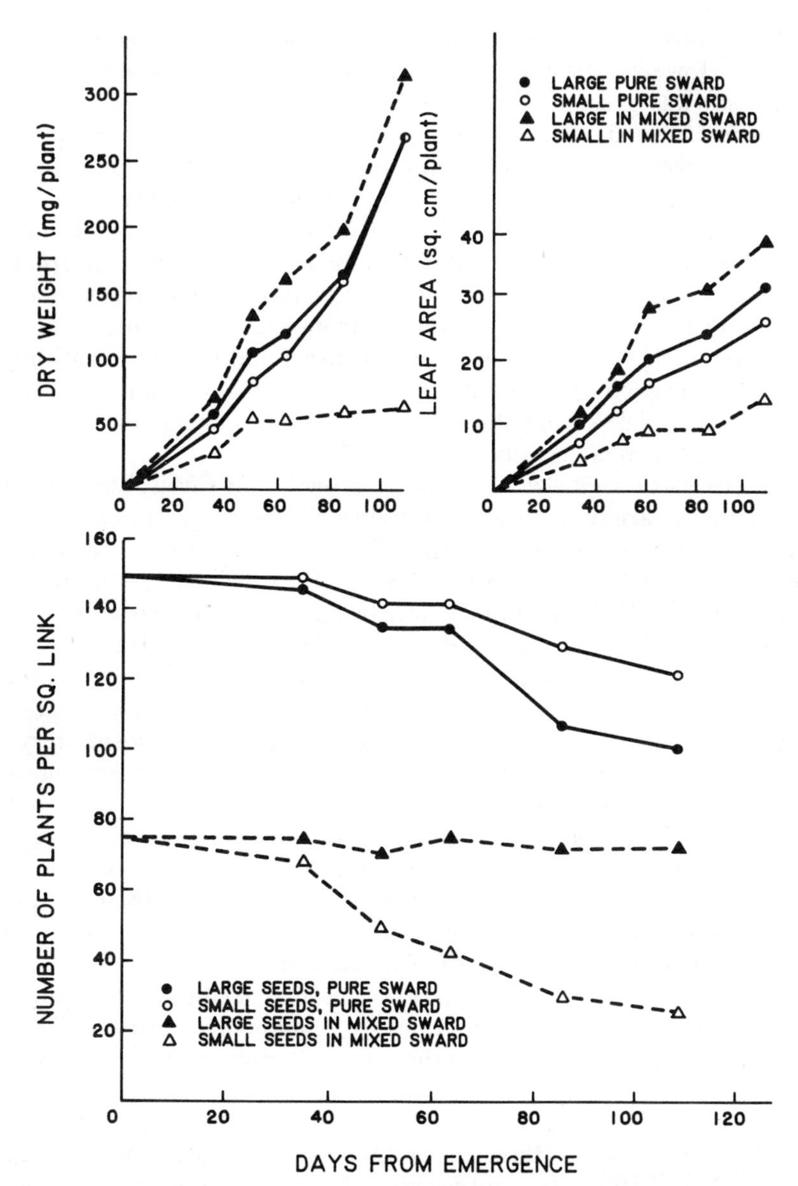

Figure 4.3 (*a*) Effect of competition between seedlings on seedling growth. Competition with seedlings from large seeds markedly depressed the growth of those from small seeds when the two types were grown together. (*b*) Mortality rates of seedlings from small seeds are higher when grown in competition with seedlings from large seeds. A "link" is about 20 cm. Redrawn from Black (1958) by permission of CSIRO.

seedling growth: the more removed, the smaller the seedling (Gould et al. 1934). However, the authors report that seedling growth and survival in all cases was greater than expected—although only hypocotyl and roots were formed, these were larger than normal. Whether such a response represents an adaptive shift initially favoring the underground parts, a response to trauma, or merely the limits of achievement with reduced resources is not known. The relative disadvantage of small propagule size was less than expected on the basis of size alone in *Helianthus tuberosus,* apparently because of faster translocation of reserves and consequent faster growth of individuals from small propagules (Kondô and Oshima 1981). In contrast, seeds of carrot and polebean that were fed upon by *Lygus* bugs were smaller and lighter than undamaged seeds, but (after an initial slowing of growth in carrot) the damaged seeds often yielded larger plants eventually (Scott 1970).

The effects of seed size on seedling growth and survival have been studied for a variety of grasses and agricultural crops, as well as for coniferous trees and a few other species. It is common to find an advantage to large seeds (e.g., Gross and Soule 1981, Schaal 1980), but the effects are not universal and may only be expressed in particular growing conditions (Baker 1972, Cideciyan and Malloch 1982, Harper et al. 1970, Howell 1981, Pollock and Roos 1972, Thomas 1966, Willson 1972a, Wulff 1973). It seems to be generally true, for within-population comparisons, that seeds containing large amounts of endosperm can produce seedlings whose cotyledons are larger and whose roots and/or shoots have a head start over those from smaller seeds. The larger seedlings frequently outcompete neighboring smaller seedlings (of the same or other species) when grown in mixed stands. The size and/or growth-rate advantages are often transient but, nevertheless, they may have great consequences for the probability of subsequent survival or the eventual size and reproductive capacity achieved (Cook 1980b, Harper 1977, Werner 1979). Survivorship of *Viola blanda* seedlings was strongly dependent on size (Cook 1980b): those 10 mm long had a probability of death greater than 80%, whereas those over 30 mm enjoyed a risk close to zero.

Sometimes seed size is related to time of seedling emergence (*Trifolium subterraneum,* Black 1956) and sometimes it is not (*Impatiens capensis,* Howell 1981). It is important to know the extent to which seed size exerts an effect independent of seedling emergence times, because time of emergence can also have a significant effect on seedling success (Howell 1981, Ross and Harper 1972). Seed size is also frequently correlated with the ability to emerge from greater soil depths, at least in some species (Harper et al. 1970, Schimpf 1977).

The importance of stored nutrition to the success of the seedling is further demonstrated by surveys of seed-size differences across a range of habitats. Salisbury (1942, 1974) long ago noted for the flora of southern England that seed weights were greater in species whose seedlings are established in shade than for those establishing in sunlit habitats, and in late successional species than in those of early succession. He suggested that large seeds give the seedling an early advantage in developing leaves so as to meet the competition from co-occurring plants. Luftensteiner (1979) indicated a similar trend in comparing two habitat types in Austria.

More recently, Baker (1972) surveyed the flora of California and found that average seed weight of herbaceous species increased with increasingly xeric conditions; the trend appeared not only in broad-scale comparisons of herbaceous communities in wet and dry habitat extremes, but also within genera whose constituent species spanned a range of habitat moisture. Because the weight of the seed coat changed little along the California wet–dry gradient, the changes in average seed weight were apparently a result of difference in stored nutrition (Baker 1972). In seeming contradiction, the seeds of desert annuals (but not the perennials) are smaller and lighter than would be expected on the basis of the correlations just described. Baker suggested that most desert annuals, in fact, live their lives in quite mesic conditions because they commonly complete the brief life cycle just after the seasonal rains. Tree seeds also exhibited an inverse correlation of size with habitat dryness, but the seeds of shrubs did not, for reasons that are not clear.

It is hardly surprising that Salisbury's early studies did not demonstrate a relationship between seed size and xericness, because Britain does not offer the extremes of xericness that California does (Baker 1972), although Salisbury did notice that the seeds of many British beach and dune plants were large, presumably in order to quickly establish a large root system. However, Baker's results for the California flora with respect to estimated relative availability of light do not parallel those of Salisbury either. Neither herbs nor trees showed an increase in seed size with decreasing light, perhaps because the importance of moisture availability was overwhelming (Baker 1972). The mean size of shrub seeds did diminish as light availability increased, but it is not clear why shrubs showed results different from the other growth forms. Levin's (1974) survey also failed to corroborate Salisbury's generalization, again perhaps because of the confounding effects of habitat moisture.

Although some within-species comparisons of seed weights parallel Baker's between-species correlations, others run counter to the correlations of dryness and seed size. *Veronica peregrina* is a predominantly self-pollinated annual, characteristic of temporary pools. Plants in the center of such pools produced larger seeds than those at the periphery, and more of them, perhaps because intense intraspecific competition at the pool center favored seedlings with a large endowment, and/or because selection on peripheral plants favored numbers at the expense of size (Linhart 1974).

Silvertown (1981) presents a further complication. Among the seeds of native plants in calcareous grasslands in Britain, small-seeded types usually germinate in the fall, whereas large-seeded species typically germinate the following spring. The author suggests that fall germination and small-seededness is a tactic for colonizing empty spaces at a time when the growth of established plants is slow and competition is slight; the price is a high mortality over the winter. In contrast, large-seeded, spring-germinating species elude high winter mortality and are equipped to grow quickly in competition with the spring growth of established plants. Just what determines the choice between these two tactics is not known; the effect is found in both annuals and perennials, although a higher proportion of annuals have small seeds (Silvertown 1981). However, small seed size is not *necessarily* associated with colonizing ability (Fenner 1978).

From these broad surveys emerges the basic proposition that larger seeds are generally adaptive where a premium is placed on rapid establishment and growth. However, deviations from the described trends are sufficient to warrant further, more detailed examination. Another problem is indicated by the studies of Schweizer and Ries (1969) with wheat; plants growing on soils with augmented minerals produced seeds whose seedlings grew faster than those whose parents received no mineral augmentation. Thus even within a species, the particular composition of the stored nutrition can influence seedling vigor even if seed size is unchanged.

Before one blindly accepts the proposition that large seeds are generally adaptive, some cautionary comments are appropriate. First, there will always be some point beyond which a further increase in size becomes ineffective (Smith and Fretwell 1974) or downright disadvantageous because of conflicting selection pressures, and not every increment in size need result in seedling advantages (Twamley 1967). We know little about such matters for any species. Second, it is much too simplistic to presume that larger size is always good. Capinera (1979) emphasizes that smaller seeds may well have advantages in some circumstances. Smaller seeds may have different dispersal capabilities, find more suitable safe sites, have greater access to water (because of their higher surface/volume ratio), be less subject to predation, and so on.

Third, even when mean seed size varies little, the variance around that mean within any population can be enormous (Janzen 1977a,c, 1978a; Marshall 1982; Schaal 1980), and that variation must be explained. Some of these points are examined in more detail below.

Predation. Seeds are consumed by many kinds of animals, especially insects, birds, and mammals. They constitute a food source rich in nutritional value; the caloric value of some seeds is higher than most of the rest of the plant (e.g., Golley 1961), and seeds are often relatively high in protein and lipids (e.g., Halls 1977). Seed predators are capable of demolishing entire seed crops in some circumstances and thus can have an important influence on plant RS. To the extent that seed predation is size-selective, we should expect evolutionary changes of seed size in response to seed predation (within the constraints on size that derive from other selection pressures).

Both increases and decreases in seed size might contribute to the escape from predation (C. C. Smith 1975, Stiles 1980). Small seeds might offer such a small food reward as to be difficult to find or uneconomical to harvest, or they could be so small that, when enclosed in an edible fruit and consumed by a dispersal agent, they pass through the digestive tract unscathed. Smallness can offer an escape from predators that consume many seeds (birds and mammals) and from predators that may need a single seed of some minimum size to complete a portion of the life cycle or to achieve maximum growth (some insects). Largeness can offer escape by being too large to crack or carry off or by having a protective covering so thick that some animals cannot penetrate it. However, it is important to remember that even size-selective predators may not always be so; a very hungry predator may settle for a less than "optimal" food source.

What evidence is there for size-selectivity by seed predators? Not much. Seed-eating birds fed on sunflower seeds in the laboratory showed some size preferences (Hespenheide 1966, Willson 1972b), the small-billed species favoring smaller seeds. But since sunflower seeds are larger than the wild seeds normally consumed by these birds, the selectivity of these birds in natural circumstances is unknown. European jays (*Garrulus glandarius*) and band-tailed pigeons (*Columba fasciata*) forage selectively on certain sizes of acorns (Bossema 1979, Fry and Vaughn 1977); in these cases the variation in seed size seems to be principally phenotypic. Tree squirrels do not select conifer seeds on the basis of size (C. C. Smith 1975), which is not surprising because the squirrels harvest the whole cone and not individual seeds; they do discriminate among the cones of certain species on the basis of numbers of seeds per cone (C. C.

Smith 1975). Ground-foraging heteromyid rodents that harvest loose seeds may select large-seeded species (Brown et al. 1979a), although the extent to which they do so is debated (Lemen 1978). Predators selected the larger seeds of *Arctostaphylos glauca* over those of *A. glandulosa* (Keeley and Hays 1976). Interisland differences in mean seed size of Galápagos *Opuntia* reside chiefly in thickness of the seedcoat and may be related to predation pressure (Racine and Downhower 1974).

Insects might be size-selective too. The lesser rice weevil (*Sitophilus oryzae*) infests the grains of sorghum. Although hardness of the grain can be one factor that influences the amount of oviposition by female weevils, when the female weevils are presented with a mixture of grain sizes, grain hardness was much less important than grain size; the oviposition rates were significantly greater on sorghum varieties with large grains (Russell 1962). Furthermore, the young weevils that emerged from the large grains were bigger and heavier than those from small grains (Russell 1962), so that the choices of the ovipositing female had potential consequences for the fitness of her offspring. Body size of insects is often associated with female fecundity, flight capacity, ability to withstand food deprivation, and perhaps other features of importance to success (e.g., Derr 1980).

Nevertheless, an evolutionary shift of seed size may only change the predator to which the seed is subject, so shifts of size are unlikely to provide complete escapes from predation. Seed-harvesting *Pheidole* ants preferred small-seeded species (Ballard and Pruess 1979). Indeed, changes in seed size might even be countered by corresponding evolutionary changes in the size of the predator (especially if an insect; Janzen 1969) or of the predator's jaw parts (especially if a vertebrate); such changes are likely only when that kind of seed is particularly important to that predator and changes of diet are uneconomical.

Janzen (1969, 1971b) argue that selection for the production of vast numbers of seeds may result in an evolutionary reduction of seed size by the principle of allocation. His argument is presented in some detail below; for present purposes it suffices that Janzen's proposition is a case in which seed size would evolve indirectly in response to selection that acts directly on seed numbers.

The risk of predation of course may affect the evolution of many seed characteristics in addition to size. These matters are addressed in later sections of this chapter.

Dispersal. Seed size is related to dispersal ability (Jackson 1981). Wind-borne seeds can be carried farther if they are small and have a large surface/volume ratio. Seeds that are dispersed by animals that ingest them (often with a surrounding edible "fruit") commonly are passed through the gastrointestinal tract if they are sufficiently small. Larger seeds often are regurgitated by birds after the fruit pulp is removed and remain in the fruit-eater for a much shorter time, on average, than do smaller seeds. As a result, smaller seeds are probably carried longer distances from the source than larger ones (Johnson et al. unpublished manuscript). If selection favors the production of numerous seeds to enhance success in dispersal and other constraints on seed size are weak, small-seededness may evolve indirectly as a means of enhancing dispersal distance. Nevertheless, dispersal by animal consumers may provide a way to have a relatively large seed without sacrificing much in dispersal distance (Webb 1966).

The basis for variation of seed size within a single seed crop has been little explored. Harper et al. (1970) have dealt briefly with the proximate means of controlling seed size, emphasizing competition among the sibling embryos for maternal resources and the predictability of resources available for seed production. Plants with indeterminate growth patterns appear to produce seeds less variable in size than those of determinate growth, because the latter commit themselves to the production of a certain number of seeds and must parcel among them whatever resources are available. Removal of flowers of fruits from some kinds of plants may lead to an increase in the weights of seeds that remain (Harper et al. 1970, Maun and Cavers 1971a,b), whereas removal of leaves or of light can sometimes lead to a decrease (e.g., Willson and Price 1980). The addition of mineral fertilizers is sometimes associated with an increase in seed weight (Willson and Price 1980). Photoperiod during seed development in *Chenopodium rubrum* affects seed weight and seed number (Cook 1975), but the ecological basis for this response seems to be unknown.

The existence of considerable phenotypic variability in seed size in many species leads to a question about the importance of the variation itself. Some of the variation in size may be related to condition of the parent; for instance, in one (but not a second) population of *Yucca whipplei,* seed size was correlated with rosette size (Aker 1982). Any proximate controls on seed size should be constrained, normally, to produce only seeds that are viable and have some prospects of success. If, for instance, severe herbivory results in much smaller-than-normal seeds, it should be advantageous for the parent to restrict further allocation of resources to those seeds to the degree that their future success is diminished; seeds too small to germinate successfully should be aborted completely, if possible. Crop plants often have been selected artificially to produce seed crops with little variation in seed size (Harper et al. 1970), demonstrating that the tightness of physiological control over seed size is regulated genetically and is subject to natural selection. That seed size is variable and that the smallest seeds are sometimes aborted suggests that the variability itself may be adaptive. Seeds of different sizes exploit different situations, as Capinera (1979) pointed out. Furthermore, Janzen (1977a,c, 1978a) documented considerable within-crop variation of seed weights of three tropical plants; the variation was not found to lie in seed-coat weights and thus resulted primarily from differences in the embryo and endosperm. He suggested that production of seeds that differ in size may result in a more homogenous dispersion of seeds around the parent, with a resulting greater probability that some seeds will arrive at suitable safe sites. Any effect on dispersion might be reflected more in the length of the seed shadow than in homogeneity within the shadow, but even that will depend greatly on whether seeds of different sizes often travel together (in the same pod, for instance) and on the behavior of the dispersal agent. This hypothesis has not yet been tested.

Composition of the Seed

Composition varies enormously among the seeds of different species. Although energy content is usually high, this is not universally true, and some seeds have little more energy per unit weight than other parts of the plant (Goldman unpublished manuscript). The energy may be stored primarily in lipids but additionally in carbohydrates (either starches or sugars) or in proteins, which are also a storehouse of nitrogen. The balance of these three major components differs greatly sometimes even within a genus (Table 4.1). Further, the protein content of seeds may vary both within a genus and within a single species, as recorded for some *Mentzelia* and *Chenopodium* species (Hill 1977). The oil content among the seeds of *Euphorbia* species ranges from 7–45% (Levin 1974). What determines this balance has never been assessed, but we can point to some of the factors that are likely to be important. Provisions for the emerging seedling are critical in determining its possible growth rates and the length of time it can depend on stored reserves before becoming self-sustaining (e.g., Schweizer and Ries 1969). The amount of energy packed into the seed depends, of course, on the form in which it is stored; lipids contain more energy per unit weight than starches (e.g., Levin 1974), for instance, but they are less quickly mobilized. Likewise, sugars are more readily available than starches. If the size of the seed is constrained by other factors such as dispersal ability, selection might favor the use of lipids as storage products in order to pack as much

Table 4.1 Protein and lipid composition of the seeds of selected species[a]

Species	% Protein	% Lipids
Pinus		
P. cembra	17	50
P. echinata	26.7–28.6	17.6–22.1
P. edulis	15	64
P. jeffreyi	33	52
P. lambertiana	27	56
P. palustris	26.7–33.9	19.3–28.9
P. ponderosa	21.3	22.0
P. sabiniana	29.6	56.6
P. sylvestris	33	28
P. taeda	15.9–17.4	6.7–13.5
Quercus		
White Oaks		
Q. alba	4.1–7.4	1.2–6.8
Q. lyrata	4.6	0.9
Q. macrocarpa	4.3	4.8
Q. michauxii	4.1	3.3
Q. prinus	3.6–8.5	1.0–14.0
Q. stellata	3.8–8.3	5.1–7.6
Q. virginiana	4.4–5.9	6.9–8.6
Black Oaks		
Q. coccinea	7.8	30.8
Q. falcata	4.2–6.5	15.6–17.0
Q. ilicifolia	6.1–10.3	19.4–20.0
Q. laurifolia	5.0	26.8
Q. marilandica	5.5–6.3	10.7–18.2
Q. nigra	3.8–5.3	20.3–23.7
Q. nuttallii	3.8–4.5	5.2–15.0
Q. palustris	3.8–7.2	15.4–17.8
Q. phellos	4.6–5.9	19.6–23.0
Q. rubra	7.2	22.5

[a] Data from Bonner (1974), Goldman (unpublished manuscript), Halls (1977), King and McClure (1944), USDA (1974). Some variation may be due to methods of determination, as well as species and provenance.

energy as possible into the smallest possible space; if rapid mobilization is necessary for a rapid spurt of initial growth, perhaps this option becomes less available. Differential foraging by seed predators could favor a shift of seed contents, upward if the consumers are also hoarders and dispersers, but otherwise downward. The necessities of chemical protection might also alter the nature of the storage products (see below). Mineral content can vary a great deal also, but for reasons that seem to be unexplored.

Sometimes seed contents vary among populations of a single species (Johnson and Robel 1968, USDA 1974), and apparently we do not know if this variation is adaptive in some way or merely a by-product of site differences. Genetic differences among individuals would be of great interest because this would allow us to investigate the action of selection on seed contents. Variation is also recorded from the same population of some species in different years (Robel 1972). Even phenotypically determined differences in the composition of seeds would also be potentially interesting because at least the consequences of the ability to change seed constituents could be explored. It would not be at all surprising to find individual differences, either genotypic or phenotypic, but there seem to be few detailed data for noncrop plants.

Snell's (1976) study of *Chamaesyce hirta* emphasizes some potentially important relationships involving seed composition: plants grown at different densities produced seeds that differed in both size and composition. High-density parents tended to produce few, lighter seeds containing less carbohydrate and more protein and lipids than low-density parents, although the relationship did not appear to be linear. Clearly, parental allocation to offspring was altered in response to parental growing conditions; the effec of this phenotypic plasticity on offspring growth and survival was not explored.

The cost to the parent of differing investments obviously will differ, too. Lipids, proteins, and carbohydrates differ in the energetic cost of synthesis (Levin 1974, Sinclair and de Wit 1975, see also Penning de Vries et al. 1974) and, furthermore, the parent may have to drain its stores of certain elements to endow its offspring properly. Although a good deal of verbiage has been addressed to the issue of quantity of seed endowment by the parent, very little has been focused on quality of endowment as it is related to offspring success and parental costs.

PROTECTION FROM PREDATION

The packaging of food reserves with the embryo inside a seed creates a convenient and tasty morsel for the many possible consumers. Whole groups of animals—genera, families, and even orders—have specialized to diets in which seeds are an important component and their adaptive radiation is very probably closely tied with their use of this food source; lygaeid bugs, bruchid beetles, finchlike birds, and rodents are obvious examples. Consumers often have highly developed means of finding and eating their food and may devour large numbers of seeds. So, from the point of view of a parent plant, it becomes important to protect its offspring from these hungry hordes. Protection first needs to be provided while the seeds are still on the parent, before dispersal. At this time, seeds are a concentrated resource for predators. And, after dispersal, protection of the seed may still be necessary if germination is not immediate. Maternal protection can be extended specifically to the young seedling after the seed germinates; toxic chemicals are translocated from seed to seedling at germination in *Canavalia,* for example (McKey 1979); the inclusion of toxic chemicals in an actively growing embryo is possible but fraught with difficulties (Orians and Janzen 1974). Protection of the seed is sometimes one feature contributing to the lowered fitness of interspecific hybrids (Drake 1981a,b).

Plants have a variety of ways to protect their seeds. These are presented here in five categories but, of course, plants may use more than one of these types of defense.

Morphological Defenses

Seeds can be protected (from at least some predators) by where and how they are borne on the parent, by hiding them, and by enclosing them in thick, hard, or prickly coverings. When seeds are carried on the tips of fragile stems, they are less available to small, climbing mammals than when they occur closer to the larger stems or to the ground (Denslow and Moermond, 1982). This is the only report on positional effects that I have found, but protection by position is likely to occur in other ways as well. Seeds may be concealed by various plant parts while on the parent, and by cryptic colors and patterns—even odors—after they leave the parent (Wiens 1978).

A common morphological protection is the enclosure of the seeds in maternal tissue: the ovary wall, the calyx and associated structures, bracts, and so on. Most conifers enclose their seeds in a fibrous cone and, in some species, the scales of the cone are much thickened and hardened and may bear a sharp spine. Such protection may function against insect predators (or fire in some instances). The hard and heavy scales of lodgepople pine (*Pinus contorta*), especially in regions where the cones are serotinous and only open to shed their seeds in response to fire, also protect the seeds against the depredations of pine squirrels (*Tamia-*

sciurus; Smith 1970). These squirrels harvest whole cones of several species of conifer in western North America and cache them for winter stores. The smaller Douglas squirrel (*T. douglasii*), west of the Cascade Mountains, has considerable difficulty opening serotinous lodgepole cones and carefully chooses the softer parts of old cones, leaving the seeds in the basal part of the cone, which are protected by the thickest, hardest bracts. Red squirrels (*T. hudsonicus*) are larger and have stronger jaws; they can consume seeds from all parts of a serotinous cone. Red squirrels that dwell in lodgepole forests are larger, with proportionately heavier jaw musculature than those living in other kinds of forest (Smith 1970). Furthermore, at least some red squirrels seem to prefer lodgepole pine cones with narrow bases (which permit easy detachment from the branch), a low proportion of protective tissue relative to the weight of the enclosed seeds, and a large number of seeds per cone (Elliott 1974). Smith and Elliott suggest that tree squirrels can serve as a powerful agent of selection on the packaging of seeds in the cones of lodgepole pine; squirrel foraging should select for wider cones with heavier scales and fewer seeds. It might prove intriguing to investigate the protective roles of the cones of other *Pinus*; the genus exhibits a prodigious array of cone size and scale structure. Cone morphology may have other functions as well, such as the capture of wind-borne pollen (Chapter 3), so multiple selective pressures may be at work.

Many other species construct protective walls around their seeds. The thick pod walls of *Asclepias syriaca* virtually prevent feeding by the three youngest instars of the seed-sucking lygaeid bug, *Oncopeltus fasciatus*, and even adult bugs seem to feed preferentially at the thinnest parts of the pod, so that many seeds escape from these predators, which are specialized exploiters of the Asclepiadaceae (Ralph 1976). However, not all species of *Asclepias* have thick pod walls and these appear to elude *O. fasciatus* in other ways (Chaplin 1980).

A cocklebur (*Xanthium strumarium*) on Long Island, New York, is attacked by two seed predators, a tortricid moth and a tephritid fly (Hare 1980, Hare and Futuyma 1978), both of which seem to be specialists on this species. Each *Xanthium* fruit envelops two seeds in a spiny burr, and burr size is controlled, in part, genetically. Smaller burrs have relatively thinner walls and shorter spines, and the seeds therein suffer significantly higher predation rates than those in larger burrs. Selection pressure from the seed predators may account for differences in burr size among populations, but within any one population much of the variation in burr size is phenotypic and selective predation, therefore, cannot cause much change in burr size within populations (Hare 1980).

Another kind of deterrent to potential predators is the presence of pubescence on the seed coverings. Plants produce a variety of types of pubescence (which may well have effects in addition to their antipredator function). There is a good deal of evidence that at least some forms of pubescence deter herbivores from foraging on leaves (e.g., Johnson 1975, Levin 1973, Pillemer and Tingey 1976, and references therein); evidence also exists that some herbivores have means of coping with such deterrents (Breedlove and Ehrlich 1972, Rathcke and Poole 1975). Although most of the existing literature on the deterrent effects of pubescence has emphasized leaves and herbivores, pubescence also occurs on the outer coverings of seed packages and may serve as a deterrent there also (Kerner 1878). Dense pubescence on the pods of *Astragalus utahensis* apparently reduces attacks by a bruchid beetle and prevents attack by a chalcid wasp that oviposits on nonhairy species of the genus (Green and Palmblad 1975), although other possible deterrent factors that might differ among these species were not investigated. However, pubescence on the pods of *A. mollissimus* seems not to deter the oviposition attack of another bruchid (Center and Johnson 1974).

The involucral bracts of several species of thistle (*Cirsium*) produce a gummy substance whose function may be the reduction of seed predation. Ants are sometimes found trapped in the gum and die there, but this may be incidental to the real function of the gum. When the gum is obliterated from the bracts of *C. discolor,* seed predation by flies and moths is significantly higher than on control inflorescences, but even the controls suffered considerable loss of seeds (Willson, Anderson and Thomas, in press).

Other morphological defenses include the production of gum by seed pods as insect larvae penetrate, and flaking and peeling of the seed-pod exterior surface. As with other defenses, some insects have evolved mechanisms of avoiding or penetrating the defense; no defense is perfect (Center and Johnson 1974, Janzen 1969). Still another probable defensive barrier may be provided by the water-filled bracts of some *Heliconia* species (Seifert 1982): both buds and immature seeds remain below the surface, emerging only when mature and ready for pollination (of the flower) or dispersal (of the seed).

Chemical Defenses

Chemical defenses of leaves against herbivores (and pathogens) are well documented; those of seeds are less well studied at the present time (Janzen 1977b). However, it is clear that many plants employ some form of chemical defense of their seeds. The protective agent may be located in the fruit wall or seed coat, the endosperm, or even the embryo itself.

Some defensive chemicals reduce digestibility of plant parts, generally by tying up proteins in unusable forms, and some are actually toxic (Feeny 1975, 1976; Rhoades and Cates 1976). Many different kinds of substances have protective effects (Bell 1978, Fox 1981, Janzen 1977b, Janzen et al. 1977, McKey 1979, Seigler 1979). But for every defense there is a possible countermeasure (Fox 1981), and the host-specificity of many seed predators depends on just this. Bruchid beetles are, in general, highly specialized seed predators, many of them attacking only a single species of plant (Center and Johnson 1974, Janzen 1977b, 1978b, 1980, Johnson 1981, Johnson and Slobodchikoff 1979). The successful attackers have obviously broken the host's defense and, in fact, may use the once-defensive chemical as an important means of locating and/or identifying the right host. These specialists all have particular methods of avoiding or detoxifying the toxin (e.g., Rosenthal et al. 1982). Although bruchids attack plants in many families (Center and Johnson 1974), Janzen (1977b) speculates that the adaptive radiation of bruchids may be founded on their evolution of countermeasures against the chemical defenses of legumes, especially; it is also likely, then, that the radiation of the legumes has been affected correspondingly by the bruchids.

The seeds of jojoba (*Simmondsia chinensis*) contain toxic compounds that deter eating by several species of heteromyid rodents and many insects (Sherbrooke 1976). However, one species of mouse (*Perognathus baileyi*) eats these seeds with impunity and a few insects have also broken the chemical defense of the seed (Sherbrooke 1976). Mourning doves (*Zenaida macroura*) selectively consumed certain morphs of *Eremocarpus setigerus* seeds, avoiding the gray, probably toxic, phenotype—which, however, had compensating disadvantages at germination (Cook et al. 1971). Cycad seeds are reported to be toxic to vertebrates but some insects can eat them (Giddy 1974).

Compounds stored in seeds can do double duty. Toxic lipids are well known in seeds (Seigler 1979) and constitute a very large proportion of the seed of *Ginkgo biloba* and *Sterculia foetida*. Their toxic properties help protect the seed and, upon germination, they are metabolized as nutrition for the young seedling (McKey 1979).

Digestibility-reducing compounds also occur in seeds—the tannins are perhaps the best known example (McKey 1979). Tannins are more concentrated in the acorns of at least some red and black oaks, for instance, than in those of the white oaks (Fox 1974, but see Wainio and Forbes 1941), and mammals (such as squirrels or mice) that eat acorns are likely to prefer those of white oaks and to digest them better.

When seeds are dispersed by animals that eat fruits, a special problem is prevention of predation and premature dispersal. Obviously, if the consumer devours the fruits before the seeds are ripe enough to survive the treatment by the animal and to germinate, the dispersal system tends to break down. Immature fruits are often protected by sour or bitter flavors (e.g., Goldstein and Swain 1963, Hulme 1970, 1971), hard consistencies, strong attachment to the parent, toxins (Herrera 1982b), and sometimes by emetic or cathartic properties that disappear when the seeds and fruits are ripe (e.g., *Rhamnus cathartica,* Sherburne 1972), or low nutritional values (e.g., Foster 1977, Sherburne 1972). Of course, it sometimes happens that immature fruits are eaten anyway (e.g., Foster 1977, Snodderly 1979).

Of considerable interest is the observation that individual plants within a population may vary in the level of chemical defense (e.g., Moore 1978). Moore's results for *Crotalaria pallida* in East Africa are enigmatic, inasmuch as the plants with the greater concentrations of certain toxins suffered higher seed predation, for reasons that are not altogether clear. Individual ponderosa pines (*Pinus ponderosa*) differ in the levels of herbivory by tassel-eared squirrels (*Sciurus aberti*; Farentinos et al. 1981), and scale insects (Homoptera) seem to discriminate clearly among individual ponderosa pines (Edmunds and Alstad 1978). It seems likely that individual variation in chemical protection will prove to be widespread. This is of fundamental importance because (if the differences are genetically controlled) it provides the material on which natural selection can work; it also makes doubly hazardous comparisons between species. Even if the variation is strictly phenotypic, differential predation could constrain phenology or spatial distribution of the population. Furthermore, if somatic mutations can produce high levels of within-plant variation, this variation itself may effectively reduce levels of predation (Whitham in press).

Animal Bodyguards

Although other defenders may be discovered in the zoological realm, ants are surely the best known. Ants are ideally suited for defensive purposes, for they are both numerous (and colonial) and pugnacious. Ants are known to defend plants against herbivores (see review by Bentley 1977) and competing vegetation (Janzen 1967); in *Catalpa speciosa* the reduction of herbivory enhances fruit production (Stephenson 1982). They have also been drawn into service to protect developing seeds. The defended plants attract their protectors by providing domiciles or, more

Figure 4.4 An inflorescence of *Costus woodsonii,* showing the single daily flower, the protective ants, and the parasitic fly. The light colored streak on the bracts is the nectary. Photo by D. W. Schemske.

often, food, which is produced in special bodies or in extrafloral nectaries.

Costus woodsonii produces a single flower per day on a single inflorescence per stem, each flower protected basally by a thick bract (Figure 4.4, Schemske 1980). The bract produces nectar as the bud reaches maturity and while the flower is open. Ants are attracted to the extrafloral nectaries. The seeds are attacked by a fly that oviposits on the immature fruit and whose larva eats the seed and fruit. During the wet season, the presence of a small, unaggressive ant significantly reduced oviposition rates and damage by the fly, and seed production was markedly greater when ants were present than when they were excluded. A large aggressive ant was present in the dry season; its presence on the inflorescence reduced ovipositions by the fly, but seed production was unimproved. Instead, protection by the large ant reduced damage to the fruit and thus enhanced success in dispersal: the fruits are eaten and the seeds dispersed by birds, which do not consume heavily damaged fruits (Schemske 1980).

Mentzelia nuda produces nectar not only during but also after flowering; nectaries are located at the base of the corolla on top of the inferior ovary. The nectar produced on the developing fruit attracts ants, which apparently defend the fruits and its seeds against various seed predators. When ants were excluded from the fruits, seed production diminished (Keeler 1981).

Nectar is produced on the involucral bracts of *Helianthella quinquenervis*. The seeds of this sunflower are attacked, before dispersal, by the larvae of several flies and at least one moth. Flowering heads with ants had fewer seed predators and greater seed production than those without ants, and ants were seen to attack flies on the inflorescence (Inouye and Taylor 1979).

Temporal Escape

Escape in time can occur on a seasonal or an annual scale. If levels of seed predation vary with the season, this is a potentially powerful agent of selection to shift the reproductive season. A lycaenid butterfly (*Glaucopsyche lygdamus*) oviposits on immature inflorescences of *Lupinus amplus* in the Colorado Rocky Mountains, and the larvae eat the corolla, stamens, and sometimes the ovary of the flower. The attacked flowers drop off, as a rule, and do not produce seeds. Breedlove and Ehrlich (1968) reasoned that the butterfly's preference for immature inflorescences means that early-flowering plants, on which the flowers mature before the butterfly is ready to oviposit, would set more seed than late-flowering individuals. The selection pressure from the herbivore may push the flowering season toward the early part of the summer; the seeds mature long before the end of the growing season. In fact, this lupine blooms so early that it sometimes suffers great damage from frost, which may place a limit on the advancement of the flowering season.

Seed predation of *Astragalus cibarius* and *A. utahensis* in Utah is strongly seasonal (Green and Palmblad 1975). Pods of *A. cibarius* and *A. utahensis* suffered 44% and 27% destruction, respectively, in mid-June but by mid-July, destruction had soared to 99% and 73%, respectively. Both species flower and disperse their seeds earlier than other species growing with them, and *A. cibarius* flowers so early that frost damage is sometimes severe. The authors suggest that seed predation may provide an important selection pressure maintaining the early reproductive season.

Another form of escape in seasonal time may be found in patterns of fruit ripening. Although some fruits exhibit a steady growth through the season, others grow in a sequence of two or three bursts (Hall and

Forsyth 1967, Kozlowski 1971, Rhodes 1980). Although these bursts can be associated with the development of particular tissues (Hall and Forsyth 1967, Kozlowski 1971, Rhodes 1980), it might also be possible that the approach to the mature size is postponed as a means of reducing attractiveness to predators. Furthermore, in *Celtis occidentalis* and in other species also (Kozlowski 1971), the fruit achieves the mature size early in the season but the seed is not filled until much later. A plausible but untested explanation might be that the lack of food reserves and the hollow space around the embryo protect the offspring from some insect predator.

Escape in time on a scale involving years is an idea that has its beginnings far back in the forestry literature. More recently, Janzen (1971b) crystallized the idea under the formal rubric of predator satiation. The hypothesis is that a plant may reduce seed predation, especially (but not exclusively) by specialized seed predators, by producing enormous seed crops at long and/or irregular intervals and making few or no seeds in between. The long and/or irregular intervals must be longer than the predator population can endure, a tactic effective only if the primary predator is dependent on the seeds of that plant and has few alternate foods, and if opportunistic predators are few. And all the individuals of the plant population must fruit in synchrony. Thus in the lean years the predator population is reduced, and, when a large seed crop is produced, there are too few predators to demolish the whole crop and some of the seeds escape. The periodic or irregular production of seeds is known as mast seeding or mast fruiting because many of the plants that do it produce large seeds or nuts called mast.

Foresters have known for a long time that exceptionally large seed crops often have relatively low predation (e.g., Lyons 1956), but Janzen's hypothesis has received formal testing only rather recently. Predicting that a positive correlation of seed predation with seed-crop size should lead to mast-fruiting, Silvertown (1980) scoured the literature and found 59 sets of data on predispersal seed predation and seed-crop size for 25 species of tree. Twenty-four of these data sets exhibited a significant positive relationship between predispersal seed survival, as a percentage of the total crop size, and the logarithm of seed-crop size for the same year; the remaining data sets showed no relationship or, in two cases, a negative one. Thus to this degree, one of the expectations of the predator satiation hypothesis seems to receive some support. That so many of the data sets do not support Silvertown's predictions, however, may be cause for concern. Although there may be many explanations for their lack of relationship, post facto "explaining away" is unsound procedure, and further, independent tests of the hypothesis are highly desirable. A potential confounding factor is the production of empty seeds by a number of tree species; empty seeds cannot be preyed upon, and a large seed-crop with many void seeds may support few predators. Empty seeds themselves may be a defense against seed predation, but their inclusion in correlation analyses may mask other relationships.

Silvertown (1980) also predicted that seed survival immediately after a mast year should decrease more than expected from the relationship between seed-crop size and same-year survivorship. If the predator population build-up in the mast year carries over to the following year, the expectation is reasonable and is a test of a basic assumption of the predator satiation hypothesis. Rates of insect attack on the cones of *Pinus edulis* were positively correlated with the size of the previous year's crop (Forcella 1980). But Silvertown found the expected association in only 2 of 14 usable data sets, so this aspect of the hypothesis receives relatively little support. Postdispersal predation rates, which are difficult to ascertain, might be important to consider in this test.

A third piece of the predator satiation hypothesis is that individuals of the host-plant population should fruit in synchrony; asynchronous individuals will endure heavy predation and lowered RS. Many of the mast-fruiting populations do indeed fruit synchronously; this is common knowledge among field biologists but is only seldom quantified. There is some anecdotal evidence that asynchronous individuals of mast-fruiters suffer unusually intense predation (Janzen 1974a). Some mast-fruiting species in North America and Europe even exhibit interspecific synchrony, but the effect is far from universal (Silvertown, 1980). Further investigation may show that different predators or levels of predation afflict these co-occurring mast-fruiting species (for instance, squirrels show clear preferences among different conifer species), or that other factors constrain the fruiting schedule.

Another approach to testing the hypothesis of predator satiation is provided by Janzen's (1975) comparison of *Hymenaea courbaril* populations in Costa Rica and in Puerto Rico. The mainland populations endure sometimes heavy seed predation by *Rhinochenus* weevils, which are absent from the island. Perhaps as a result, the Puerto Rican individuals of *H. courbaril* produce softer pods with less resin, produce small seed crops while still young, and continue producing seed crops annually. In contrast, Costa Rican trees have harder pods, delayed female reproduction, and produce offspring at three to five year intervals. However, regional synchrony of fruiting is imperfect and *Rhinochenus* is not eliminated (Janzen 1978c).

One constraint on the schedule of masting is the dispersal mechanism. Silvertown (1980) provided evidence that species with seeds enclosed in edible dispersal units, which are consumed by animals, are less likely to be mast-fruiters than are those with other means of dispersal.

When a plant relies on animals for seed dispersal, a pattern of mast-fruiting might satiate dispersers and lower dispersal success. Nevertheless, several animal-dispersed species are mast-fruiters, and so it would be of interest to know more about the foraging habits and opportunism of the animal dispersal agents involved. In the case of the large-seeded *Pinus* that are dispersed by birds, the avian dispersal agents are known for great vagility (see below).

Longevity is another potential constraint in the evolution of mast-fruiting (Silvertown, 1980, Waller 1979). Short-lived plants cannot adopt a pattern of mast-fruiting without drastically lowering their lifetime offspring production. Periodic reproduction entails a delay of maturation and, for iteroparous species, reduced frequency of crop production with the attendant risks of mortality before the next fruiting and of seedling failure if not every year is equally good for seedling survival. (Note, however, that these plants may still function as males; Janzen 1975.)

The energetics of seed and fruit production impose still other constraints. The cost of producing a crop of offspring is often considerable, involving a diversion of resources from other activities. Production of an extraordinarily large seed crop may require the acquisition (and storage) of resources over an extended period. The bigger the crop required to satiate seed predators, the more resources must be stored. Thus the interval between mast years may be controlled, in part, by the time required to accumulate enough resources to build the next massive crop. Janzen (1974a) noted that, in habitats where soils are poor in nutrients, it may take a long time to store sufficient resources for reproduction. It seems reasonable to expect that different species can do this at different rates and that, therefore, not all sympatric mast-fruiting species may be able to adopt the same schedule.

Synchrony of reproduction across a population of iteroparous plants is likely to require a strong, unambiguous, environmental cue to which all individuals can respond (Janzen 1974a). However, there are a number of semelparous perennials that are mast-fruiters using internal rather than external cues for flowering (Janzen 1976a); bamboos are the most famous representatives of this group. The factors determining whether external or internal cues are used are not known.

It is important to emphasize that predator satiation is not thought to be the only factor driving irregular or cyclical fruiting schedules. Nonannual fruiting is often linked with climatic unpredictability and loss of the critical proximate cues (Janzen 1974a, 1978c); in fact, such irregularities probably preceded and were reinforced by predator pressures (Silvertown 1980). The accumulation of resources for fruiting might necessitate an interval between seed crops of more than a year, even in the absence of predators. Foul weather may kill flowers or reduce the activity of insect pollinators (Janzen 1978c). Nevertheless, Janzen suggests that long intervals between crops coupled with reproductive synchrony in the population are probably an evolutionary response to seed predators. Tests of this hypothesis are beginning to come in, but clearly much of the story is yet to be told.

Spatial Escape

Escape in space has two aspects. The first affects the dispersion of seed over the body of the parent; the second occurs on a larger scale.

Many plants bear seeds as single units; oaks and hickories are two examples. But many other, perhaps most, package their seeds in groups, enclosing the clustered seeds in maternal tissues while they remain on the parent (the clustering of seeds when they leave the parent is discussed later). The aggregation of multiple seeds within a fruit may be economical in terms, for example, of the supply of nutrients through the vascular system or of the materials required for defenses against predators. The investment per seed can diminish if seeds are clustered: as the number of seeds enclosed in a pod or capsule increases, pod or capsule weight commonly increases more slowly than total seed weight (e.g., Mitchell 1977, C. C. Smith 1975). At least sometimes, seed size diminishes with increasing numbers of seeds per pod (Mitchell 1977, C. C. Smith 1975), but this is not always true (Janzen 1977a).

However, clumps of seeds may have the disadvantage of providing a concentrated resource for seed predators. The lygaeid bug, *Oncopeltus fasciatus*, is a specialized feeder on milkweed seeds and, in Missouri, forages preferentially on two species of milkweed. *Asclepias syriaca* is the only one of several co-occurring milkweeds to provide enough seeds per stem for the development of an entire clutch of *Oncopeltus* nymphs, and *A. verticillata* (whose pods and seeds are much smaller, but which clones extensively and develops dense patches of stems) grows densely enough to provide enough seeds within the short distance that nymphs can move (Chaplin 1980).

The pods of the desert legume *Cercidium floridum* are thin-walled and offer little protection against seed predators (Mitchell 1977). As many as eight seeds are sometimes set in a pod, but the usual number is one or two. Rodents consume many seeds but do not discriminate among pods containing different numbers of seeds. A bruchid beetle, *Mimosestes amicus*, attacks pods in proportion to the number of seeds therein, so the probability of seed destruction is unrelated to the number of siblings found together in a pod. However, another bruchid, *Stator limbatus,* differentially destroys seeds in multiseeded pods. *Stator* enters the pod through the exit hole of *M. amicus* and can consume all the seeds in the pod, once entered. Over 80% of all multiseeded pods are infested by *S. limbatus*, but only 50% of one-seeded pods are attacked. *Stator* does not discriminate among the multiseeded pods, as might be expected, perhaps because many-seeded pods are too rare to be worth searching for or because all multiseeded pods are the same length and thus there are no external cues indicating the number of seeds inside (Mitchell 1977). As a result of *Stator* attacks, seeds in one-seeded pods have a marked advantage, which may account, in part, for their predominance—over 67% of all pods have only one seed, and only 6% have more than two seeds. There is, no doubt, more to the story. There is some evidence for interaction among the pods produced at a single node on the stem. Pods from nodes bearing two pods contain significantly fewer seeds than those from single-podded nodes, which implies a local "competition" for resources used in pod and seed development (Mitchell 1977).

Polemonium foliosissimum seeds fall prey to the larvae of an anthomyiid fly (Zimmerman 1980). The fly larvae eat all the seeds in an ovary before they emerge. The same individual plants were consistently attacked in the two years of the study. The fly differentially decimated the seeds in ovaries containing many seeds but did this only toward the end of the flowering season. Attack rates were high but not differential during the early parts of the season.

A contrasting and more complex situation is found for *Scheelea* palms in Panama and Costa Rica (Bradford and Smith 1977, C. C. Smith 1975). Rodents consume the fleshy fruit around the palm nut and then scatter-hoard the nuts. The nuts are subject to heavy predation both by rodents and by bruchids. These palm nuts contain one to three, or rarely four, seeds, but the frequency distribution of seeds per nut varies geographically. The Panama population produced mostly one-seeded nuts (> 94%), a few two-seeded nuts (5%), and a very few three-seeded nuts (< 1%). The Costa Rican population produced 73%, 23%, and 4% in each of these categories (Bradford and Smith 1977, C. C. Smith 1975). There is individual variation in the frequency of multiseededness (Janzen 1971c). Rodent (squirrel, agouti, paca, spiny rat) predation predominates in Panama but also occurred in Costa Rica. Rodents did not discriminate among nuts containing different numbers of seeds but often consumed more than one seed from each multiseeded nut, especially in Costa Rica where multiseeded nuts were relatively common. Bruchids accounted for most of the seed predation in the Costa Rican population.

They exhibited a possible slight and inconsistent preference for one-seeded nuts, perhaps because single seeds are larger. Generally, the bruchids attack only one seed per nut and, therefore, seeds in multiseeded nuts have a higher probability of escape from bruchid predation. This advantage to multiseededness may contribute to the higher frequency of multiseeded pods in Costa Rica. A difficulty with the interpretation is that Panamanian rodents seldom extracted more than one seed per nut, which should result in a corresponding advantage to multiseededness in Panama too. Perhaps when multiseeded nuts are rare it is uneconomical for the rodents to extract more than one seed (Bradford and Smith 1977, C. C. Smith 1975). This argument, however, seems to remove the very selection pressure that is proposed to account for the frequency distribution of nut contents. In addition, Smith (1975) emphasized that there are other components of fitness besides escape from predation, and that the nutrient endowment of the seed and its success in dispersal and establishment must contribute to the overall balance.

Two species of Darwin's finches in the Galápagos Islands feed on the seeds of *Croton scouleri* before they are dispersed (Downhower and Racine 1976). Three seeds are contained in a capsule, which is cracked by the foraging birds. When the capsule cracks, one or two seeds drop to the ground. Rather than retrieve them, birds of these two species continue to forage from the tree on which virtually all capsules ripen simultaneously so that the available crop is abundant. Although other birds may scavenge the loose seeds from the ground, some are likely to fall into crevices or be carried off by ants.

Overall, then, we find that sometimes the clustering of seeds brings attendant high predation, and escape is enhanced by "not putting all the eggs in one basket." In other cases, clustering seems to improve the likelihood of escape. Thus no general rules for seed clustering on the parent have emerged yet.

At a slightly larger scale, but still within the parent plant, seed predation might increase with increasing seed-crop size. This is an underlying presumption for the evolution of predator satiation tactics, as previously discussed. However, many species are not mast-fruiters, despite the possibility that seed predators of these species might also respond to a large concentration of food resources. The number of *Dysdercus bimaculatus* nymphs increased with the size of the fruit crop of *Sterculia apetala* in Panama and Costa Rica (Derr 1980), but we have no information regarding seed survival. *Mabea occidentalis* in Panama is subject to heavy predation of seeds by a moth (De Steven 1981). Female moths oviposit on the outside of the fruit, and the larvae burrow in and consume all three seeds. In this case large seed crops did attract more predators, but not enough more to reduce the female RS of fecund parents. The percentage of the seed crop attacked by the moth did not change with the size of the seed crop, with the result that large seed crops produced more surviving seeds than small crops. However, if the intensity of predation on groups of plants can be any indication of what would happen on individual seed crops, it may be noted that weevils wreaked greater destruction on *Astragalus canadensis* seeds at low plant densities than at higher densities (Platt et al. 1974).

Escape in space may also occur through dispersal away from the parent (Janzen 1970). There is considerable evidence that many seed predators forage in a density-dependent way, selectively consuming seed resources when they are at high density. When this is true, then clearly there is an advantage to spreading seeds around. A common way to accomplish this is through dispersal. Not all plants that are subject to density-dependent seed predation have particularly good means of dispersal, however, and they presumably use some of the other types of defense. Dispersal is a sufficiently broad topic that it is discussed in a separate section (following), despite its relevance to escape from predation.

DISPERSAL

The first question to be answered is: Why disperse at all (see also Howe and Smallwood 1982)? It is easy to see that the ability to disperse away from the parent enhances the probability of that at least one dispersing propagule will find and occupy a space, when suitable habitats are ephemeral and patchy. For example, dispersal increases the probability that the seeds of two tropical wind-dispersed trees (*Lonchocarpus velutinus, Platypodium elegans*) will reach a light gap in the forest, where a treefall creates an opening in the canopy; seedlings of these species survive better in gaps than under the canopy (Augspurger in press, Augspurger and Hogan unpublished manuscript). The colonizing of ephemeral sites may be an important advantage for a number of tropical trees (Augspurger unpublished manuscript a). It is less easy to see, but nevertheless true, that dispersal is also advantageous in predictable and uniform environments, even when the mortality of the propagules is high (Hamilton and May 1977). A parent whose offspring (or some of them, at least) can claim a space where another individual has died, in addition to replacing the parent itself, will have an advantage over a parent whose offspring never leave the parental site. Some sort of dispersal is found for almost every seed plant, although the efficiency and the distance vary enormously. However, selection pressures on seed protection or other characteristics may constrain dispersibility, and the importance of dispersal relative to other, conflicting pressures may vary in different ecological circumstances (Ellner and Shmida 1981).

An additional impetus to disperse may be provided by density-dependent predation of seeds (or seedlings; Janzen 1970, 1971b). It is very common to observe that the distribution of seeds around a parent exhibit a high concentration immediately below the mother and diminishing numbers with increasing distance (e.g., Auspurger in press, Boyer 1958, Ender 1977, Harper 1977, Liew and Wong 1973, Platt and Weis 1977). Therefore, predators responsive to seed densities would be expected to have a greater impact on the high densities near the parent. However, such effects may not be universal, Denslow (1980a) observed that seedling damage by certain herbivores varied with seedling density but by others did not, and there may be great variation with habitat (Janzen 1971c).

Scattered seeds of *Andira inermis* endured far less attack by *Cleogonus* weevils than seeds remaining beneath the parent (Janzen et al. 1976). Seeds of this species are dispersed by bats, which roost in other trees, and the defecated seeds are deposited below the roost. Seed predation beneath bat roosts appeared to vary with the size of the seed deposit to some extent, although predation rates were assayed before the end of the fruiting season and the ultimate predation rates might have been different (Janzen et al. 1976).

Platt (1976) recorded density-dependent seed predation by ants and mice on the seeds of *Mirabilis hirsuta* such that seedlings were most common at some distance from the parent. Preference of rodents for clumped seeds has been reported several times (references in Heithaus 1981). Bruchid beetles attacked single seeds of *Scheelea rostrata* less frequently (2–5%) than seeds placed in 50-seed piles (34–36%; Wilson and Janzen 1972). Seeds of *Sterculia apetala* were found very rapidly by *Dysdercus bimaculatus* (a seed-sucking bug) when they lay beneath a tree of that species (Derr 1980, Janzen 1972). However, even seeds many meters outside the area of heaviest seed concentrations were found by many bugs within three to four hours (Janzen 1972), so the real extent of escape was not established in this case. Percent predation on seeds of *Prunus serotina* and *Vitis riparia* by mice (especially *Peromyscus leucopus*) increased at greater seed densities (A. J. Smith 1975). The intensity of predation need not be always responsive to the concentration of resources (e.g., Futuyma and Wasserman 1980), and we need to know in precisely what circumstances it is the rule.

However, even without the density dependence, seed predation can be reduced by dispersal. Heithaus' (1981) study of three ant-dispersed species (*Asarum canadense, Jeffersonia diphylla,* and *Sanguinaria canadensis*) is illustrative. Rodents, especially *Peromyscus leucopus,* removed many more seeds when dispersal by ants was prevented.

Mere proximity to a parent may also increase the risk of predation. Seedlings of *Dioclea megacarpa* suffered more damage from caterpillars when they were close to the parent vine (Janzen 1971a). Seedlings of the introduced *Ulmus parvifolia* were much more subject to beetle attack when directly beneath the parent canopy than when they were not directly under the adult (Lemen 1981). However, the distance response was not observed for *Scheelea* seeds (Wilson and Janzen 1972). The spread of a fungus among seedlings of *Lepidium sativum* was slower when the seedlings were dispersed among several clusters than when the seedlings were evenly spaced (Burdon and Chilvers 1976); apparently the greater distance among the clusters slowed the rate of spread. Seedlings of *Platypodium elegans* suffered high rates of mortality from fungal infection at high densities near their parent and survivorship was markedly greater at greater distances from the source (Augspurger in press). Density and distance are closely correlated, but experiments indicate that density effects are probably more important than distance (Augspurger unpublished manuscript b).

Similarly, competitive effects among densely growing seedlings may select for better dispersal. A possible example is drawn from a study of the annual *Cakile edulenta* on a Nova Scotia beach (Keddy 1980). Keddy found dense clusters of seedlings around the remnants of the putative parent. In two different beach habitats, the density of seedlings decreased sharply with increasing distance from the parent. Survivorship of seedlings increased with distance from the parent in one of the habitats; in both habitats the eventual seed output of the offspring increased strongly with greater distance from the parent. Keddy did not ascribe a cause for the difference in offspring success but discussed the possible relationship between predation and survivorship. It seems possible that the effect on the size of the seed crop might have been related to decreasing competition among seedlings—but this possibility was not explored experimentally in this study. Nevertheless, the literature is replete with studies of density and the intensity of competition (see e.g., Harper 1977), so the suggestion seems plausible. The survival of certain species of dipterocarp seedlings increased with greater distances from the parent, but this effect was not found in all species investigated (Liew and Wong 1973), and again no specific cause was ascribed.

It is not always the eventual dispersion but sometimes the process of being dispersed that saves some seeds. If *Acacia* seeds in Africa are eaten by large mammals soon after they fall to the ground, the digestive juices kill the bruchid beetle already in the seed and the still-germinable seeds are passed from the herbivore's digestive tract (Lamprey et al. 1974).

One cannot suppose, however, that dispersal necessarily saves a seed from predation. Some species of seed-harvesting ants in the desert capitalized primarily on clumped seeds, but others collected mainly scattered seeds (Davidson 1977). Some heteromyid rodents also seem to collect seeds equally well or even primarily from scattered sources (Price 1978, Reichman and Oberstein 1977). Price's study was done with captive mice in the laboratory with no confounding effects of interspecific aggression or of microhabitat preferences, which are known to affect predation rates of some seeds (Hay and Fuller 1981). The large kangaroo rat (*Dipodomys merriami*) favored clumped seed sources, but two species of *Perognathus* (*P. penicillatus* and *P. amplus*) preferred more scattered seeds.

Having some suggestions for the *why* of dispersal, we next ask *how* dispersal is accomplished. It is simplest to categorize the general types of dispersal mechanisms recognizing that some may intergrade, or occur together on the same plant. Ridley (1930) provided the classic compendium, still a gold mine of interesting lore, and van der Pijl (1972) a more concise survey, unfortunately cluttered with unattractive jargon. A brief review of anatomical devices is found in Fahn and Werker (1972).

No Special Dispersal

A number of species bear seeds that have no means of dispersal. The subterranean seeds of *Gymnarrhena micrantha* and *Amphicarpa bracteata* have no particular means of dispersal, although van der Pijl (1972) claims that squirrels may carry off the underground seeds of *A. bracteata*. The giant seed of the double coconut (*Lodoicea maldivica*) can only fall off the tree and, if the tree grows on a slope, roll downhill. It does not survive transport by the sea, but the ordinary coconut (*Cocos*) does. The curious two-part fruit of *Cakile edentula* has one detachable part, presumably for dispersal by wind or water, and another part that remains attached to the parent and probably drops to the ground in the spot occupied by the annual parent (Keddy 1980). Some dispersal may occur fortuitously if the seeds of any species are stuck on dead leaves moved by the wind, carried in the mud on birds' feet, by sheet floods after unusually heavy rains, and so on. The lack of special dispersal mechanisms and occasional dispersal by happenstance may have important consequences for population and community structure, but at the level of the individual, there is little story to unfold. Still, given the generally accepted advantages of dispersal, the apparent lack of it or the serendipitous nature of it foster the question of how these plants succeed. It is possible that some of these seeds are indeed dispersed by special means; for instance, Zimmerman (1977) and Janzen (unpublished manuscript) note that small hard seeds are often successfully dispersed in the interiors of birds and large herbivores, and a glance at the *Chenopodium* population atop an old cow-manure pile will attest to this.

Ballistic Dispersal

The dispersing entities are released explosively and thrown from the parent. In some cases the explosive release is triggered by a passing animal or by rain, in others it is controlled by the plant itself. The ejection mechanism takes many forms. Distribution of seeds around the parent depend on initial velocity, angle of projection, and air resistance of the seed (Beer and Swaine 1977). Dispersal distances are not enormous in most instances, but generally represent an improvement over simple seed-dropping. *Hura crepitans*, a tropical tree, has a fruit that shatters with a bang, scattering seeds up to 14 m, according to Ridley, and I have seen them flung at least 6 m with great force.

Wind Dispersal

The devices for dispersal by wind are many. One of the simplest means of gaining access to the wind is to be extremely tiny, with a large surface/volume ratio. Seeds this small can be blown about like dust particles and no special devices are needed. Production of such minuscule seeds permits the production of enormous numbers. This tactic is characteristic of most orchids and of many parasitic and saprophytic plants. Salisbury (1975) remarked that parasitic plants are far more fecund than their hosts; a reasonable interpretation is that safe sites are many fewer for the parasites and that the probability of finding them is low. Then production of increased numbers of seeds, perhaps at the cost of making smaller seeds, would enhance the probability that some seed would land in a suitable spot. Extreme reductions of seed size are generally associated with the dependence on another organism for nutrition: orchids often require mycorrhizal fungi and many parasites (but not all) are parasitic from the time they germinate (those that initially support themselves have somewhat larger seeds).

A common device for becoming air-borne is the production of a plume; the plume is derived from a variety of anatomical structures,

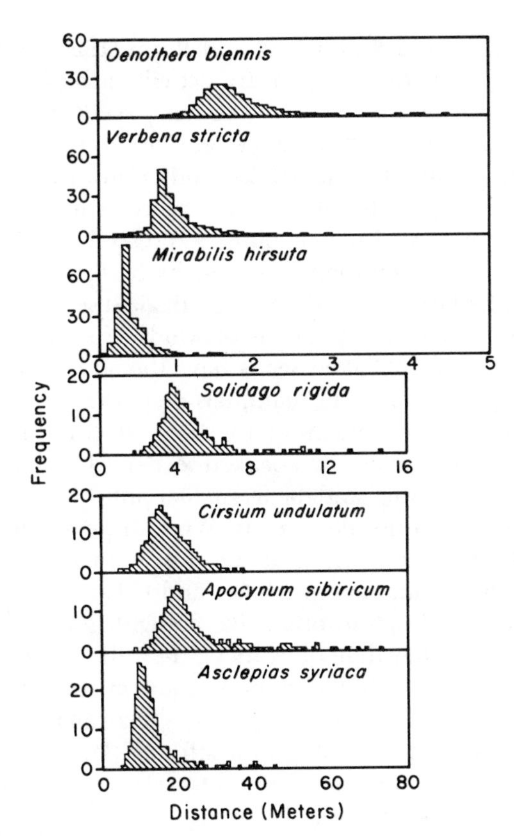

Figure 4.5 Frequency distributions of dispersal distances, from a release point, of seeds of several prairie herbs. The top three species have no special dispersal devices on the seed, the center one has a small plume, the bottom three have large plumes on the seed. Note the difference scales for distance. The wind velocity for this experiment was 10–15 km/hr. Redrawn from Platt and Weiss (1977) by permission of The University of Chicago Press, © 1977 by The University of Chicago.

depending on the taxon. Sometimes the seed or fruit bears many plumes or hairs and has a fuzzy or woolly appearance. Plumed seeds are more common in open habitats than in forests. Seeds dispersed in this way usually travel singly; as a rule, they can travel much farther than winged seeds. Patterns of seed dispersion around the parent are characteristically leptokurtic (Figure 4.5; e.g., Platt and Weis 1977). If everything else is equal (seed weight, height of release, environmental conditions, etc.), we would expect seeds with larger plumes to travel farther (Sheldon and Burrows 1973). For interspecific comparisons, everything else is almost never equal, and Rabinowitz and Rapp (1981) showed that the small seeds of grasses of small stature actually achieved dispersal distances as great or even greater than the hairy seeds of other species.

Hypochoeris glabra is one of many species that produce propagules with two different dispersal capacities (Baker and O'Dowd 1982). Seeds (technically achenes or "pseudonuts") from the central flowers of the inflorescence are "beaked," with the plumes and seed separated by a stalk-like "beak" that is longer than the seed itself. Seeds from the outer flowers are attached directly to the plume. Beaked seeds have potentially greater dispersal distances by means of wind because they can remain air-borne longer, but unbeaked seeds have the additional possibility of transport on the coats of small mammals or perhaps by harvester ants that drop them before returning to the nest (the ants have more difficulty carrying unbeaked than beaked seeds). At high plant densities, parent plants are shorter and bear smaller inflorescences producing proportionately more unbeaked seeds. Wind dispersal may be less successful from short parents and thus the shift in proportions of the two seed types may be advantageous (Baker and O'Dowd 1982).

Plumes, hairs, bristles, and other like appendages on a seed or fruit may have functions other than buoyancy. They may orient the fall of the propagule so that it lands in a certain position, anchor the seed in position, deter certain predators, or even cause short-range movement after landing that enhances the finding of a safe site (Peart 1979, 1981).

The umbels of *Daucus carota* respond to relative humidity by spreading the pedicels supporting the seeds; there are both individual and seasonal differences in the rate of response (Lacey 1980). An umbel that responds rapidly to relative humidity drops its seeds more quickly and disperses them over shorter distances than an umbel with a slower response: the seeds on a wide-spread umbel are more easily removed by a breeze and so become detached even in a light wind, whereas the seeds on less open umbels require a stronger wind for detachment and hence are released under conditions enhancing greater dispersal. In this case the response of the parent plant has considerable influence over the dispersion pattern of the offspring.

The whole plant or a large segment of it may become detached and blow in the wind. These "tumbleweeds" are found in many taxa but are restricted in habitat to open, wind-swept steppes, short-grass prairie, and certain deserts. *Agrostis hiemalis* uses this type of dispersal in addition to simple dropping of single seeds. Not surprisingly, the distances achieved by tumbling were markedly greater than for seed dropping: 95% of the dropped seeds fell within 2 m of the parent, but the same fraction extended to a distance of 9 m when carried by the tumbling inflorescence (Rabinowitz and Rapp 1979).

Wings are another common means of moving with the wind; they derive from a diversity of anatomical structures—the fruit, the flower, and the seed itself. As for plumes, wings of some species are important in orienting the seed upon landing. Some wings are impressively large: *Dipterocarpus grandiflorus* seeds are about 6 cm long, with a wing of around 14 cm; *Macrozanonia macrocarpa* seeds have lateral wings with a total "wing-spread" of about 15 cm (Ridley 1930). Wings may be placed symmetrically with respect to the seed as in *Fraxinus*, or asymmetrically as in *Acer*, which may affect behavior of the dispersal unit in different wind conditions and the achievable dispersal distances (Green 1980, unpublished manuscript). Asymmetrical wing placement enhances dispersal distances in light winds, but the results are likely to be different in strong, gusty winds (Green 1980, unpublished manuscript). Dispersal distances for winged seeds are often quite short and many seeds can land near the parent. Nevertheless, if the wing-loading (weight/area of wing) is low, considerable distances may be achieved in some cases (Ridley 1930). Regrettably little seems to be known, however, about the frequency of dispersal over long distances. If *Pinus alba* seeds can travel 880 yards, as Ridley mentions, and a Malayan windstorm can carry dipterocarp seeds in large numbers as far as half a mile (Webber 1934), how often is this achieved, in what circumstances, and how do seedling establishment and survival vary with distance?

Spergularia media and *S. marina* are intriguing because their seeds are polymorphic, and the development of wings on the seeds is, in part, genetically controlled (Sterk 1969, Sterk and Dijkhuizen 1972). The frequency of individuals producing winged seeds varies among habitats for *S. media*. Plants growing in closed stands of vegetation produce more winged seeds than plants growing in sparsely populated stands. In contrast, *S. marina* occupies a smaller array of habitats, being found almost exclusively in disturbed and open stands, and almost never produces winged seeds (the trait in this species is recessive, which itself is probably an evolved characteristic). The degree of wing development is presumably adaptive in terms of habitat condition (Sterk 1969, Sterk and Dijkhuizen 1972), and, although the authors do not specify it, the facility of dispersal seems a probable factor in determining the frequency of wings. The simplest suggestion is that crowded growing conditions and the ability to grow and reproduce in a variety of habitats favor the capacity to disperse over longer distances, and that broad wings make this possible. The interpretation is complicated, however, by the observation

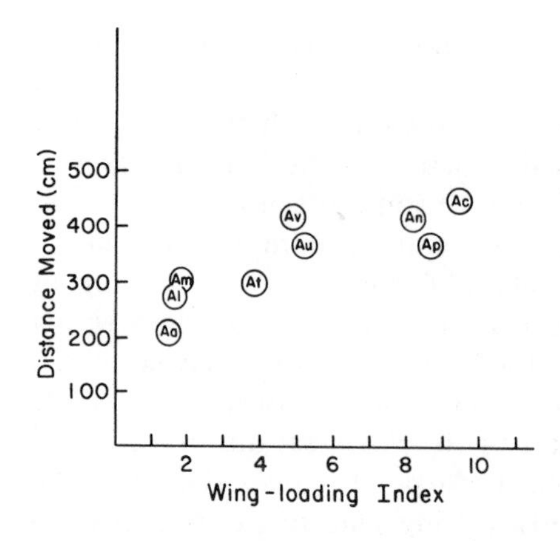

Figure 4.6 Dispersal distances of the seeds of several species of *Abronia* as a function of wing-loading index (surface area/weight). Note that this index is the inverse of the usual measure of wing-loading. The correlation is statistically significant. The species are indicated by their initials; those referred to in the text are *A. maritima, A. alpina, A. nanacovillei,* and *A. crux-maltae.*

(Sterk 1969, Sterk and Dijkhuizen 1972) that seed (and seedling) size vary concomitantly with wingedness in both species. Therefore, we need to know if the effective wing-loading and potential dispersal distances are really as different as the differing wing breadths would indicate. Wing polymorphism and variation in morph frequencies among populations is also known in *Plectritis congesta* (Ganders et al. 1977).

Abronia also produces winged propagules, with both within-plant polymorphism and considerable differences among species (Wilson 1976). Both surface areas of the winged dispersing units and the distance traveled (in the laboratory) were measured for nine species; from weights of the dispersal units and the surface area, an index of wing-loading was calculated. The distance moved was clearly correlated with the wing-loading index (calculated by Wilson as surface area/unit weight rather than the more customary weight/area): the larger the surface area relative to weight of the propagule, the farther it could be blown by the wind (Figure 4.6). The correlation is not perfect, reflecting interspecific differences in details of wing structure, the absolute weight of the dispersal unit, and probably other factors (Wilson 1976). From the discussion presented by Wilson, it is difficult to relate estimated dispersal capacity to habitat conditions in any simple way: both montane and dune species are found at both ends of the spectrum (i.e., *A. alpina* and *A. nana-covillei* are montane; *A. maritima* and *A. crux-maltae* live on dunes).

Dispersal distances of the fruits of the tropical leguminous tree *Lonchocarpus velutinus* were inversely related to the (square root of) wing-loading (Augspurger and Hogan unpublished manuscript). Lighter fruits traveled farther from the parent and their seedlings had a higher probability of survival. Seed number per fruit, seed weight, and wing-loading all varied considerably, even among the fruits of a single tree.

Ballooning is yet another means of traveling by wind. A layer of the seed coat may be loose, or the seed may be surrounded by inflated parts of the flower. Such arrangements increase buoyancy and, even if the unit is too heavy to be air-borne for long, rolling over the ground is facilitated.

Finally, a number of plants achieve some dispersal by swaying with a passing breeze and dumping their seeds as they swing around. This may occur in serendipitous fashion in many species, but some plants contain their seeds within walls from which the seeds can escape only when the stem flexes. Poppies (*Papaver*) are a classic example; the seeds are shaken from tiny windows at the top of the capsule.

Water Dispersal

Dispersal by water ranges from relatively short-distance movement by rain-wash or sheet flooding over the ground and within-pond movement (e.g., Jain 1978) to very long-distance travel by river or ocean currents. In many cases, seed movement is a casual consequence of sudden availability of water and no special adaptations are involved. (This does not imply that population consequences of such movement are trivial, merely that selection is not discriminating among parents whose genotypes specify traits related to movement by water.) However, a number of plants seem to rely on water for important dispersal movements; in some cases water transport may be supplemented by other means as well. *Anastatica hierochuntica* in the Negev desert depends on raindrops for detachment of the seeds and on run-off for longer distance dispersal; birds and possibly mammals may also contribute to longer movements (Friedman and Stein 1980).

Most seeds that move long distances by water have a flotation device of some sort, which obviously increases the amount of time they can be carried along. Very tiny seeds have high surface/volume ratios and surface tension keeps them atop the water. Others trap air in corky or fibrous layers on the surface, contain internal air bubbles, or lower their specific gravity by storing lipids or other components that are lighter than water. Not only must the seeds float, they must survive their treatment by moving water, and so they must be resistant to prolonged soaking and in some cases to extended immersion in salt water. Sometimes reproductive units other than seeds (e.g., bulbils, seedlings) are transported by water. Movement of seeds by water currents typically must be unidirectional—downstream (e.g., Waser et al. 1982). In the absence of a second means of dispersal, the distribution of seeds is necessarily restricted.

Dispersal of seeds by water has been even less studied, perhaps, than other types of dispersal. Staniforth and Cavers (1976) showed that some species of *Polygonum* living on gravel bars in Ontario rivers released their seeds at a time appropriate for the annual floods to carry them off; the seeds achieved sufficient buoyancy by means of an inflated perianth around them and survived their immersion very well. Seeds of one species failed to germinate, however, after their deposition by the river current, and the species presumably owes its presence on the gravel bars to some other dispersal agent.

Mangroves in four families are found on Panamanian seashores. The seeds have no dormancy and germinate before dispersal; the propagules are dispersed by water. The species characteristic of higher ground on the landward edge of the intertidal zone has small propagules that can become established only if they are stranded above the tide mark for several days: the propagules of the Panamanian species of *Avicennia* do not sink, whereas those of *Laguncularia racemosa* lose vigor before they lose flotation, so in both cases establishment can only occur on (temporarily) dry land. *Rhizophora* spp. and *Pelliciera rhizophorae* have larger propagules that sink and establish below the waterline; their greater weight may prevent dislodging by the tides and they normally grow on the seaward edge of the intertidal. And if initial establishment fails, they may regain buoyancy and float away to try again elsewhere (Rabinowitz 1978).

Dispersal on Animal Exteriors

The dispersal units of some plants are equipped with hooks, barbs, bristles, or spines that catch in the fur of passing mammals or, perhaps, the feathers of birds. Animals that frequently groom themselves will not carry such seeds very far, but at least sometimes mammal fur can carry quite a load of seeds. For instance, Kenyan hares (*Lepus capensis*)—especially the females—frequently carried burrs of several types in their fur (Agnew and Flux 1970). As measured by their adhesion to human clothing, some attached propagules can travel for long distances (Bullock and Primack 1977).

The other major means of attaching to animal exteriors is glue. Many plants have gummy parts that stick a seed or fruit to a mammal or bird (e.g., Kerner 1895). Ridley (1930) described some *Pisonia* fruits so sticky that small birds get caught accidentally and can't get away. The dispersal agents for the seeds of these fruits may be larger birds, but even for these a large accumulation of gooey fruits can prove fatal. The seeds of mistletoes (Loranthaceae) are generally covered with "stickum." The fruits of southeast Asian species are ingested by flower-peckers (Dicaeidae) and the seeds passed through the gut. After eating a few fruits, the birds typically sit sluggishly on a nearby branch while digesting the meal (Docters van Leeuwen 1954), and most seeds are probably deposited close to the parent. However, some may become stuck to the feathers in the vent region and be carried to greater distances. The birds also sometimes rub the vent against a twig as they void the seeds (Docters van Leeuwen 1954). The dwarf mistletoes (*Arceuthobium*) in North America are dispersed principally by explosive mechanisms, but in some cases they are carried stuck to the exteriors of birds and mammals (Hudler et al. 1979, Ostry and Nicholls 1977). Some neotropical epiphytic bromeliads produce fruits that are eaten by tanagers, which squeeze the sticky seed out before swallowing the pulp (Snow 1981); the effectiveness of dispersal clearly depends on how far the bird carries the fruit.

This mode of dispersal has been studied very little in any detail. It is important to recall that both hooking devices and sticky materials may function as protection or as establishment mechanisms in addition to, or instead of, dispersal methods.

Dispersal by Animal Harvesters

Ants. These tiny, busy creatures are very important dispersers of seeds in many kinds of habitats; their involvement in seed dispersal has been known for a long time but only recently has serious study begun. Seed-eating ants in North American deserts constitute an important component of the community; they harvest vast quantities of seeds and cache this booty in their nests (e.g., Brown et al. 1979b). Some of these seeds escape consumption and eventually germinate.

Most seeds that depend on ants for some phase of their dispersal bear an oily (sometimes starchy) appendage or elaiosome on the outside of the dispersal unit. At least in Australia, the kinds of ants particularly attracted to elaiosomes are somewhat different from the set of general harvesters: general harvesters sometimes may take elaiosome-bearing seeds, but the elaiosome-collectors usually specialize on elaiosome-bearing seeds (Berg 1975). All ants typically eat the elaiosome; the general harvesters commonly (but not always) eat the seed as well. Berg found ant-dispersal surprisingly common in Australia (see also Rice and Westoby 1981). Most of the Australian ant-dispersed species and genera are endemic to the region and, therefore, must have evolved locally. Berg asked why ant-dispersal should be so common in Australia, especially in these particular habitats, when it is known to be effective only over short distances (although Westoby and Rice 1981 suggest that even these rather short distances may be critical). Historical factors probably played a part, namely the nature of the ancestors of present-day plants and the community structure that features a rich and diversified ant fauna. Escape from seed predation might be involved, as in Heithaus' (1981) and O'Dowd and Hay's (1980) studies in North America, but we do not yet have a reason to expect seed predation to be a more serious problem in Australia than elsewhere. Berg suggested that fire is a major environmental factor in these Australian habitats, and that removal and storage of seeds by ants in underground chambers might well protect them from fires and place them in a site suitable for postfire germination. Some evidence exists that ant-dispersed plants (and maybe others also) grow larger on the microsite of an ant mound, as a result of higher concentrations of nutrients in the soil of ant mounds than in the surrounding soil (Davidson and Morton 1981a,b). As with predation, however, it remains to be seen if this advantage is better developed in Australia than in other regions.

At the time of Berg's writing, ant dispersal was best described from Eurasian and North American systems, and relatively little was known from elsewhere in the world. That may still be true, but information is beginning to emerge from some tropical areas (Horvitz and Beattie 1980, Lu and Mesler 1981). Two species of *Calathea* in tropical forests of Mexico are reported to be dispersed by ants, and several other species probably are (Horvitz and Beattie 1980). The flowering stem of *Calathea* curves toward the ground at maturity, and seeds are released at ground level. The seeds are collected by carnivorous ants (also important dispersers in Australia), which may be less dependent on the plants than vice versa. Most dispersal distances were extremely short, the longest being just over 10 m, but seeds are carried farther than if the only dispersal occurred by means of the down-curving stem. Treatment of the seeds by the ants affects seed germination in various ways, depending on when the aril is removed (Horvitz 1981).

The most intricate and captivating ant-dispersal systems of the tropics may be those involving arboreally nesting ants that establish mutualistic relationships with certain plants (e.g., Huxley 1978, Janzen 1974b, Uphof 1942). Some of these ants establish "ant gardens" by planting the seeds of their host in the walls of the nest, which appears to be a far better site for germination and establishment than most other sites (Kleinfeldt 1978). However, if the seeds of a particular plant are deposited in an ant garden located on that individual, we can hardly claim that dispersal has occurred. In some cases the seeds become stuck to the ants and are deposited along arboreal ant trails, rather than in the nest (Macedo and Prance 1978).

Pachycentria tuberosa is a shrub parasitic on *Hydnophytum formicarium* in Sarawak. Ants living in the swollen stems of *H. formicarium* collect the seeds of *P. tuberosa* and plant them in the nest debris stored in the stem. There the seeds germinate, the roots ramify from one debris deposit to another, and the shoots emerge through holes in the stem of the host. Distances over which the seeds of this parasite are carried remain unknown. *Dischidia gaudichaudii* is a sometime parasite on *D. rafflesiana.* Its seeds are planted in the walls of ant nests in the leaves of *D. rafflesiana,* the roots spread through the debris dumps, and the stems grow out the entrance holes of the ant nest (Janzen 1974b). The seeds of *D. rafflesiana* itself are treated in the same way, in which case the degree of dispersal achieved is questionable. Janzen (1974b) suggested that the use by plants of ant nests as a source of nutrients is particularly well developed in otherwise nutrient-poor habitats (see also Thompson 1981b). We seem to have just enough information to perceive that some fascinating relationships must be there, but not enough to explain the workings of the system.

Most of the more detailed work on ant dispersal of seeds has been done in North America. Berg (1966) described the dispersal mechanism in two species of tree poppy, *Dendromecon,* in California. Dispersal of seeds occurs in two stages: the mature capsule dehisces explosively, shooting the seeds to at least 5 m and perhaps 10 m (Berg 1966). Ants avidly collect the seeds and carry them off to the nest. Unfortunately, the fates of these seeds are not recorded.

The seeds of *Datura discolor* in the Sonoran Desert are dispersed by harvester ants, which eat the elaiosome but not the seed (in contrast to Berg's statements about Australian harvesters). Seed dispersal distances are rather short, averaging less than 3 m from the parent, but dispersal reduces predation on the seeds by rodents (O'Dowd and Hay 1980). These authors argued that escape from predation was the most likely advantage to be obtained by dispersing. Competition with parent is nil because the plants are desert ephemerals, and competition with siblings would occur at high densities both under the parent (predispersal) and in ant middens (postdispersal). Furthermore, rain-wash and flash floods move the seeds from their deposition site in ant middens and potentially mitigate density-dependent competitive effects. Since rains and floods move the seeds away from the possibly nutrient-rich middens and to the eventual site of establishment, the effect of removal by ants is not likely to be related to either colonization of new sites or exploitation of unusually rich patches in this habitat (O'Dowd and Hay 1980). Just how much the ants contribute to seed dispersal, given the apparent importance of surface water, is not clear, however. The situation may be different in other habitats, such as the eastern deciduous forest of North America, where ants often deposit seeds near fallen logs, where much colonization occurs (Thompson 1980).

At least one species of *Carex, C. pedunculata,* in eastern North America, is known to be dispersed by ants; a few European species are also (Handel 1976). The elaiosome of *C. pedunculata* is large and attractive to ants, which consume it and dump the seed in their refuse heaps. This species may depend on ants for dispersal to rotting logs (often the site of ants' nests), where the seeds germinate and the seedlings grow in the absence of the depressing effects of competition from other species of *Carex* (Handel 1978). The possible effects of intraspecific competition are undescribed. More recently, several other species of *Carex* are reported to be ant-dispersed (Beattie and Culver 1981).

Several species of the eastern deciduous forest of North America are dispersed by ants (e.g., Handel et al. 1981). Most are herbaceous and most ripen their seeds in late spring or summer, a time when dispersal by birds is less probable (Thompson 1981a). The ants involved in these dispersal systems are not very specialized to particular plant species (Beattie et al. 1979). Unlike some of the other ant-dispersed species, bloodroot (*Sanguinaria canadensis*) has no alternative means of dispersal and thus is entirely dependent on ants (Pudlo et al. 1980). In undisturbed habitat, relatively few seeds were produced but these were carried as far as 12 m from the parent. However, in disturbed sites, seed production per clone was high, although removal by ants was infrequent and only for short distances—seldom beyond the borders of the parent clone (Pudlo et al. 1980). The authors emphasize that when this plant favors sexual reproduction in dense populations, in accordance with one of the models of Williams (1975), the failure of the dispersal system renders maladaptive conformance with the model.

The genus *Viola* exhibits three means of dispersal: ant, ballistic, and the combination of the two. Beattie and Lyons (1975) suggest that predation on *Viola* seeds may be severe; ballistic dispersal then represents escape in space and ant dispersal represents escape in time, inasmuch as ants remove the seeds very rapidly. Ballistic dispersal scatters the seeds to an average distance of 1–2 m with a maximum of about 5 m for certain species; in some cases, ants will then find the seeds and carry them to the nest, extending the maximum potential dispersal distance (see Westoby and Rice 1981). If predation is a significant problem and ants are not entirely reliable, either the ballistic or the combination mechanism may be advantageous. Obligate reliance on ants necessitates that the ants be dependable if seed predation is a serious risk; this may limit the habitats that can be occupied by obligately ant-dispersed plants.

Direct evidence that dispersal can aid escape of some types of seeds from predation was provided by Heithaus (1981) and O'Dowd and Hay (1980). Deposition in ant nests significantly increased seed germination and seedling emergence of two European species of *Viola,* and the first adult leaves were larger (Culver and Beattie 1978, 1980).

Clearly, many questions remain. Why do ants carry both seed and elaiosome back to the nest if only the elaiosome is eaten? The probable energetic cost of carrying a larger item is likely to be countered by some benefit, as yet unknown. Perhaps the time required to gnaw an elaiosome from the seed too greatly increases the risk of predation to the ant or the chance that another ant, from another nest, will commandeer the seed. Variation in the success of dispersal from place to place may affect other aspects of the life history: clutch size, frequency of reproduction, alternate forms of reproduction, and so on. As is true for other dispersal modes, the nexus of inter- and intraspecific competition, predation of different sorts, and success in establishment needs to be unraveled for most cases. We now have a series of anecdotes, developed to various stages, but no emerging generalities.

Birds. When birds, such as the acorn woodpecker (*Melanerpes formicivorus*) and the red-headed woodpecker (*M. erythrocephalus*) hoard seeds and acorns in pits drilled in tree trunks, they are acting as seed-predators. However, other birds cache seeds and acorns in the ground, and these are dispersal agents in addition to seed predators. The only cases described in any detail are from temperate regions; I do not know of any tropical examples of birds that store seeds (see Roberts 1979).

Blue jays (*Cyanocitta cristata*) in North America transport and bury acorns as far as 5 km from the source; several acorns can be carried on a single trip but they are buried singly (Darley-Hill and Johnson 1981). Numerous oak seedlings were found in the cache areas, indicating that retrieval by jays is imperfect and that germination conditions were good.

European jays eat a wide variety of foods, both animal and vegetable, but in the autumn they collect and stash large numbers of acorns (Bossema 1979). Acorns form a great portion of the jays' diet through the year, and even the nestlings are fed on acorns. When caching acorns of *Quercus robur,* jays prefer medium sized acorns and those of long, slim form. Jays carry multiple acorns by swallowing all but the last one; this last one, or single acorns, are carried in the bill and are often larger and fatter than those carried internally. Jays can carry as many as five acorns at a time, although such a large load is rare. When only one acorn is carried, the distance carried is usually less than 20 m, but when three to five acorns constitute the load, trips over 100 m are most frequent.

Acorns are usually stored singly, usually near some conspicuous local landmark. They are buried in the soil and covered with litter or small stones. The birds retrieve some portion of their buried stores through the seasons and, in addition, detach the acorns from emerged seedlings in the spring and eat the enclosed cotyledons. Scatter-hoarding of food reserves probably reduces the risk that other seed-eaters, including other jays, will find and consume the entire stock. Many acorns are not retrieved by their hoarders and constitute potentially successful dispersals. The jays' habit of stripping cotyledons from seedlings must be commonly detrimental but seems not to be so in all circumstances (Bossema 1979). If jays discriminate against small acorns, they exert simultaneous selection in favor of larger acorns (assuming some genetic control of acorn size), but only at the price of sacrificing some of these large offspring to the predators.

Clark's nutcracker (*Nucifraga columbiana*) inhabits montane coniferous forests in western North America and specializes on the seeds of pines (Tomback 1977, 1982; vander Wall and Balda 1977). They carry the seeds in a pouch under the tongue. Nutcrackers feed on white-bark pine (*P. albicaulis*) seeds in the Sierra Nevada of California; a bird carries an average of 77 whitebark pine seeds per trip, with a maximum of 150 recorded, but places only about 4 in each cache (Tomback 1977, 1982). As many as 95 seeds of pinyon pine (*P. edulis*) can be carried at a time, but the average is 55; it takes about 45 min to extract that many seeds from the cones (vander Wall and Balda 1977). In northern Arizona, the nutcrackers may travel as far as 22 km from the harvest site to the caching site. Each bird caches two to three times as many seeds as are needed for winter food and uses these stores to attain breeding conditions and to feed nestlings. Nutcrackers begin to harvest pine seeds before the cones open, thus getting a head start over some of the other harvesters. Other species of birds and rodents also eat pinyon seeds, either from the cones, once opened, or from the caches. Nutcrackers use landmarks to retrieve up to about 78% of their caches (in an experimental situation); interestingly, other birds can successfully locate these caches some of the time by searching near appropriate landmarks (vander Wall 1982). Losses to other seed eaters probably necessitates the deposition of "extra" caches. Several seeds are stored in each cache and may germinate if not eaten. Intense seedling competition usually means that only one will survive; if they are siblings and form root grafts (Balda 1980), there may be kin-selected "cooperation." Nutcrackers commonly store seeds on south-facing slopes at the bases of trees or shrubs, sites that are both relatively easy to locate and favorable for germination. Curiously, in northern Arizona, the caching areas are located largely in the vegetation zone above where pinyon pine is most common. Although this diminishes the threat of thievery from some of the other seed-eating birds (vander Wall and Balda 1977), the post-dispersal success of these seeds must be reduced. *Pinus edulis* is a mast-fruiting species, and failure of the seed crop, if simultaneous with low supplies of other pine seeds, results in huge population irruptions as the nutcrackers move out in search of better foraging (see also vander Wall and Balda 1981).

The Eurasian nutcracker (*N. caryocatactes*) harvests pine seeds and hazel nuts. It differs from the North American species in that it is at least sometimes territorial; the seeds are collected on a communal feeding ground and cached on each bird's territory where they are defended against other nutcrackers (references in vander Wall and Balda 1977).

The pinyon jay (*Gymnorhinus cyanocephalus*) is another bird whose biology is closely tied to that of pinyon pine in some regions. The availability of pinyon seeds determines the ability to breed and the timing of the breeding effort (Balda and Bateman 1971, 1973; Ligon 1978). Pinyon seeds are stored on the home range of each flock, usually close to the nesting area. Like the nutcrackers, pinyon jays readily discriminate among good and bad seeds (Ligon and Martin 1974) and commonly place their caches in sites favorable for germination. The jays, however, favor a habitat where pinyon pine is common, and perhaps may be better dispersal agents than the nutcrackers.

Pinyon pines have several features that appear to be adaptive to dispersal by birds. The seeds are large, cone-opening is spread out over about two months, and the cones are oriented upward, so that seeds are held in the cone for some time. Healthy and diseased seeds, or those with aborted embryos, can be distinguished by color, and the birds can clearly discriminate among the seeds (Smith and Balda 1980, vander Wall and Balda 1977). Similarly, whitebark pine cones do not open when ripe, ripening is somewhat asynchronous, and the seeds are large. Cones are borne in conspicuous positions at the tips of branches (Tomback 1977). In contrast, Jeffrey pine (*P. jeffreyi*) has winged seeds, rapidly shed from open cones. Although nutcrackers harvest many seeds of this species, adaptations of seeds or cones to avian foragers are not apparent (Tomback 1977). Likewise, limber pine (*P. flexilis*) sheds its seeds from the open cones, which ripen quite synchronously. Seeds are unwinged, but there are no easily detected color differences between edible and inedible seeds (vander Wall and Balda 1977). Differences in the ways that harvesting birds exploit these food resources might account for the differences in the apparent degree of adaptation to bird dispersal.

Mammals. The best hoarders of seeds are among the rodents. One thinks immediately of tree squirrels, perhaps, but there are also the chipmunks, ground squirrels, the neotropical agoutis (*Dasyprocta*) and acouchis (*Myoprocta*), and mice in several families. The best-studied are the North American tree squirrels. Some of these (*Tamiasciurus*) commonly create a large central cache in which they store conifer cones for winter food (Smith 1968). How many of the seeds from these stored cones escape from predation is not recorded; in this system the squirrels may be predators more than dispersers. Other tree squirrels (*Sciurus*) are scatter-hoarders, burying seeds singly or in small clusters. They collect the seeds of walnut (*Juglans*), hickory (*Carya*), oak (*Quercus*), beech (*Fagus*), chestnut (*Castanea*), as well as maples (*Acer*), whose seeds have wings. Where *Tamiasciurus* coexists with *Sciurus* in deciduous forest, it too scatter-hoards nuts. Tropical *Sciurus* are also scatter-hoarders (Heaney and Thorington 1978). It seems clear that scatter-hoarders often fail to retrieve seeds from some of their caches, as is true for birds of similar habit, and that the remaining seeds (if not destroyed by other predators or by pathogens) can germinate. Thus as with scatter-hoarding birds, the dispersal system is expensive to the parent plant because many offspring are sacrificed so that some survive.

Many of the temperate-zone plant species dispersed by this means are mast-fruiters, and the effects of the mammalian harvest system may differ among years. *Carya glabra* seeds in a year of low seed production were almost entirely destroyed by weevils or carried off by gray squirrels (*Sciurus carolinensis*) (Sork and Boucher 1977). The squirrels removed nuts in proportion to their availability at different times in the season: the more nuts lay on the ground, the higher the probability of their removal by squirrels (Sork and Boucher 1977). If the squirrels are more likely to retrieve cached seeds when the crop is small, then predation will outweigh dispersal. On the other hand, when crops are large, then squirrels won't collect the whole crop, which is then subject to destruction by other predators and, furthermore, fails to be dispersed completely (Barnett 1977). But large crops may lead to longer dispersal distances because squirrels tend to spread out their caches rather than increasing the cache density as the number increases (Stapanian and Smith 1978). The squirrels' common habit of "notching" the fall-germinating acorns of white oaks before burying them could make dispersal irrelevant (Barnett 1977, Fox 1982). Fall germination of these species might have originated, in part, as a means of eluding predation by squirrels (a form of escape in time), but the squirrels' behavior can be interpreted as a countertactic (Fox 1982). Nevertheless, many of these acorns are cached without notching, especially by young squirrels, so dispersal may still occur. The tradeoffs between predation and dispersal in such a system as a function of seed-crop size have not been adequately studied.

Although seeds are known to germinate from caches, and some studies emphasize the central importance of rodent caches to the establishment of certain plant species (e.g., *Purshia tridentata,* Sherman and Chilcote 1972), there seem to be very few data that register variations (temporal, spatial, taxonomic) in the success of this system. Burial of seeds affects seeds in differing ways. White oak (*Quercus alba*) acorns escaped predation more successfully after burial, but predation on pignut hickory (*Carya glabra*) was unaffected (Barnett 1977). The effects of seed burial vary with conditions: sometimes it is virtually essential, perhaps especially in hot, dry habitats, but other times it is not (Barnett 1977, Harper 1977).

Both mammalian and avian foragers may harvest a fruit, carry it off some distance in jaws or feet, and consume some seeds whereas others drop to the ground without being eaten. This may happen fortuitously (as in the cycads, probably, in which baboons and vervets often carry away whole cones; Giddy 1974), but in some instances the fruit seems to be designed to encourage the rejection of part of the fruit and seeds by the would-be predator. The pods of *Sterculia apetala* are lined with short, stiff hairs. Several types of mammals feed on the seeds. Some of these consume the seeds at the parent tree, but certain arboreal species (squirrels, monkeys) are likely to carry away some pods to eat the seeds elsewhere. When they open the pod, the hairs accumulate in lips and paws and eventually the animal becomes so irritated by them that it drops the fruit and concentrates on getting rid of the hairs (Janzen 1972). Any seeds remaining in the pod have been dispersed. Some tropical oak (*Quercus, Pasania*) and chestnut (*Castonopsis*) seeds are harvested by squirrels, but they are protected somewhat by spines or hairs, and the squirrels often drop an acorn or a cluster of chestnuts without much damage (Ridley 1930).

Dispersal in Animal Interiors

Seeds to be dispersed by animals that consume them are enclosed in or attached to edible material that is the food reward for the consumer. The edible portion derives from various morphological parts but is here called a "fruit" regardless of its anatomical origins, in order to emphasize its ecological function. Fruits come in an astounding array of sizes, colors, aromas, flavors, and nutritional values. To some extent, this variation can be related to particular kinds of dispersal agents, although there are perhaps no hard-and-fast rules. Not only must these plants offer a food reward (or an apparent reward, Wiens 1978) to the dispersers of their offspring, those offspring must either evade or resist the potentially harsh handling of the disperser. The seeds may somehow spend little time in the digestive tract, either by being regurgitated (as by some birds) or by passing quickly through the digestive tract, perhaps aided by a cathartic chemical in the fruit. Alternatively, they may become resistant to the grinding and chemical treatment received in the animal guts. The most common form of resistance is a hard seed-coat (McKey 1975). The fruits also must be presented so that the appropriate dispersal agents can reach them readily; the importance of difficulty of access to the fruits appears to vary among frugivores and with the nutritional value of the fruit (Denslow and Moermond in press).

There must be several costs to adopting this type of dispersal. Fruits may be expensive to build: many are rich in nutritional value and some are large. Hard seed-coats or cathartic agents may be required, and immature fruits may need to be protected (probably chemically) from premature dispersal. Some frugivorous animals destroy some portion of the crop, and such losses must be compensated for by producing more fruits or changing the schedule of fruit production. Nevertheless, dispersal in the innards of animals is very widespread among the seed plants, both gymnosperms and angiosperms, so presumably the benefits have outweighed the costs. Precisely what those benefits are is not known, regrettably. It is likely that dispersal distances are frequently longer (predictably so) than for other dispersal systems, and it may be that the probability of deposition in a favorable site is higher. The effectiveness of the manure often applied at the time of deposition apparently has not been ascertained.

Regular internal dispersal is known to occur by modern representatives of four of the seven major groups of vertebrates, which are by far the most common internal packers of seeds (earthworms notwithstanding, see McRill and Sagar 1973). Only the jawless fishes, the cartilaginous fishes, and the amphibians seem to be left out. Strange as it may seem, certain fishes commonly eat fruits and some seeds are germinable after passage through the guts. Frugivory is best known among fishes of the riparian floodplains and swamps in Amazonia, although it is not exclusive to this region. Gottsberger (1978) recorded a dozen species of larger fishes whose stomachs contained the seeds of 33 plant species; of these, 16 species were in germinable condition. This list is far from exhaustive, of course, because it derives from a single locale and neglects both smaller seed-eating fishes and possible effects of passage through the entire gastro-intestinal tract. Goulding (1981) surveyed a number of Amazonian fishes (primarily characins but also some catfishes) and showed conclusively that several are regular and moderately specialized seed predators. However, the seeds of certain species can pass unharmed through the guts of some of these predators, and a few fishes consume fleshy fruits, in addition to hard seeds, and may serve as important dispersal agents. The fish-dispersal (and predation) system may be best developed where the terrestrial habitats are only seasonally flooded and seed deposition in suitable habitat has a high probability compared to that of riverbank plants whose seeds may be eaten by fishes that release the seeds into flowing water, to be deposited who knows where. Furthermore, the development of a frugivorous fish fauna may be encouraged in river systems with low levels of nutrients and low primary productivity, so that many fishes depend heavily on food material derived from adjacent terrestrial habitats (Goulding 1981).

Perhaps better known are the dispersal agencies provided by lizards and tortoises, the only extant reptiles in which some form of herbivory is a regular occurrence, although Ridley (1930) does mention the possibility that alligators eat (and disperse?) the alligator apple (*Annona palustris*). Various types of tropical iguanas include fruit in their diets and at least sometimes may pass viable seeds. The fruits of several species are eaten by tortoises and viable seeds emerge in the feces. The Galápagos tomato (*Lycopersicon esculentum* var. *minor*) is dispersed by the Galápagos tortoise (*Testudo elephantopus*); the seeds take 12–20 days to pass through the digestive tract but germination is more successful by the seeds taking shorter times (Rick and Bowman 1961). This is a variety of the domestic tomato species, but the seeds of domestic tomatoes are damaged by passage through tortoise guts (Rick and Bowman 1961). Box turtles (*Terrapene carolina*) may be among the major dispersers of mayapples (*Podophyllum peltatum*) in eastern North America (Rust and Roth 1981); could they disperse wild strawberries (*Fragaria*) and raspberries (*Rubus*) as well? Rust and Roth found that seedlings from seeds deposited within the parent clone of mayapples had much lower survival than those dispersed to greater distances. Box turtles can see colors very well, especially in the red–orange wave lengths, and eat a variety of fruits (Klimstra and Newsome 1960). Gopher tortoises (*Gopherus* spp.) in the southeastern United States and northeastern Mexico reportedly eat fruits of *Asimina pygmaea* (Kral 1960) and *Opuntia,* among others, and may disperse the seeds of some of these. To the best of my knowledge, virtually nothing is known about the ecology of most of these interactions—dispersion patterns, effectiveness of deposition, possible influence of simultaneously deposited manure, seedling success, the roles of predation, competition, site preemption remain unstudied. Although it might seem silly to recount information at such an anecdotal and preliminary level, perhaps it will be justified if it spurs someone to inspect the systems more closely.

Whatever role these "lower" vertebrates may play, the mammals and birds constitute the great majority of internal seed-carriers. Representatives of both classes have been known to be seed-dispersers for a long time, but little serious ecological investigation has been attempted until rather recently. Ridley (1930) provides examples of fruit-eating and possible seed dispersal for some present-day species of most (12 of 18) of the extant orders of mammals, including not only those usually classed as herbivores but also many usually thought of as carnivores, even including a single species of Insectivora (the Malay tree-shrew, *Tupaia ferruginea*). Since Ridley's writing, at least two additional orders may be added to the list (Dermoptera, Edentata). Among the birds, frugivory and a possible role in seed dispersal is also widespread, occurring in at least 16 or 17 of the 29 orders. Just how commonly each of the frugivores in these taxa serve as successful agents of seed dispersal is seldom known; many of them may be more important as predators.

Although mammals ranging in size from mice to elephants are known to consume fruits and sometimes to pass the seeds in germinable condition (Hladik and Hladik 1967, Krefting and Roe 1949, Lieberman et al. 1979, Ridley 1930, van der Pijl 1972), very little has been studied of the ecology of these interactions. A neotropical tapir (*Tapirus bairdi*) destroyed most of the seeds of *Enterolobium cyclocarpum* it ingested (Janzen 1981a). About 20% of the seeds emerged from the tapir's digestive tract in a viable state, taking from 4 to 23 days to pass through, even though all were consumed in a single meal. A number of seeds were spat out when the tapir ate the fruits. Clearly, if this single animal is a fair representative, it exacts a high toll of offspring from the parent tree: some are not dispersed, others are digested. However, the dispersal potential for the remainder may be rather good, because the seeds are deposited over a considerable span of time, during which a wild tapir might range widely.

Janzen (1981b) also fed the seeds of *E. cyclocarpum* to range horses, under the hypothesis that wild horses may have been important dispersers during the Pleistocene. Individual horses retained seeds for differing lengths of time; most seeds had emerged after 14 days but two horses passed the last seed after 30 and 33 days and one retained some seeds for as long as 60 days. Furthermore, the survivorship of seeds subject to treatment by the different horses' guts differed markedly, ranging from 17–56%. The fates of seeds carried by horses are not known. Janzen and Martin (1981) have suggested that the large fruits of several neotropical plants may have been consumed principally by now-extinct large mammals.

Elephants may be important but largely overlooked agents of dispersal, at least in Africa (Alexandre 1978). They are the chief dispersers of perhaps 30% of the forest trees in the Ivory Coast and seeds germinate rapidly from elephant dung (Alexandre 1978). Some fruits never accumulate under the parent trees when elephants are present (Janzen and Martin 1981).

Bats are common agents of seed dispersal in many tropical regions. Why frugivory by bats is so poorly developed in temperate regions is unclear; seasonal exploitation of food resources is certainly possible for insectivorous bats that migrate to temperate zone breeding grounds. The large Megachiroptera are found in Africa, southern Asia, and Australia; they eat rather large fruits (up to 250 g), squeezing out and swallowing the juice, then spitting out the seeds and pressed pulp (van der Pijl 1957). The neotropical phyllostomatid bats are smaller; about half the members of the family are frugivores. These bats both spit out seeds and pass them through the digestive tract (van der Pijl 1957). Bat fruits have distinctive aromas, little conspicuous coloration (bats are chiefly nocturnal and colorblind), and they are borne in exposed positions outside the crown of foliage. The dispersion of seeds from bat-consumed fruits is highly uneven (Fleming and Heithaus 1981, Janzen et al. 1976) since many seeds are deposited beneath special roosting trees. The most successful dispersal may be accomplished by seeds that fall elsewhere—between roosts or under occasional perches, although August (1981) suggested that dispersal of seeds to sites beneath roosts was often successful in his Venezuelan study area.

Primates are unusual among mammals in that color vision is well developed in many species, especially in the Old World taxa, but also to a lesser degree in New World forms. Snodderly (1979) argued that neotropical frugivorous primates must discriminate their foods from a generally green background of foliage so that the important discriminations are among shades of green or between green and some other color. In contrast, many more paleotropical primates forage in habitats subject to intense dry seasons that cause the main background colors to be the browns, and selection may then have favored visual systems capable of discriminating hues at the red end of the color spectrum, in addition to the various greens, perhaps. A neotropical primate, *Callicebus torquatus*, in Peruvian Amazonia eats large quantities of certain types of fruits. Several are consumed before ripening, while they are green, and the monkeys are mainly seed predators (Snodderly 1979). When the monkeys eat fruits of colors other than green, Snodderly suspects that usually only the fruit is digested and the seeds are dispersed. The green color of some fruits may reduce their conspicuousness to visually searching foragers, but in some cases these monkeys obviously circumvent this attempted camouflage, perhaps by using other senses in the search for food.

Monkeys are major dispersal agents for *Tetragastris panamensis* on Barro Colorado Island in Panama (Howe 1980): two species account for 88% of the viable seeds removed from the parent. But their services were terribly wasteful, because many of the seeds were simply dropped and even more, although they survived passage through the digestive tract, germinated in dense clumps of siblings. Howe argued that only one sib is likely to survive from such a compact cluster. These two species of monkey may not contribute regularly to the dissemination of *T. panamensis* seeds, however, inasmuch as they may capitalize extensively on the fruits of this species principally in years of unusually high fruit production (Howe 1980).

The dispersion pattern or seed shadow produced by mammals (ants and probably birds as well) is likely to be very patchy. Deposition of seeds in ant middens is clearly localized, as is true in bat roosts. Baboons deposit dung on rocky outcrops and seeds are washed down to the ground by rains (Lieberman et al. 1979). The seeds of an Australian cycad, *Macrozamia riedlei,* are dispersed frequently by brush-tailed possums (*Trichosurus vulpecula*), chiefly along straight trails from the parent plant to large trees climbed by the [illegible] sums (Burbidge and Whelan in press). The consequences of such patchiness include possible effects of predation and competition but have been unexplored in any systematic way. We do not seem to know whether these kinds of patchiness have consequences different from the patchiness created by the usual leptokurtic seed shadow around the parent.

The dispersal of seeds by frugivorous birds has been studied a good deal more than mammalian dispersal and yet, even then, our knowledge remains quite elementary. Most plants whose seeds are dispersed predominantly by fruit-eating birds produce fruits in a colorful display; such plants are especially common in forest understory (e.g., Hilty 1980) but also occur in canopy and nonforest habitats. They may be particularly frequent in certain habitat patches (e.g., Macedo and Prance 1978 for scrub patches in Brazilian rain forest; Chapter 2 for light gaps in North American deciduous forest). Snow (1976) made the interesting point that, whereas insects are selected to avoid predation by birds and are often inconspicuous so that there are many ways to hunt for these specialists at hiding, fruits may be selected to be consumed and are often very conspicuous. He argued that there are a limited number of ways to be conspicuous to a visually searching forager and that, therefore, there are fewer ways for birds to specialize to eating fruits. As support for his suggestion, he noted that among the neotropical frugivores there are 79 species of cotinga (Cotingidae) and 59 kinds of manakin (Pipridae), many fewer than the purely insectivorous ovenbirds (Furnariidae, 215 spp.) and antbirds (Formicariidae, 222 spp.).

The fruit itself is often brightly colored; reds, blues, and black are common colors (Ridley 1930, Turcek 1963, van der Pijl 1972). This is generally true in all regions; however, the frequency of green or yellow fruits is greater in the tropics, perhaps because specialized frugivores there are better at locating or can remember the location of fruiting plants and do not need special "flags." Conspicuous colors are frequently used in combination. Birds generally have excellent color vision and it is likely that the colors of many fruiting displays have evolved in relation to this capacity of the avian visual system. This is not to say that color vision in primates or tortoises is irrelevant to the evolution of colored displays, but merely to emphasize that birds are likely to be central actors in most cases, though they may be preempted by primates in some instances (Boucher 1981) and that the direction of the interaction is one in which fruit-display color is probably the dependent variable.

Bird-consumed fruits are usually rather small; Snow (1981) estimated an upper limit of about 4 × 7 cm for an oval fruit. Fruits of that size can be eaten only by the very largest avian frugivores, such as the toucans in the neotropics, the hornbills in Africa and southeast Asia, fruit pigeons and cassowaries (Crome 1976) in Australasia, and the now-extinct dodo of Mauritius Island (Temple 1977). Morphological correlations of fruit size and bird size are anything but simple, however—bill size and body size may be less important than one might think at first glance (Ricklefs 1977, Crome 1975).

Frugivorous birds have three basic means of dealing with the seeds: they pick or squeeze the pulp free from the seed (in which case they are probably poor dispersal agents), they swallow the entire fruit, and either regurgitate or defecate the seed. Many tropical frugivores handle fruits with large seeds and their ability to regurgitate these seeds is well-known (McKey 1975). Less noticed is that many temperate birds regurgitate seeds of a relatively small size: seeds as small as those of *Phytolacca americana* (0.009 g) were frequently regurgitated by captive red-eyed vireos (*Vireo olivaceus,* 18 g) and sometimes even by the larger thrushes (*Catharus* spp.). The still larger (77 g) robin (*Turdus migratorius*) usually defecated smaller seeds and voided larger seeds, such as those of *Prunus serotina* (0.1 g) and *Cornus racemosa* (0.04 g), in both directions (Johnson et al. unpublished manuscript, see also Kerner 1895). Contrastingly, the 81 g *Turdus merula* and 17 g blackcap (*Sylvia atricapilla*) both apparently pass the 0.1 g seeds of *Prunus mahaleb* all the way through the digestive tract (Herrera and Jordano 1981).

The ability to regurgitate seeds means the birds can unload useless ballast rather quickly; unloading time for regurgitated seeds in Johnson et al.'s laboratory study was perhaps a third of unloading time for defecated seeds. This method of unloading means that, on average, regurgitated seeds will be deposited closer to the source than if they passed through the gut. It also means that parent plants can afford to increase their seed size, if it is advantageous, without diminishing the profitability of their fruits for foraging birds with good ejection mechanisms (McKey 1975). In most cases, regurgitation of seeds occurred in a matter of minutes (Johnson et al. unpublished manuscript), whereas defecation of seeds may take almost an hour or even several hours (Johnson et al. unpublished manuscript, Kerner 1895, Walsberg 1975).

Passage through the digestive tract of some birds destroys seeds, and fails to affect others, but in others, seed germination usually becomes more rapid and more immediate (Howe and vande Kerckhove 1981, Krefting and Roe 1949, McKey 1975; but see Kerner 1895). The dodo may have been the chief disperser of *Calvaria major* on Mauritius, for it was large enough to swallow the fruit and to scarify the seed on passage through the gut. The extinction of the dodo in the seventeenth century may have doomed natural reproduction by *C. major,* for no other Mauritian animals are large enough to swallow the fruit and defecate the seed (Temple 1977). Emus (*Dromaius novaehollandiae*) in Australia devour the fruits of *Nitraria billardieri,* and so do certain mammals. Seeds passed through emu guts germinate more quickly and in larger proportions than seeds passed through mammals or through no animal at all (Noble 1975). Noble makes the interesting observation that on clay soils, emu treatment may enhance the likelihood that *N. billardieri* can germinate quickly when the rains come and thus get a head start over other species, but that on sandy soils, where seeds become buried by wind-blown sand, germination occurs readily even without emus.

It seems to be generally accepted that rapid and complete germination of a cohort of seeds is advantageous to the plants (either the parents, or the offspring themselves, or both). However, this is not necessarily so (see also Janzen 1981b). Prolonged periods of dormancy in seeds are a means of waiting for the arrival of suitable conditions for germination and establishment—a form of dispersal in time (see next section). Thus removal of the germination retardants by digestive processes means that dispersal in time is sacrificed for more immediate growth. For this to be advantageous, the conditions surrounding rapid germination must foster greater fitness than delays of germination. To the best of my knowledge, we do not have the faintest idea if that is true.

Tropical avian frugivores have been arrayed on a spectrum between two extreme categories (McKey 1975, Snow 1981): specialists, which depend heavily on fruits for proteins and lipids as well as carbohydrates and water, and opportunists, which use fruits mainly for carbohydrates and water and derive proteins and lipids primarily from other sources. Some tropical fruit specialists and semispecialists feed large quantities of fruit to their nestlings (Foster 1978, Morton 1973, Snow 1971). The seeds of forest trees are often large and, therefore, if they are to be bird-dispersed, the amount of fruit around the seed(s) must be correspondingly large or very nutritious to attract avian foragers. Such plants, in the tropics, commonly rely on rather specialized frugivores for seed dispersal. The large seeds are often unloaded by regurgitation soon after the fruit is eaten, but in some cases they are passed through the digestive tract (and must, therefore, be protected by a hard seed-coat). Plants of second-growth, forest openings, and forest edges more often have small seeds and often pack many of them in a single fruit. These fruits are generally exploited by more opportunistic frugivores, often by a considerable diversity of opportunists, typically of smaller body size than the specialists. The seeds of these species are more often defecated by the birds than are larger seeds. It is important to remember, however, that the specialist–opportunist spectrum is just that; it is not a sharp dichotomy. There are degrees of specialization of both birds and fruits (e.g., Frost 1980, Wheelwright and Orians 1982) and exceptions to almost any rule. Extreme specialization of a single disperser to a particular plant species is uncommon and could be disadvantageous to the plant anyway, if the offspring of one parent were removed only to be deposited beneath another conspecific parent (Frost 1980, Wheelwright and Orians 1982). It is by no means clear that the seed shadows created by specialists are any better than those generated by generalists.

Although the nutrient content of fruits for specialist frugivores is often high, Herrera (1981a) is quick to point out that mere percentages of proteins and lipids are inadequate indicators of the reward to the forager (see also Sorensen 1981). The total amount of pulp per fruit, the relative amount of (undigested) seed material, the digestibility of the pulp by the particular bird, and the handling time required must all affect the level of profitability achieved by the foraging bird. However, these matters have not been studied in any detail. When the seed is large (for success in seedling establishment), the addition of a large volume of fruit pulp around the seed is likely to make the fruit too large for the available consumers. Thus large-seeded tropical plants may be able to increase the food reward to dispersers chiefly by improving the nutrient content more than by increasing the overall size of the fruit (Herrera 1981a).

Snow (1980, 1981) has reviewed the available information on the diets of tropical frugivorous birds. Some intriguing biogeographical patterns begin to emerge. The African flora is poor in species, compared to other tropical areas, and the African avifauna has relatively few specialized frugivores. Nor are there any groups of frugivorous birds that exhibit adaptive radiation on a scale similar to that achieved by the cotingas in the neotropics or the birds-of-paradise in Australasia. Snow suggested that this may not always have been so, however, for there is evidence that the African forest flora and perhaps the avifauna as well were once considerably richer.

Some plant families are fairly well-represented by bird-dispersed species in all four major tropical regions (Africa usually with the shortest list); the Lauraceae, Palmae, Sapindaceae, and especially the Euphorbiaceae are examples. Other families have many bird-dispersed representatives in one area and few in others: for example, there are 14 species of Rubiaceae in the neotropics but only one in southeast Asia. It would be interesting to know to what extent these distributions parallel the distributions of the entire family and if the radiation of some taxa in certain regions is related to the historically present frugivore fauna.

Even the fruits of a single genus may be treated differently in different regions. Figs (*Ficus*) are exploited particularly by specialized frugivores in the Old World, but in the neotropics they are exploited by many opportunists as well (Snow 1981). Persimmons (*Diospyros*) are consumed by frugivorous birds in most regions except the neotropics. And in some families that are dispersed principally by nonavian means, a single genus or group of species has shifted to bird-dispersal: neotropical grasses of the genus *Lasiacis,* and some Old World bamboos; a single genus of Asteraceae in each of three tropical regions (Snow 1981). Little can be made of such broad-scale patterns at this point for the data are still too fragmentary. However, it is clear that further study will be rewarding.

Not only do different kinds of birds differ in their effectiveness as dispersal agents, but the relative effectiveness differs among various types of fruit. The neotropical masked tityra (*Tityra semifasciata*) is a good disperser of *Casearia corymbosa* and *Virola sebifera* in rain forest, but a poor disperser of *Virola surinamensis,* of which it eats the fruit but drops the seed beneath the parent (Howe 1977, 1981; Howe and vande Kerckhove 1981). And the best dispersers for two species differ with habitat (*C. corymbosa,* Howe and vande Kerckhove 1979) and geographic range (*V. surinamensis,* Howe and vande Kerckhove 1981). Individual trees of

V. sebifera were served, to some extent, by different avian foragers (Howe 1981).

Fruits turn out to be important sources of food for many birds that migrate between tropic and temperate areas. Thrushes, flycatchers, orioles, tanagers, vireos, and warblers often consume large quantities of fruit in their neotropical wintering grounds (Fitzpatrick 1980, Greenberg 1981, Howe and De Steven 1979, Morton 1980). Swainson's thrush (*Catharus ustulatus*), red-eyed vireos (*Vireo olivaceus*), Tennessee warblers (*Vermivora peregrina*), and great crested flycatchers (*Myiarchus crinitus*) all breed in North America and winter in the neotropics. These four species accounted for about 60% of the seeds removed from *Guarea glabra* (Howe and De Steven 1979). Bay-breasted warblers (*Dendroica castanea*), Tennessee warblers, and, to a lesser extent, chestnut-sided warblers (*D. pensylvanica*) ate many fruits while wintering in Panama; the bay-breasted warbler appeared to be the principal consumer of the fruits of *Lindackeria laurina* and *Miconia argentea* (Greenberg 1981). Blackcaps (*Sylvia atricapilla*) wintering in Spain consume many kinds of fruits (Jordano and Herrera 1981) and are potential dispersal agents.

Migrant frugivores are potentially important dispersal agents in their winter residences, and, in addition, many of them also consume large volumes of fruit as they move between winter and summer ranges (e.g., Skutch 1980). These observations are particularly noteworthy because many, perhaps most, of the migrant frugivores are chiefly insectivorous on the breeding ground, and they tend to be stereotyped as such by north-temperate bird-watchers. Northbound eastern kingbirds (*Tyrannus tyrannus*), moving from their South American wintering grounds to the breeding grounds in North America, are major consumers of *Didymopanax morototoni* fruits in Panama (Morton 1971). Figs (*Ficus carica,* a domesticated species) are an important food source for garden warblers (*Sylvia borin*) in passage from Europe to Africa south of the Sahara Desert (Thomas 1979). Likewise, the fruits of *Salvadora persica* are a principal food of whitethroats (*Sylvia communis*) on their way north from central Africa to Europe (Fry et al. 1970). Note, however, that we have no information regarding possible dispersal success of the seeds of these fruits.

Snow (1971) emphasized the strongly seasonal pattern of fruit-ripening in Europe and correlated this with the southward passage of frugivorous birds in fall migration. A similar pattern, documented in more detail, exists in eastern North America (Thompson and Willson 1979, Stiles 1980). Most of the fleshy-fruited species in Illinois bear mature fruits during August, September, and early October; a few retain some dried fruits through the winter if they are not consumed earlier. Waves of bird migration pass through central Illinois beginning usually in late August and continuing through mid or late October. The diversity of fruiting plants and of migrating birds is highest in September. An October wave is composed primarily of yellow-rumped warblers (*Dendroica coronata*) feeding heavily on poison ivy berries (*Rhus radicans*), hermit thrushes (*Catharus guttatus*) feeding on greenbriars (*Smilax* spp.) and wahoo (*Euonymus atropurpurea*), and robins, which become particularly abundant in years when hackberry (*Celtis occidentalis*) produces large fruit crops. Fruit removal rates are high during the peaks of migration and usually lower at other times. Stapanian (1982a) noted that, in Kansas, the variation of frugivore numbers is also highest in fall and winter, indicating that fruit removals are probably very patchy in space, as observed also by Willson and Melampy (in press).

Heavy consumption of fruits especially after the breeding season is also characteristic of European thrushes (Simms 1978, Sorensen 1981). *Prunus mahaleb* bears fruits mainly during July in southern Spain at mid elevations. The black fruits are consumed and dispersed chiefly by post-breeding *Turdus merula* and *Sylvia atricapilla*. These two species consume the most fruit and seem to transport the seeds to the most favorable sites (Herrera and Jordano 1981). Most of the seeds that fell beneath the parent were eaten by deer or mice. The fruit is composed mostly of carbohydrate, with little protein or lipid; fruit crops are large and the fruiting period is short. Dispersal is accomplished primarily by two species that seem to have highly frugivorous diets at this time of year (and in the case of *T. merula*, other times as well). At least in this study, the tree is quite dependent on these dispersal agents which, in turn, are locally and temporarily dependent on the fruits. The authors suggest that both tree and birds are relatively specialized and, by Howe and Estabrook's (1977) model for specialized tropical frugivores, we would expect *P. mahaleb* to have nutritious fruits, small crops, and a long fruiting season. Clearly, this species does not fit the model. Whether the fault lies in the model, its application to a temperate system, or in the definitions of specialist is uncertain.

Dispersal by winter residents in north temperate zones can be important for species with late-ripening and persistent fruits (Karasawa 1978, Snow 1971). Both migrant and wintering birds seem to be important for dispersal of *Juniperus* spp. Migrating robins in New England are the principal disseminators of *J. communis* and *J. virginiana*. Passage of the seeds through avian digestive tracts reduces germinability, and seeds are frequently deposited in sites not optimal for seedling growth. Nevertheless, robins are good dispersal agents because they consume vast numbers of juniper "berries" and deposit seeds on stones in pastures; rain washes these seeds into cracks in the soil, and the seeds and seedlings occupy sites with favorable moisture conditions (the stones provide micro-watersheds) and with some protection from trampling (Livingston 1972). Townsend's solitaires (*Myadestes townsendi*) in western North America defend feeding territories in patches of western juniper (*J. occidentalis*) and subsist largely on the juniper berries (Lederer 1977a,b). Seeds are probably dispersed, at least within the confines of the territory, by this species. Other birds (robins, jays) are great berry-eaters also, and range freely through solitaire territories; they may be responsible for dispersal over larger distances. Some solitaires also defend winter territories in stands of *J. monosperma,* although some individuals are free-ranging (Salomonson 1978). The effect on juniper seeds of passage through the solitaire's digestive tract varied with age of the seed: six months after ripening, avian treatment reduced germination, but 12 months after ripening, the same treatment enhanced germination. Despite the presence of solitaires and other frugivorous birds, the density of seeds was far higher below parent trees than elsewhere. However, seedling density was higher in the spaces between tree canopies. Before we jump to concluding that this is a case documenting the importance of dispersal, it must be noted that in another habitat, seedling survival between parent trees was reduced drastically by competition with grasses, and seedlings were more numerous directly below their parents (references in Salomonson 1978).

A few temperate-zone breeding birds are frugivorous. The cedar waxwing (*Bombycilla cedrorum*) comes the closest to being a complete frugivore (Martin et al. 1951, Putnam 1949), consuming many types of fruits. The phainopepla (*Phainopepla nitens*) feeds predominantly on mistletoe berries in desert scrub habitats (Walsberg 1975). A number of thrushes and mimids feed some fruits to their nestlings, as well as eating fruits themselves during the breeding season.

Fruiting displays must be attractive to the dispersal agents. Fruiting seasons may be timed to correspond to seasons of frugivore abundance, as seems to be the case for some species at some temperate latitudes. Fruits may offer large nutritional rewards. Although the quality of fruit pulp is no doubt important, the quantity is at least as critical. Herrera (1982) provided evidence that the nutritional quality of bird-consumed fruits in southern Spain is probably adjusted to the metabolic needs of the avian seed-dispersers. Summer fruits contain large proportions of water, costly to the parent plant but needed by birds at a time when water is scarce; winter fruits contain high proportions of energy-rich lipids. The ratio of energy in fruit to energy in the seed for bird-dispersed fruits in Kansas riparian forest declined through the summer and leveled off in late summer and fall (Stapanian 1982a). The maximum ratio and hence peak energy reward per consumed fruit occurred in June, when the estimated abundance of caterpillars (a major avian food resource) was also highest. Stapanian suggested that it might be advantageous for bird-dispersed plants fruiting in June to produce high-reward fruits to attract birds that might otherwise forage only on abundant caterpillars.

Some tropical frugivorous birds seem to be quite selective of the fruits they consume (see also Denslow and Moermond in press), even among individuals of the same species (Howe and vande Kerckhove 1980, 1981). Although removal rates from *Virola surinamensis* trees were not correlated with variation in nutrient composition of the fruits, with the concentration of a presumably protective set of compounds, or with fruit crop size (as seen below) a clear and significant positive correlation appeared between the fraction of a seed crop removed by frugivores and the ratio between fruit weight and seed weight: the higher the weight of fruit relative to the enclosed seed—and the smaller the seed—the greater the consumption by frugivores. The birds (and spider monkeys) effectively select fruits with smaller seeds and often drop larger ones with fruit attached; the weight of the fruit itself is not critical. The authors suggest that the birds are avoiding the accumulation of excessive indigestible bulk in the form of seeds. How can we then account for the continued presence of large-seeded individuals in the population? Disruptive selection is one possibility (Howe and vande Kerckhove 1981): when the effectiveness of dispersal is sometimes low, selection may favor large seeds to enhance seedling survival, but at other times effective dispersal is favored. It might also be possible that selection is indeed directional favoring smaller seeds, and still in progress, perhaps as a result of a shift in the available dispersal agents, or that the dispersers of *V. surinamensis* in other regions are not size-selective and gene flow among populations prevents elimination of large-seeded forms. There are still other unknowns, including rates and selectivity of seed predation by other animals, so the end of the story is not yet told.

The apparent ability of frugivores to discriminate among similar food items raises a general question about how seeds should be packed into an edible fruit. The pulp/seed ratio in several fleshy-fruited species decreases with increasing numbers of seeds (*Asimina triloba,* Willson and Schemske 1980; *Vaccinium macrocarpon,*Hall and Alders 1965; but not *V. angustifolium,* Aalders and Hall 1961; *Smilax aspera,* Herrera 1981), even though the total amount of fruit increases with seed number. This indicates that, from the parent's point of view, many-seeded fruits are cheaper on a per-seed basis to construct than are few-seeded fruits. If they are equally attractive to dispersal agents, selection might favor parents that economically pack numerous seeds into a fruit. However, Herrera (1981b) provided some evidence that frugivorous birds may prefer fruits with relatively more pulp, thus providing counterpressures favoring few-seededness. Other factors might also contribute to low seed numbers in a fruit: seed predators might differentially attack multiseeded fruits (not known in *A. triloba*), limited success in pollination might favor the maturation even of uneconomical one-seeded fruits (possible for *A. triloba*), and dispersal agents might tend to deposit all the seeds from a single meal in one pile, so that seeds might be subject to attack by density-responsive pathogens or seedling competition might be intense. We have seen that large seeds, simultaneously consumed, often emerge from a consumer's gut asynchronously, but this is probably not so common with small seeds.

Whether bird-dispersed plants compete for the attentions of dispersal agents or the foraging animals compete for fruits is uncertain. Howe (1981) observed aggression by chestnut-mandibled toucans (*Rhamphastos swainsonii*) against both conspecifics and other avian visitors to fruiting crowns of *Virola sebifera*, a phenomenon seldom recorded among the visitors to tropical fruiting trees. To the extent that fruit supplies are small in certain habitats or at certain times of the year, competition among frugivores may occur, although direct evidence is generally lacking (Crome 1975, Fleming 1979). The situation among the fruit-bearing plants is equally unestablished. *Spondias mombin*, in Panama, produces colorful, sweet, and flavorful fruits that are eaten by vertebrates, including the scatterhoarding agouti. It fruits principally in August and September, at a time when other edible fruits are rather abundant. A close relative, *S. radlkoferi*, fruits chiefly in October and November when the abundance of other fruits is very low. The pulp of *S. radlkoferi* fruits is edible but reportedly less tasty than that of its congener, and the fruit itself is an inconspicuous green. Croat (1974) proposed that a shift in fruiting season to a time when frugivores have few alternative food resources might facilitate the evolution of a fruit that is (perhaps) cheaper to make. The implication is that perhaps the phenological shift reduced competition for dispersers sufficiently to permit the building of less expensive fruits. The suggestion is interesting, but documentation is still lacking. Snow (1965) suggested that the distribution of fruiting seasons for various *Miconia* species in Trinidad more or less evenly throughout the year might be evidence of competitive displacement to increase the probability of successful dispersal by the local populations of frugivorous birds. However, in the absence of statistical analysis of the data demonstrating a more-even-than-random spread of fruiting seasons (Poole and Rathcke 1979, Rabinowitz et al. 1981), or of any demonstration that fruit resources are indeed limiting, the suggestion remains unsubstantiated. A correlation between pulp-seed ratio of *Smilax aspera* and the number of other readily available, coexisting, fruit-species suggested to Herrera (1981b) that competition for dispersal agents might favor large rewards. Nevertheless, other factors need to be considered also, including tradeoffs between seed weight and seed number, which may differ among habitats, and levels of successful pollination.

Color is commonly a major means of advertising a ripe fruit crop, and it is not uncommon for more than one color to be used in a fruiting display. A high proportion of the summer-fruiting species in eastern North America have fruits that ripen in a color sequence from green to red to black (or sometimes blue; Willson and Thompson 1982). Ripening is generally asynchronous (in contrast to fall-fruiting species), thus each plant bears a two-colored fruiting display until the end of the season when all fruits are ripe (see also Ridley 1930). Fruiting displays of red and black together are more attractive to birds than single-color displays or larger displays of all-black fruits (in which *all* the fruits are fully edible, red ones are usually scorned), and fruits usually are removed more rapidly from the bicolored displays (Morden-Moore and Willson 1982, Willson and Melampy in press).

Asynchronous ripening and bicolored displays may be adaptations to the low levels of frugivore populations during the summer (Willson and Thompson 1982). At that time the few fruit-eaters present are often still breeding and tied, to some extent, to their breeding territories, which further reduces their availability as dispersal agents. Many invertebrates nibble on or suck the juices from ripe fruits; if the damage is severe enough, the remaining fruit will not be eaten by birds and those seeds are not dispersed. If summer-fruiting plants ripened their fruit crops synchronously and removal rates by dispersers were low, many ripe fruits might be damaged by invertebrates and rendered unsuitable for later dispersal by birds. Asynchronous ripening may adjust the size of the crop of ripe fruits to the possible removal rates, and bicolored displays may serve as a flag to local birds, identifying a suitable food source and thus enhancing removal rates.

The story is complicated by the existence of a number of species that produce bicolored fruiting displays by means of a color contrast between the fruit and some associated structure, such as a pedicel, bract, or capsule (Ridley 1930, Willson and Thompson 1982) or between fruit and leaves (Kerner 1895). Usually one of the colors is red or pink, but the other color may be white, blue, black, purple, or orange. These displays show no known marked seasonal patterns, nor is the conspicuous display associated with fruits of unusually small nutritional value or small fruit crops as a possible compensation for the small rewards offered. The color contrasts cannot, in most cases, be associated with attraction of pollinators earlier in the season (because so often the fruit itself provides the second contrasting color), and their adaptive value is unknown.

Furthermore, many bird-dispersed fruits are exhibited in monochromatic displays, the color differing among the species. We know nothing about the factors determining the evolutionary choice among the colors. Why should some be red, but others blue? Several factors need to be weighed in the estimation of adaptive value: the costs of different pigments, the need to protect fruits from predators (apparently few insects can see red; Is red a cryptic color with respect to insect foragers?; Do nonred fruits have other forms of protection?), contrast with the background at the time of fruit maturation, and the spectral sensitivities of the desirable dispersal agents.

It is of interest to know if plants with large fruit crops attract more dispersal agents, or a greater diversity of them, than conspecific individuals with small crops and if, as a result, individuals with large crops enjoy more successful dispersal. Further, these relationships might differ with the nature of the frugivores. Howe and Estabrook (1977) suggested that, for species dependent on opportunistic frugivores, large crops might be more attractive to the many available frugivores and that dispersal rates of seeds should increase with the size of the fruit crop. Such plants should exhibit relatively large fruit crops and short fruiting seasons. In contrast, species using more specialized dispersers should exhibit peak success in dispersal at some intermediate crop size, because when vast numbers of fruits are available at any one time, the population of specialists will be too small to consume them all and the remainder will be consumed by less reliable opportunists or simply drop to the ground. These species generally are predicted to have smaller fruit crops and longer fruiting seasons (Howe and Estabrook 1977). Observations of fruit dispersal for several tropical trees can be used to assay at least parts of the Howe–Estabrook model.

Casearia corymbosa growing in dry forest in Costa Rica is dispersed by several bird species, but primarily by the yellow-green vireo (*Vireo flavoviridis*). Visits by this species plateau at intermediate crop sizes and the proportions of seed removed were highest at intermediate crop sizes, whereas the total number of seeds dispersed increased directly with increasing crop sizes (Howe and vande Kerckhove 1979). These observations are in accord with the expectations of the model for fruits consumed by specialists.

Virola sebifera and *V. surinamensis* in Panama each are served by a relatively small array of avian dispersers: three bird species account for over 50% of the (estimated) dispersal of *V. surinamensis,* and two do the same for *V. sebifera*; only 7 species of birds were seen feeding on the fruits of each species (Howe 1981, Howe and vande Kerckhove 1981). The seeds are rather large (18 × 15 mm for *V. surinamensis* and 14 × 10 mm for *V. sebifera*) and are surrounded by a thin layer of very oily fruit (an average of 54–63% for both species, with considerable variation among individuals of *V. surinamensis*). The number of frugivore visits increased with size of the daily fruit crop for *V. sebifera,* but the number of species visiting the trees reached a plateau at intermediate crop sizes. For *V. surinamensis,* the numbers of fruits removed increased linearly with crop size, although the *percentage* of the fruit crop removed (an index of efficiency of dispersal) was unrelated to crop size. Both *Virola* species may be considered to be dispersed by rather specialized frugivores, but neither seems to fit the expectations of Howe and Estabrook (1977).

Two other species provide a contrast with *Virola. Tetragastris panamensis* and *Guarea glabra* trees are visited by many species of frugivorous birds (at least 23 and 19, respectively; Howe 1980, Howe and De Steven 1979), but *T. panamensis* is dispersed largely by monkeys (Howe 1980). The number of *T. panamensis* seeds removed increased with size of the fruit crop, as did the number of visits by foraging animals and their diversity (Howe 1980). However, the proportion of seed wasted by chewing or dropping also increased with crop size, with the result that very large fruit crop sizes may be at a disadvantage (Howe 1980). Likewise, *G. glabra* individuals with large seed crops attracted more birds in higher diversity and had larger numbers of seeds removed (Howe and De Steven 1979). Both of these species seem to be served by a collection of more opportunistic avian frugivores, but in neither case did individuals with large fruit crops enjoy a *disproportionate* advantage in seed removal.

Thus, regardless of the putative specialization of the consumers, the more fruits produced by an individual the more seeds were removed, but the proportion of the seed crop removed usually did not vary with seed crop size in the predicted manner. In addition, large fruit crops may receive a greater diversity of visitors or more effective visitors than small fruit crops; large *Didymopanax morototoni* crops were exploited differentially by large, wide-ranging birds such as aracaris in Panama (T. Martin, unpublished manuscript). Assuming that removal is a reasonably accurate index of successful dispersal, we can say that there is an advantage, in terms of dispersal, to high fecundity. Possible counters to this advantage are the diminished efficiency (or higher cost) for those species in which the proportion of seeds removed decreased, or wastage and predation increased, at high fecundity. At this point, we lack the information needed to assess the effects of these pressures. In addition, the effect of fruit-crop size may vary within a fruiting season, as the abundance of alternate foods changes (Stapanian 1982b) or as birds learn the location of fruit sources. Furthermore, any advantage of high fecundity must be balanced within the entire life history (Howe 1980): if high fecundity results in overwhelming dispersal success, selection might go so far as to favor semelparity. But production of small seed crops at early stages in the life history and of intermediate ones in middle age may have advantages too, if success of establishment varies greatly from year to year, if success at an early age carries with it a large increase in fitness relative to the cost in poor dispersal, or if years of consistent production of medium seed crops and intermediate dispersal success have cumulative effects that outweigh a huge one-shot effort. The possible effects of seed-crop size on effectiveness of dispersal are still not well understood in any detail.

HIDING AND WAITING

The seeds of many plants delay germination, sometimes for decades or more. Huge numbers of seeds can accumulate in the soil as a seedbank (Harper 1977, Nelson and Chew 1977, Rabinowitz 1981, and others), with important consequences for the evolution of the population (Templeton and Levin 1979). The seeds of successional species are best known for this ability (Cheke et al. 1979, Hall and Swaine 1980, Harper 1977, Kellman 1974, Whipple 1978) but the tendency toward delayed germination is not equally developed in all areas, even among pioneer species (Grime et al. 1981, Ng 1978) and the seeds of some mature tropical forest plants can also be found in the soil (Denslow 1980b, Hall and Swaine 1980).

Seed dormancy may be genetically controlled (Hilu and de Wet 1980, Naylor and Fedec 1978, other references in Harper 1977), but many cases are known of polymorphism among the offspring of a single parent (Harper 1977, Venable and Lawlor 1980). Furthermore, environmental conditions (both biotic and abiotic) can influence the expression of dormancy among the seeds of numerous species; such effects can be elicited both by mutilation (defoliation, removing inflorescences) of the mother plant (Maun and Cavers 1971a,b) and by direct effects on the seeds themselves (e.g., Koller 1972).

The actual longevity of dormant seeds in the soil is seldom known under natural conditions, although there are some records for germinable seeds known to be over 1000 years old (Cook 1980, Harper 1977). Less spectacular estimates are in the neighborhood of 30 to 40 years for some species (Hill and Stevens 1981, Thurston 1960). The shapes of the survivorship curves for dormant seeds seem to be largely unknown (Hill and Stevens 1981, Thurston 1960), although there is some evidence that there may be a constant probability of mortality at least over the first few years, such that percentage loss each year is the same (Cook 1980, Harper 1977). This, of course, means that only a small part of an original cohort survives for any length of time. However, when seed predation is severe, survivorship curves may be steeper and nonlinear (e.g., Sarukhan 1980).

The ability to delay germination is commonly referred to as dormancy; an enormous literature addresses chiefly the physiological mechanisms involved (references in Harper 1977, Villiers 1975). There are three main types of dormancy, classified by Harper (1959, in 1977). Innate dormancy is inherent in the seed itself while the seed is on the parent and during dispersal and for various periods thereafter. Induced dormancy is an acquired inability to germinate, engendered by some environmental factor. Enforced dormancy is one imposed continuously by environmental conditions unfavorable for germination; when the restraint is removed, the seed will germinate. Both induced and enforced dormancy allow seeds to germinate opportunistically when appropriate cues occur.

Westoby (1981) raised the fascinating possibility that selection may operate on the fitness of mother plants and of their offspring in different, potentially conflicting ways. However, he argued, the conflict seems to have been resolved principally in favor of the mother. Almost all known cases of germination polymorphism involve proximate, physiological, control of germination by the seed coat, which is purely maternal tissue (op. cit.). Even when germinability is regulated by size of embryo or its degree of development (Harper 1977, Westoby 1981), this could easily be subject to maternal control. Westoby, in fact, argued that, by surrounding embryos with maternal tissues, the seed plants have taken control of offspring behavior until germination.

Delayed germination is dispersal in time (Cook 1980, Harper 1977, Venable and Lawlor 1980). Polymorphisms in dormancy, however controlled, spread out the seed shadow in time. Presumably, the ability to remain dormant for various lengths of time increases the probability that seeds will encounter conditions suitable for germination and successful establishment and reproduction. If the environment is predictably suitable, delays are far less valuable than when conditions are unpredictable (Cohen 1968, Thompson and Grime 1979); nevertheless, a lack of delaying ability does not mean that survivorship will be good (Augspurger 1979). Clearly, delayed germination can only succeed when seed survivorship is relatively high, and thus the evolution of seeds that wait should be associated with enhanced longevity (Cohen 1968) and the ability to evade predators and pathogens.

Venable and Lawlor (1980) and Angevine and Chabot (1979) suggested that increased dispersibility should be associated with lower degrees of innate dormancy. Venable and Lawlor proposed specifically that dispersal in space and in time are inversely correlated, especially for annuals, which reproduce only by seed and have only one chance to do so. Dispersal in space improves the probability of reaching a site that is presently suitable and thus diminishes the value of delayed germination. Dormancy, especially when expressed to varying degrees among the offspring, allows the seeds to wait until suitable conditions may appear. A survey of certain species of Asteraceae and Brassicaceae that produce two kinds of seeds indicates that the seeds with greater dispersal ability usually germinate quickly, but those with less usually have delayed germination (Venable and Lawlor 1980). However, there are several exceptions to the trend and these warrant further investigation.

While a seed is waiting to germinate, it is subject to attack by predators and pathogens. Chemical protection and some forms of morphological protection are still important. Burial may reduce predation by some animals (e.g., Barnett 1977, A. J. Smith 1975) but simultaneously increase exposure to microorganisms and fungi. So far as I can determine, we know very little about this aspect of the dormancy period. For *Vulpia fasciculata,* Watkinson (1978) estimates that about 5% of the seeds died of unknown causes before germination, seed predation ranged between 1–9%, on average, and the seedling phase was the most vulnerable. However, the relative importance of postdispersal seed predation no doubt varies greatly among species.

When a seed germinates, it enters a period when the risks of mortality are usually very high (Baskin and Baskin 1979, Connell 1971, Cook 1979, Harper 1977, Hett 1971, Phillips 1927, Sarukhan 1980). The seedling itself may encounter one or more periods of waiting before it embarks on a subsequent episode of growth (Grime 1979, Sarukhan 1980). The sources of mortality are legion. Although seedling morphology has received a good deal of descriptive attention (e.g., Vogel 1980), I have found little information of an ecological nature. It seems highly probable that the site of storage for nutritional reserves, the degree of exposure and form of the cotyledons, the protective devices for various seedling parts, and so on can be associated with selection to improve competitive ability, to reduce predation, and to facilitate establishment in particular habitats (Ng 1978, Sarukhan 1980). Physiological studies of seedlings demonstrate that the role of the cotyledons in seedling development may be substantial, and that the photosynthetic capacity of the cotyledons and their placement have significant consequences for rates of seedling growth (Ampofo et al. 1976, Lovell and Moore 1971). These tantalizing results indicate a profitable field of investigation.

There is a large element of chance in the success of seedling establishment. Whether a seed escapes predation may be quite independent of the characteristics of the seed itself (e.g., Racine and Downhower 1974). Furthermore, very localized differences in soil density, texture, nutrient levels, and the like can have profound effects on seedling growth and eventual fecundity (e.g., Harper et al. 1965, Hartgerink 1980). Such size differences need not reflect genetically controlled fitness differentials (Gottlieb 1977) and can slow the rates of response to natural selection by the population.

LITERATURE CITED

Aalders, L. E. and I. V. Hall. 1961. Pollen incompatibility and fruit set in lowbush blueberries. *Can. J. Genet. Cytol.* **3:** 300–307.

Agnew, A. D. Q. and J. E. C. Flux. 1970. Plant dispersal by hares (*Lepus capensis* L.) in Kenya. *Ecology* **57:** 735–737.

Aker, C. L. 1982. Regulation of flower, fruit and seed production by a monocarpic perennial, *Yucca whipplei*. *J. Ecol.* **70:** 357–372.

Alexandre, D. Y. 1978. Le rôle disséminateur des éléphants en forêt de Taï, Cote-d'Ivoire. *La Terre et La Vie* **32:** 47–72.

Ampofo, S. F., K. G. Moore, and P. H. Lovell. 1976. Cotyledon photosynthesis during seedling development in *Acer*. *New Phytol.* **76:** 41–52.

Angevine, M. W. and B. F. Chabot. 1979. Seed germination syndromes in higher plants. *In* Solbrig, O. T., S. Jain, G. B. Johnson, and P. H. Raven (eds.), *Topics in Plant Population Biology*. Columbia University Press, New York.

Augspurger, C. K. 1979. Irregular rain cues and the germination and seedling survival of a Panamanian shrub (*Hybanthus prunifolius*). *Oecologia* **44:** 53–59.

Augspurger, C. K. (in press). Seed dispersal of the tropical tree, *Platypodium elegans*, and the escape of its seedlings from fungal pathogens. *J. Ecol.*

Augspurger, C. K. (unpublished manuscript a). Comparative patterns of seedling survival among tropical tree species: interactions of dispersal distance, light gaps, and fungal pathogens.

Augspurger, C. K. (unpublished manuscript b). Experimental studies of the effects of light conditions, dispersal distance, and seedling density on fungal mortality.

Augspurger, C. K. and K. P. Hogan. (submitted for publication). Wind dispersal of fruits with variable seed number in a tropical tree, *Lonchocarpus velutinus*:(Leguminosae). *Am. J. Bot.*

August, P. V. 1981. Fig fruit consumption and seed dispersal by *Artibeus jamaicensis* in the llanos of Venezuela. *Biotropica* **13** (Suppl.): 70–76.

Baker, G. A. and D. J. O'Dowd. 1982. Effects of parent plant density on the production of achene types in the annual *Hypochoeris glabra*. *J. Ecol.* **70:** 201–215.

Baker, H. G. 1972. Seed weight in relation to environmental conditions in California. *Ecology* **53:** 997–1010.

Balda, R. P. 1980. Are seed caching systems co-evolved? *Proc. Int. Ornithol. Congr.* **17:** 1185–1191.

Balda, R. P. and G. C. Bateman. 1971. Flocking and annual cycle of the piñon jay, *Gymnorhinus cyanocephalus*. *Condor* **73:** 287–302.

Balda, R. P. and G. C. Bateman. 1973. The breeding biology of the piñon jay. *Living Bird* **11:** 5–41.

Ballard, J. B. and K. P. Pruess. 1979. Seed selection by an ant *Pheidole bicarinata longula* Emery (Hymenoptera: Formicidae). *J. Kansas Entomol. Soc.* **52:** 550–552.

Barnett, R. J. 1977. The effect of burial by squirrels on germination and survival of oak and hickory nuts. *Am. Midl. Nat.* **98:** 319–330.

Baskin, J. M. and C. M. Baskin. 1979. Studies on the antecology and population biology of the weedy monocarpic perennial, *Pastinaca sativa*. *J. Ecol.* **67:** 601–610.

Beattie, A. J. and D. C. Culver. 1981. The guild of myrmecochores in the herbaceous flora of West Virginia forests. *Ecology* **62:** 107–115.

Beattie, A. J. and N. Lyons. 1975. Seed dispersal in *Viola* (Violaceae): Adaptations and strategies. *Am. J. Bot.* **62:** 714–722.

Beattie, A. J., D. C. Culver, and R. J. Pudlo. 1979. Interactions between ants and the diaspores of some common spring flowering herbs in West Virginia. *Castanea* **44:** 177–186.

Beer, T. and M. D. Swaine. 1977. On the theory of explosively dispersed seeds. *New Phytol.* **78:** 681–694.

Bell, E. A. 1978. Toxins in seeds. *In* J. B. Harborne (ed.), *Biochemical Aspects of Plant and Animal Coevolution*. Academic, New York.

Bentley, B. L. 1977. Extrafloral nectaries and protection by pugnacious bodyguards. *Annu. Rev. Ecol. Syst.* **8:** 407–427.

Berg, R. Y. 1966. Seed dispersal of *Dendromecon*: Its ecologic, evolutionary, and taxonomic significance. *Am. J. Bot.* **53:** 61–73.

Berg, R. Y. 1975. Myrmecochorous plants in Australia and their dispersal by ants. *Aust. J. Bot.* **23:** 475–508.

Black, J. N. 1956. The influence of seed size and depth of sowing on preemergence and early vegetative growth of subterranean clover (*Trifolium subterraneum* L.). *Aust. J. Agric. Res.* **7:** 98–109.

Black, J. N. 1957a. The early vegetative growth of three strains of subterranean clover (*Trifolium subterraneum* L.) in relation to size of seed. *Aust. J. Agric. Res.* **8:** 1–14.

Black, J. N. 1957b. Seed size as a factor in the growth of subterranean clover (*Trifolium subterraneum* L.) under spaced and sward conditions. *Aust. J. Agric. Res.* **8:** 335–351.

Black, J. N. 1958. Competition between plants of different initial seed sizes in swards of subterranean clover (*Trifolium subterraneum* L.) with particular reference to leaf area and the light microclimate. *Aust. J. Agric. Res.* **9:** 299–318.

Blom, C. W. P. M. 1978. Germination, seedling emergence and establishment of some *Plantago* species under laboratory and field conditions. *Acta Bot. Neerl.* **27:** 257–271.

Bonner, F. T. 1974. Chemical components of some southern fruits and seeds. *USDA, For. Serv. Res. Note* **SO-183.**

Bossema, I. 1979. Jays and oaks: An ecoethological study of a symbiosis. *Behaviour* **70:** 1–117.

Boucher, D. H. 1981. The "real" disperser of *Swartzia cubensis. Biotropica* **13** (Suppl.): 77–78.

Boyer, W. D. 1958. Longleaf pine seed dispersal in south Alabama. *J. For.* **56:** 265–268.

Bradford, D. F. and C. C. Smith. 1977. Seed predation and seed number in *Scheelea* palm fruits. *Ecology* **58:** 667–673.

Breedlove, D. E. and P. R. Ehrlich. 1968. Plant-herbivore coevolution: Lupines and lycaenids. *Science* **162:** 671–672.

Breedlove, D. E. and P. R. Ehrlich. 1972. Coevolution: Patterns of legume predation by a lycaenid butterfly. *Oecologia* **10:** 99–104.

Brown, J. H., D. W. Davidson, and O. J. Reichman. 1979a. An experimental study of competition between seed-eating desert rodents and ants. *Am. Zool.* **19:** 1129–1143.

Brown, J. H., O. J. Reichman, and D. W. Davidson. 1979b. Granivory in desert ecosystems. *Annu. Rev. Ecol. Syst.* **10:** 201–227.

Bullock, S. H. and R. B. Primack. 1977. Comparative experimental study of seed dispersal on animals. *Ecology* **58:** 681–686.

Burbidge, A. H. and R. J. Whelan. (in press). Seed dispersal in a cycad, *Macrozamia riedlei. Aust. J. Ecol.*

Burdon, J. J. and G. A. Chilvers. 1976. The effect of clumped planting patterns on epidemics of damping-off disease in cress seedlings. *Oecologia* **23:** 17–29.

Capinera, J. L. 1979. Qualitative variation in plants and insects: Effect of propagule size on ecological plasticity. *Am. Nat.* **114:** 350–361.

Center, T. D. and C. D. Johnson. 1974. Coevolution of some seed beetles (Coleoptera: Bruchidae) and their hosts. *Ecology* **55:** 1096–1103.

Chaplin, S. J. 1980. An energetic analysis of host plant selection by the large milkweed bug, *Oncopeltus fasciatus. Oecologia* **46:** 254–261.

Cheke, A. S., W. Nanakorn, and C. Yankoses. 1979. Dormancy and dispersal of seeds of secondary forest species under the canopy of a primary tropical rain forest in northern Thailand. *Biotropica* **11:** 88–95.

Cideciyan, M. A. and A. J. C. Malloch. 1982. Effects of seed size on the germination, growth and competitive ability of *Rumex crispus* and *Rumex obtusifolius. J. Ecol.* **70:** 227–232.

Cohen, D. 1968. A general model of optimal reproduction in a randomly varying environment. *J. Ecol.* **56:** 219–228.

Connell, J. H. 1971. On the role of natural enemies in preventing competitive exclusion in some marine animals and in rain forest trees. *In* P. J. den Boen and G. R. Gradwell (eds.), *Dynamics of Populations.* Centre for Agricultural Publishing and Documentation, Wageningen, Netherlands.

Cook, A. D. 1972. Polymorphic and continuous variation in the seeds of dove weed, *Eremocarpus setigerus* (Hook.) Benth. *Am. Midl. Nat.* **87:** 366–376.

Cook, A. D., P. R. Atsatt, and C. A. Simon. 1971. Doves and dove weed: Multiple defenses against avian predation. *Bioscience* **21:** 277–281.

Cook, R. E. 1975. The photoinductive control of seed weight in *Chenopodium rubrum* L. *Am. J. Bot.* **62:** 427–431.

Cook, R. E. 1979. Patterns of juvenile mortality and recruitment in plants. *In* O. T. Solbrig, S. Jain, G. B. Johnson, and P. H. Raven (eds.), *Topics in Plant Population Biology.* Columbia University Press, New York.

Cook, R. E. 1980a. The biology of seeds in soil. *Bot. Monogr.* **15:** 107–129.

Cook, R. E. 1980b. Germination and size dependent mortality in *Viola blanda. Oecologia* **47:** 115–117.

Corner, E. J. H. 1966. *The Natural History of Palms.* University of California Press, Berkeley.

Croat, T. B. 1974. A case for selection for delayed fruit maturation in *Spondias* (Anacardiaceae). *Biotropica* **6:** 135–137.

Crome, F. H. J. 1975. The ecology of fruit pigeons in tropical north Queensland. *Aust. Wildl. Res.* **2:** 155–185.

Crome, F. H. J. 1976. Some observations in the biology of the cassowary in northern Queensland. *Emu* **76:** 8–14.

Crouch, P. A. and S. P. Vander Kloet. 1980. Variation of seed characters in populations of *Vaccinium* § *cyanococcus* (the blueberries) in relation to latitude. *Can. J. Bot.* **58:** 84–90.

Culver, D. C. and A. J. Beattie. 1978. Myrmecochory in *Viola:* Dynamics of seed-ant interactions in some West Virginia species. *J. Ecol.* **66:** 53–72.

Culver, D. C. and A. J. Beattie. 1980. The fate of *Viola* seeds dispersed by ants. *Am. J. Bot.* **67:** 710–714.

Darley-Hill, S. and W. C. Johnson. 1981. Acorn dispersal by the bluejay (*Cyonocitta cristata*). *Oecologia* **50:** 231–232.

Davidson, D. W. 1977. Foraging ecology and community organization in desert seed-eating ants. *Ecology* **58:** 725–737.

Davidson, D. W. and S. R. Morton. 1981a. Competition for dispersal in ant-dispersed plants. *Science* **213:** 1259–1261.

Davidson, D. W. and S. R. Morton. 1981b. Myrmechory in some plants (F. Chenopodiaceae) of the Australian arid zone. *Oecologia* **50:** 357–366.

Denslow, J. S. 1980a. Notes on the seedling ecology of a large-seeded species of Bombacaceae. *Biotropica* **12:** 220–221.

Denslow, J. S. 1980b. Gap partitioning among tropical rainforest trees. *Biotropica* **12** (Suppl.): 47–55.

Denslow, J. S. and T. C. Moermond. 1982. The effect of accessibility on rates of fruit removal from tropical shrubs: An experimental study. *Oecologia* (in press).

Derr, J. A. 1980. Coevolution of the life history of a tropical seed-feeding insect and its food plants. *Ecology* **61:** 881–892.

De Steven, D. 1981. Predispersal seed predation in a tropical shrub (*Mabea occidentalis,* Euphorbiaceae). *Biotropica* **13:** 146–150.

Docters van Leeuwen, W. G. 1954. On the biology of some Javanese Loranthaceae and the role birds play in their life-historie. *Beaufortia* **41:** 105–207.

Downhower, J. F. and C. H. Racine. 1976. Darwin's finches and *Croton scouleri*: An analysis of the consequences of seed predation. *Biotropica* **8:** 66–70.

Drake, D. W. 1981a. Reproductive success of two *Eucalyptus* hybrid populations, I: Generalized seed output model and comparison of fruit parameters. *Aust. J. Bot.* **29:** 25–35.

Drake, D. W. 1981b. Reproductive success of two *Eucalyptus* hybrid populations, II: Comparison of predispersal seed parameters. *Aust. J. Bot.* **29:** 37–48.

Edmunds, G. F. and D. N. Alstad. 1978. Coevolution in insect herbivores and conifers. *Science* **199:** 941–945.

Elliott, P. F. 1974. Evolutionary responses of plants to seed-eaters: Pine squirrel predation on lodgepole pine. *Evolution* **28:** 221–231.

Ellner, S. and A. Shmida. 1981. Why are adaptations for long-range seed dispersal rare in desert plants? *Oecologia* **51:** 133–144.

Endler, J. A. 1977. *Geographic Variation, Speciation, and Clines.* Princeton University Press, Princeton, N.J.

Fahn, A. and E. Werker. 1972. Anatomical mechanisms of seed dispersal. *In* T. T. Kozlowski (ed.), *Seed Biology,* Vol. 1. Academic, New York.

Farentinos, R. C., P. J. Capretta, R. E. Kepner, V. M. Littlefield. 1981. Selective herbivory in tassel-eared squirrels: Role of monoterpenes in ponderosa pines chosen as feeding trees. *Science* **213:** 1273–1275.

Feeny, P. 1975. Biochemical coevolution between plants and their insect herbivores. *In* L. E. Gilbert and P. H. Raven (eds.), *Coevolution of Animals and Plants.* University of Texas Press, Austin.

Feeny, P. 1976. Plant apparency and chemical defense. *Adv. Phytochem.* **10:** 1–40.

Fenner, M. 1978. A comparison of the ability of colonizers and closed-turf species to establish from seed in artificial swards. *J. Ecol.* **66:** 953–963.

Fitzpatrick, J. W. 1980. Wintering of North American tyrant flycatchers in the neotropics. *In* A. Keast and E. G. Morton (eds.), *Migrant Birds in the Neotropics.* Smithsonian Institution Press, Washington, D.C.

Fleming, T. H. 1979. Do tropical frugivores compete for food? *Am. Zool.* **19:** 1157–1172.

Fleming, T. H. 1981. Fecundity, fruiting pattern, and seed dispersal in *Piper amalago* (Piperaceae), a bat-dispersed tropical shrub. *Oecologia* **51:** 42–46.

Fleming, T. H. and E. R. Heithaus. 1981. Frugivorous bats, seed shadows, and the structure of tropical forests. *Biotropica* **13** (Suppl.): 45–53.

Forcella, F. 1980. Cone predation by pinyon cone beetle (*Conopthorus edulis*: Scolytidae): Dependence on frequency and magnitude of cone production. *Am. Nat.* **116:** 594–598.

Foster, M. S. 1977. Ecological and nutritional effects of food scarcity on a tropical frugivorous bird and its fruit source. *Ecology* **58:** 73–85.

Foster, M. S. 1978. Total frugivory in tropical passerines: A reappraisal. *Trop. Ecol.* **19:** 131–154.

Fox, J. F. 1974. Coevolution of white oak and its vertebrate seed predators. Ph.D. thesis, Univeristy of Chicago, Chicago.

Fox, J. F. 1982. Adaptation of gray squirrel behavior to autumn germination by white oak acorns. *Evolution* **36:** 800–809.

Fox, L. R. 1981. Defense and dynamics in plant-herbivore systems. *Am. Zool.* **21:** 853–864.

Friedman, J. and Z. Stein. 1980. The influence of seed-dispersal mechanisms on the dispersion of *Anastatica hierochuntica* (Cruciferae) in the Negev Desert, Israel. *J. Ecol.* **68:** 43–50.

Frost, P. G. H. 1980. Fruit-frugivore interactions in a South African coastal dune forest. *Proc. Int. Ornithol. Congr.* **17:** 1179–1184.

Fry, C. H., J. S. Ash, and I. J. Ferguson-Lees. 1970. Spring weights of some Palearctic migrants at Lake Chad. *Ibis* **112:** 58–82.

Fry, M. E. and C. E. Vaughn. 1977. Acorn selection by band-tailed pigeons. *Calif. Fish Game* **63**(1): 59–60.

Futuyma, D. J. and S. S. Wasserman. 1980. Resource concentration and herbivory in oak forests. *Science* **210:** 920–922.

Ganders, F. R., K. Carey, and A. J. F. Griffiths. 1977. Natural selection for a fruit dimorphism in *Plectritis congesta* (Valerianaceae). *Evolution* **31:** 873–881.

Giddy, C. 1974. *Cycads of South Africa.* Purnell, Cape Town.

Goldstein, J. L. and T. Swain. 1963. Changes in tannins in ripening fruits. *Phytochemistry* **2:** 371–383.

Golley, F. B. 1961. Energy values of ecological materials. *Ecology* **42:** 581–584.

Gottlieb, L. D. 1977. Genotypic similarity of large and small individuals in a natural population of the annual plant *Stephanomeria exigua* ssp. *coronaria* (Compositae). *J. Ecol.* **65:** 127–134.

Gottsberger, G. 1978. Seed dispersal by fish in the inundated regions of Humaitá, Amazonia. *Biotropica* **10:** 170–183.

Gould, S. A., R. Pearl, and T. I. Edwards. 1934. On the effects of partial removal of the cotyledons upon the growth and duration of life of canteloupe seedlings without exogenous food. *Ann. Bot.* **48:** 575–599.

Goulding, M. 1981. *The Fishes and the Forest.* University of California Press, Berkeley.

Green, D. S. 1980. The terminal velocity and dispersal of spinning samaras. *Am. J. Bot.* **67:** 1218–1224.

Green, T. W. and I. G. Palmblad. 1975. Effects of insect seed predators on *Astragalus cibarius* and *Astragalus utahensis* (Leguminosae). *Ecology* **56:** 1435–1440.

Greenberg, R. 1981. Frugivory in some migrant tropical forest wood warblers. *Biotropica* **13:** 215–222.

Grime, J. P. 1979. *Plant Strategies and Vegetation Processes.* Wiley, New York.

Grime, J. P., G. Mason, A. V. Curtis, J. Radman, S. R. Band, M. A. G. Mowforth, A. M. Neal, and S. Shaw. 1981. A comparative study of germination characteristics in a local flora. *J. Ecol.* **69:** 1017–1059.

Gross, K. L. and J. D. Soule. 1981. Differences in biomass allocation to reproductive and vegetative structures of male and female plants of a dioecious, perennial herb, *Silene alba* (Miller) Krause. *Am. J. Bot.* **68:** 801–807.

Hall, I. V. and L. E. Aalders. 1965. The relation between seed numbers and berry weight in the cranberry. *Can. J. Plant Sci.* **45:** 292.

Hall, I. V. and F. R. Forsyth. 1967. Respiration rates of developing fruits of the lowbushblueberry. *Can. J. Plant Sci.* **47:** 157–159.

Hall, J. B. and M. D. Swaine. 1980. Seed stocks in Ghanaian forest soils. *Biotropica* **12:** 256–263.

Halls, L. K. 1977. Southern fruit-producing woody plants used by wildlife. *USDA, For. Serv. Tech. Rep.* **SO-16.**

Hamilton, W. D. and R. M. May. 1977. Dispersal in stable habitats. *Nature* **269:** 578–581.

Handel, S. N. 1976. Dispersal ecology of *Carex pedunculata* (Cyperaceae), a new North American myrmecochore. *Am. J. Bot.* **63:** 1071–1079.

Handel, S. N. 1978. The competitive relationship of three woodland sedges and its bearing on the evolution of ant-dispersal of *Carex pedunculata. Evolution* **32:** 151–163.

Handel, S. N., S. B. Fisch, and G. E. Schate. 1981. Ants disperse a majority of herbs in a mesic forest community in New York State. *Bull. Torrey Bot. Club* **108:** 430–437.

Hannan, G. L. 1980. Heteromericarpy and dual seed germination modes in *Platystemon californicus* (Papaveraceae). *Madroño* **27:** 164–170.

Hare, J. D. 1980. Variation in fruit size and susceptibility to seed predation among and within populations of the cocklebur, *Xanthium strumarium* L. *Oecologia* **46:** 217–222.

Hare, J. D. and D. J. Futuyma. 1978. Different effects of variation in *Xanthium strumarium* L. (Compositae) on two insect seed predators. *Oecologia* **37:** 109–120.

Harper, J. L. 1977. *Population Biology of Plants.* Academic, New York.

Harper, J. L., P. H. Lovell, and K. G. Moore. 1970. The shapes and sizes of seeds. *Annu. Rev. Ecol. Syst.* **1:** 327–356.

Harper, J. L., J. T. Williams, and G. R. Sagar. 1965. The behaviour of seeds in soil, I: The heterogeneity of soil surfaces and its role in determining the establishment of plants from seed. *J. Ecol.* **53:** 273–286.

Hartgerink, A. P. 1980. Effects of some events experienced by seedlings in competition. Ph.D. thesis, University of Illinois.

Hay, M. E. and P. J. Fuller. 1981. Seed escape from heteromyid rodents: The importance of microhabitat and seed preference. *Ecology* **62:** 1395–1399.

Heaney, L. R. and R. W. Thorington. 1978. Ecology of neotropical red-tailed squirrels, *Sciurus granatensis,* in the Panama Canal Zone. *J. Mammal.* **59:** 846–851.

Heithaus, E. R. 1981. Seed predation by rodents on three ant-dispersed plants. *Ecology* **62:** 136–145.

Herrera, C. M. 1981a. Are tropical fruits more rewarding to dispersers than temperate ones? *Am. Nat.* **118:** 896–907.

Herrera, C. M. 1981b. Fruit variation and competition for dispersers in natural populations of *Smilax aspera. Oikos* **36:** 51–58.

Herrera, C. M. 1982a. Seasonal variation in the quality of fruits and diffuse coevolution between plants and avian dispersers. *Ecology* **63:** 773–785.

Herrera, C. M. 1982b. Defense of ripe fruit from pests: Its significance in relation to plant-disperser interactions. *Am. Nat.* **120:** 218–241.

Herrera, C. M. and P. Jordano. 1981. *Prunus mahaleb* and birds: The high-efficiency seed dispersal system of a temperate fruiting tree. *Ecol. Monogr.* **51:** 203–218.

Hespenheide, H. A. 1966. The selection of seed size by finches. *Wilson Bull.* **78:** 191–197.

Hett, J. M. 1971. A dynamic analysis of age in sugar maple seedlings. *Ecology* **52:** 1071–1074.

Hill, M. O. and P. A. Stevens. 1981. The density of viable seed in soils of forest plantations in upland Britain. *J. Ecol.* **69:** 693–709.

Hill, R. J. 1977. Variability of soluble seed proteins in populations of *Mentzelia* L. (Loasaceae) from Wyoming and adjacent states. *Bull. Torrey Bot. Club* **104:** 93–101.

Hilty, S. L. 1980. Flowering and fruiting periodicity in a premontane rain forest in Pacific Colombia. *Biotropica* **12:** 292–306.

Hilu, K. W. and J. M. J. de Wet. 1980. Effect of artificial selection on grain dormancy in *Eleasine* (Gramineae). *Syst. Bot.* **5:** 54–60.

Hladik, C. M. and A. Hladik. 1967. Observations sur le rôle des primates dans la dissémination des végétaux de la forêt gabonaise. *Biol. Gabon.* **3:** 43–58.

Horvitz, C. C. 1981. Analysis of how ant behaviors affect germination in a tropical myrmecochore *Calathea microcephala* (P. E.) Koernicke (Marantaceae): Microsite selection and aril removal by neotropical ants, *Odontomachus, Pachycondyla,* and *Solenopsis* (Formicidae). *Oecologia* **51:** 47–52.

Horvitz, C. C. and A. J. Beattie. 1980. Ant dispersal of *Calathea* (Marantaceae) seeds by carnivorous ponerines (Formicidae) in a tropical rain forest. *Am. J. Bot.* **67:** 321–326.

Howe, H. F. 1977. Bird activity and seed dispersal of a tropical wet forest tree. *Ecology* **58:** 539–550.

Howe, H. F. 1980. Monkey dispersal and waste of a neotropical fruit. *Ecology* **61:** 944–959.

Howe, H. F. 1981. Dispersal of a neotropical nutmeg (*Virola sebifera*) by birds. *Auk* **98:** 88–98.

Howe, H. F. and D. De Steven. 1979. Fruit production, migrant bird visitation, and seed dispersal of *Guarea glabra* in Panama. *Oecologia* **39:** 185–196.

Howe, H. F. and G. F. Estabrook. 1977. On intraspecific competition for avian dispersers in tropical trees. *Am. Nat.* **111:** 817–832.

Howe, H. F. and J. Smallwood. 1982. Ecology of seed dispersal. *Annu. Rev. Ecol. Syst.* **13:** 201–228.

Howe, H. F. and G. A. Vande Kerckhove. 1979. Fecundity and seed dispersal of a tropical tree. *Ecology* **60:** 180–189.

Howe, H. F. and G. A. Vande Kerckhove. 1980. Nutmeg dispersal by birds. *Science* **210:** 925–927.

Howe, H. F. and G. A. Vande Kerckhove. 1981. Removal of wild nutmeg (*Virola surinamensis*) crops by birds. *Ecology* **62:** 1093–1106.

Howell, N. 1981. The effect of seed size and relative emergence time on fitness in a natural population of *Impatiens capensis* Meerb. (Balsaminaceae). *Am. Midl. Nat.* **105:** 312–320.

Hudler, G. W., N. Oshima, and F. G. Hawkesworth. 1979. Bird dissemination of dwarf mistletoe on ponderosa pine in Colorado. *Am. Midl. Nat.* **102:** 273–280.

Hulme, A. C. (ed.). 1970–1971. *The Biochemistry of Fruits and Their Products.* 2 vols. Academic, London.

Huxley, C. R. 1978. The ant-plants *Myrmecodia* and *Hydnophytum* (Rubiaceae) and the relationships between their morphology, ant occupants, physiology and ecology. *New Phytol.* **80:** 231–268.

Inouye, D. W. and O. R. Taylor. 1979. A temperate region plant-ant-seed predator system: Consequences of extra floral nectar secretion by *Helianthella quinquenervis. Ecology* **60:** 1–7.

Jackson, J. F. 1981. Seed size as a correlate of temporal and spatial patterns of seed fall in a neotropical forest. *Biotropica* **13:** 121–130.

Jain, S. K. 1978. Local dispersal of *Limnanthes* nutlets: An experiment with artificial vernal pools. *Can. J. Bot.* **56:** 1995–1997.

Janzen, D. H. 1967. Fire, vegetation structure, and the ant x acacia interaction in Central America. *Ecology* **48:** 26–35.

Janzen, D. H. 1969. Seed-eaters vs. seed size, number, toxicity and dispersal. *Evolution* **23:** 1–27.

Janzen, D. H. 1970. Herbivores and the number of tree species in tropical forests. *Am. Nat.* **104:** 501–528.

Janzen, D. H. 1971a. Escape of juvenile *Dioclea megacarpa* (Leguminosae) vines from predators in a deciduous tropical forest. *Am. Nat.* **105:** 97–112.

Janzen, D. H. 1971b. Seed predation by animals. *Annu. Rev. Ecol. Syst.* **2:** 465–492.

Janzen, D. H. 1971c. The fate of *Scheelea rostrata* fruits beneath the parent tree: Predispersal attack by bruchids. *Principes* **15:** 89–101.

Janzen, D. H. 1972. Escape in space by *Sterculia apetala* seeds from the bug *Dysdercus fasciatus* in a Costa Rican deciduous forest. *Ecology* **53:** 350–361.

Janzen, D. H. 1974a. Tropical blackwater rivers, animals, and mast-fruiting by the Dipterocarpaceae. *Biotropica* **6:** 69–103.

Janzen, D. H. 1974b. Epiphytic myrmecophytes in Sarawak: Mutualism through the feeding of plants by ants. *Biotropica* **6:** 237–259.

Janzen, D. H. 1975. Behavior of *Hymenaea courbaril* when its predispersal seed predator is absent. *Science* **189:** 145–147.

Janzen, D. H. 1976a. Why bamboos wait so long to flower. *Annu. Rev. Ecol. Syst.* **7:** 347–391.

Janzen, D. H. 1976b. Reduction of *Mucuna andreana* (Leguminosae) seedling fitness by artificial seed damage. *Ecology* **57:** 826–828.

Janzen, D. H. 1977a. Variation in seed size within a crop of a Costa Rican *Mucuna andreana* (Leguminosae). *Am. J. Bot.* **64:** 347–349.

Janzen, D. H. 1977b. The interaction of seed predators and seed chemistry. *In* V. Labeyrie (ed.), *Comportement des Insectes et Milieu Trophique.* Colloques Internationaux du C.N.R.S., Paris.

Janzen, D. H. 1977c. Variation in seed weight in Costa Rican *Cassia grandis* (Leguminosae). *Trop. Ecol.* **18:** 177–186.

Janzen, D. H. 1978a. Inter- and intra-crop variation in seed weight of Costa Rican *Ateleia herbert-smithii* Pitt. (Leguminosae). *Brenesia* **14–15:** 311–323.

Janzen, D. H. 1978b. The ecology and evolutionary biology of seed chemistry as relates to seed predation. *In* J. B. Harborne (ed.), *Biochemical Aspects of Plant and Animal Coevolution.* Academic, New York.

Janzen, D. H. 1978c. Seeding patterns of tropical trees. *In* P. B. Tomlinson and M. H. Zimmerman (eds.), *Tropical Trees as Living Systems.* Cambridge University Press, Cambridge.

Janzen, D. H. 1980. Specificity of seed-attacking beetles in a Costa Rican deciduous forest. *J. Ecol.* **68:** 929–952.

Janzen, D. H. 1981a. Digestive seed predation by a Costa Rican Baird's tapir. *Biotropica* **13** (Suppl.): 59–63.

Janzen, D. H. 1981b. *Enterolobium cyclocarpum* seed passage rate and survival in horses, Costa Rican Pleistocene seed dispersal agents. *Ecology* **62:** 593–601.

Janzen, D. H. and P. S. Martin. 1981. Neotropical anachronisms: The fruits the gomphotheres ate. *Science* **215:** 19–27.

Janzen, D. H., H. B. Juster, and E. A. Bell. 1977. Toxicity of secondary compounds to the seed-eating larvae of the bruchid beetle *Callosobruchus maculatus. Phytochemistry* **16:** 223–227.

Janzen, D. H., G. A. Miller, J. Hackforth-Jones, C. M. Pond, K. Hooper, and D. P. Janos. 1976. Two Costa Rican bat-generated seed shadows of *Andira inermis* (Leguminosae). *Ecology* **57:** 1068–1075.

Johnson, C. D. 1981. Interactions between bruchid (Coleoptera) feeding guilds and behavioral patterns of pods of the Leguminosae. *Environ. Entomol.* **10:** 249–253.

Johnson, C. D. and C. N. Slobodchikoff. 1979. Coevolution of *Cassia* (Leguminosae) and its seed beetle predators (Bruchidae). *Environ. Entomol.* **8:** 1059–1064.

Johnson, H. B. 1975. Plant pubescence: An ecological perspective. *Bot. Rev.* **41:** 233–258.

Johnson, R. A., M. F. Willson, J. N. Thompson, and R. R. Bertin. (unpublished). Nutritional value of wild fruits and consumption by migrant frugivorous birds.

Johnson, S. R. and R. J. Robel. 1968. Caloric values of seeds from four range sites in northeastern Kansas. *Ecology* **49:** 956–961.

Jordano, P. and C. M. Herrera. 1981. The frugivorous diet of blackcap populations *Sylvia atricapilla* wintering in southern Spain. *Ibis* **123:** 502–507.

Karasawa, K. 1978. Relationships between fruit-eating birds and seed dispersal in urbanized areas (English summary). *Tori* **27:** 1–20.

Keddy, P. A. 1980. Population ecology in an environmental mosaic: *Cakile edentula* on a gravel bar. *Can. J. Bot.* **58:** 1095–1100.

Keeler, K. H. 1981. Function of *Mentzelia nuda* (Loasaceae) postfloral nectaries in seed defense. *Am. J. Bot.* **81:** 295–299.

Keeley, J. E. and R. L. Hays. 1976. Differential seed predation on two species of *Arctostaphylos* (Ericaceae). *Oecologia* **24:** 71–78.

Kellman, M. 1974. Preliminary seed budgets for two plant communities in coastal British Columbia. *J. Biogeogr.* **1:** 123–133.

Kerner, A. 1878. *Flowers and Their Unbidden Guests* (Translated and edited by W. Ogle). Paul, London.

Kerner von Marilaun, A. 1895. *The Natural History of Plants: Their Forms, Growth, Reproduction, and Distribution* (English translation by F. W. Oliver). Holt, New York.

King, T. R. and H. E. McClure. 1944. Chemical composition of some American wild feedstuffs. *J. Agric. Res.* **69:** 33–46.

Kleinfeldt, S. E. 1978. Ant-gardens: The interaction of *Codonanthe crassifolia* (Gesneriaceae) and *Crematogaster longispina* (Formicidae). *Ecology* **59:** 449–456.

Klimstra, W. D. and F. Newsome. 1960. Some observations on the food coactions of the common box turtle, *Terrapene carolina. Ecology* **41:** 639–647.

Koller, D. 1972. Environmental control of seed germination. *In* T. T. Kozlowski (ed.), *Seed Biology*, Vol. 2. Academic, New York.

Kondo, Y. and Y. Oshima. 1981. Propagule size and growth of plant—phenomenon of gaining in growth by plants handicapped by small initial size. *Jap. J. Ecol.* **31:** 217–219.

Kozlowski, T. T. 1971. *Growth and Development of Trees*, Vol. 2. Academic, New York.

Kral, R. 1960. A revision of *Asimina* and *Deeringothamnus* (Annonaceae). *Brittonia* **12:** 233–278.

Krefting, L. W. and E. I Roe. 1949. The role of some birds and mammals in seed germination. *Ecol. Monogr.* **19:** 269–286.

Lacey, E. P. 1980. The influence of hygroscopic movement on seed dispersal in *Daucus carota* (Apiaceae). *Oecologia* **47:** 110–114.

Lamprey, H. F., G. Halevy, and S. Makacha. 1974. Interactions between *Acacia*, bruchid seed beetles and large herbivores. *East Afr. Wildl. J.* **12:** 81–85.

Lederer, R. J. 1977a. Winter territoriality and foraging behavior of the Townsend's solitaire. *Am. Midl. Nat.* **97:** 101–109.

Lederer, R. J. 1977b. Winter feeding territories in the Townsend's solitaire. *Bird-Band* **48:** 11–18.

Lemen, C. A. 1978. Seed size selection in heteromyids. *Oecologia* **35:** 13–19.

Lemen, C. A. 1981. Elm trees and elm leaf beetles: Patterns of herbivory. *Oikos* **36:** 65–67.

Levin, D. A. 1973. The role of trichomes in plant defense. *Q. Rev. Biol.* **48:** 3–15.

Levin, D. A. 1974. The oil content of seeds: An ecological perspective. *Am. Nat.* **108:** 193–206.

Lieberman, D., J. B. Hall, M. D. Swaine, and M. Lieberman. 1979. Seed dispersal by baboons in the Shai Hills, Ghana. *Ecology* **60:** 65–75.

Liew, T. C. and F. O. Wong. 1973. Density, recruitment, mortality and growth of dipterocarp seedlings in virgin and logged-over forests in Sabah. *Malay. Forester* **36:** 3–15.

Ligon, J. D. 1978. Reproductive interdependence of piñon jays and piñon pines. *Ecol. Monogr.* **48:** 111–126.

Ligon, J. D. and D. J. Martin. 1974. Piñon seed assessment by the piñon jay, *Gymnorhinus cyanocephalus*. *Anim. Behav.* **22:** 421–429.

Linhart, Y. B. 1974. Intra-population differentiation in annual plants, I: *Veronica peregrina* L. raised under non-competitive conditions. *Evolution* **28:** 232–243.

Livingston, R. B. 1972. Influence of birds, stones and soil on the establishment of pasture juniper, *Juniperus communis*, and red cedar, *J. virginiana* in New England pastures. *Ecology* **53:** 1141–1147.

Lovell, P. and K. Moore. 1971. A comparative study of the role of the cotyledon in seedling development. *J. Exp. Bot.* **22:** 153–162.

Lu, K. L. and M. R. Mesler. 1981. Ant dispersal of a neotropical forest floor gesneriad. *Biotropica* **13:** 159–160.

Luftensteiner, H. W. 1979. The eco-sociological value of dispersal spectra of two plant communities. *Vegetatio* **41:** 61–67.

Lyons, L. A. 1956. Insects affecting seed predation in red pine, Part I: *Conophthorus resinosae* Hopk. (Coleoptera: Scolytidae). *Can. Entomol.* **88:** 599–608.

Macedo, M. and G. T. Prance. 1978. Notes on the vegetation of Amazonia, II: The dispersal of plants in Amazonian white sand campinas: The campinas as functional islands. *Brittonia* **30:** 203–215.

Marshall, D. L. 1982. The nature and consequences of variation in the components of yield in *Sesbania macrocarpa, S. drummondii,* and *S. vesicaria*. Ph.D. thesis, University of Texas.

Martin, A. C., H. S. Zim, and A. L. Nelson. 1951. *American Wildlife and Plants*. Dover, New York.

Maun, M. A. and P. B. Cavers. 1971a. Seed production and dormancy in *Rumex crispus*, I: The effects of removal of cauline leaves at anthesis. *Can. J. Bot.* **49:** 1123–1130.

Maun, M. A. and P. B. Cavers. 1971b. Seed production and dormancy in *Rumex crispus*, II: The effects of removal of various proportions of flowers at anthesis. *Can. J. Bot.* **49:** 1841–1848.

McCurrach, J. C. 1960. *Palms of the World.* Harper, New York.

McKey, D. 1975. The ecology of coevolved seed dispersal systems. *In* L. E. Gilbert and P. H. Raven (eds.), *The Coevolution of Animals and Plants*. University of Texas Press, Austin.

McKey, D. 1979. The distribution of secondary compounds within plants. *In* S. A. Rosenthal and D. H. Janzen (eds.). *Herbivores: Their Interaction with Secondary Plant Metabolites.* Academic, New York.

McRill, M. and G. R. Sagar. 1973. Earthworms and seeds. *Nature* **243:** 482.

Mitchell, R. 1977. Bruchid beetles and seed packaging by palo verde. *Ecology* **58:** 644–651.

Moermond, T. C. and J. S. Denslow. (in press). Fruit choice in neotropical birds. *J. Anim. Ecol.*

Moore, L. R. 1978. Seed predation in the legume *Crotalaria*, II: Correlates of interplant variability in predation intensity. *Oecologia* **34:** 203–223.

Morden-Moore, A. L. and M. F. Willson. 1982. On the ecological significance of fruit color in *Prunus* and *Rubus*: Field experiments. *Can. J. Bot.* **60:** 1154–1160.

Morton, E. S. 1971. Food and migration habits of the eastern kingbird in Panama. *Auk* **88:** 925–926.

Morton, E. S. 1973. On the evolutionary advantages and disadvantages of fruit eating in tropical birds. *Am. Nat.* **107:** 8–22.

Morton, E. S. 1980. Adaptations to seasonal changes by migrant land birds in the Panama Canal Zone. *In* A. Keast and E. S. Morton (eds.), *Migrant Birds in the Neotropics.* Smithsonian Institution Press, Washington, D.C.

Naylor, J. M. and P. Fedec. 1978. Dormancy studies in seed of *Avena fatua*, 8: Genetic diversity affecting response to temperature. *Can. J. Bot.* **56:** 2224–2229.

Nelson, J. F. and R. M. Chew. 1977. Factors affecting seed reserves in the soil of a Mohave Desert ecosystem, Rock Valley, Nye County, Nevada. *Am. Midl. Nat.* **97:** 300–320.

Ng, F. S. P. 1978. Strategies of establishment in Malayan forest trees. *In* P. B. Tomlinson and M. H. Zimmerman (eds.), *Tropical Trees as Living Systems.* Cambridge University Press, Cambridge.

Noble, J. C. 1975. The effects of emus (*Dromaius novaehollandiae* Latham) on the distribution of the nitre bush (*Nitraria billardieri* DC.). *J. Ecol.* **63:** 979–984.

O'Dowd, D. J. and M. E. Hay. 1980. Mutualism between harvester ants and a desert ephemeral: Seed escape from rodents. *Ecology* **61:** 531–540.

Orians, G. H. and D. H. Janzen. 1974. Why are embryos so tasty? *Am. Nat.* **108:** 581–592.

Ostry, M. E. and T. H. Nicholls. 1977. Animal vectors of the eastern dwarf mistletoe of black spruce. *Proc. Am. Phytopathol. Soc.* **4:** 110.

Peart, M. H. 1979. Experiments on the biological significance of the morphology of seed-dispersal units in grasses. *J. Ecol.* **67:** 843–864.

Peart, M. H. 1981. Further experiments on the biological significance of the morphology of seed-dispersal units in grasses. *J. Ecol.* **69:** 425–436.

Penning de Vries, F. W. T., A. H. M. Brunsting, and H. H. van Laar. 1974. Products, requirements and efficiency of biosynthesis: A quantitative approach. *J. Theor. Biol.* **45:** 339–377.

Phillips, J. 1927. Mortality in the flowers, fruits and young regeneration of trees in the Knysna forests of South Africa. *Ecology* **8:** 435–444.

Pillemer, E. A. and W. M. Tingey. 1976. Hooked trichomes: A physical plant barrier to a major agricultural pest. *Science* **193:** 482–484.

Platt, W. J. 1976. The natural history of fugitive prairie plant (*Mirabilis hirsuta* (Pursh) MacN.). *Oecologia* **22:** 399–409.

Platt, W. J. and I. M. Weis. 1977. Resource partitioning and competition within a guild of fugitive prairie plants. *Am. Nat.* **111:** 479–513.

Platt, W. J., G. R. Hill, and S. Clark. 1974. Seed production in a prairie legume (*Astragalus canadensis* L.). *Oecologia* **17:** 55–63.

Pollock, B. M. and E. E. Roos. 1972. Seed and seedling vigor. *In* T. T. Kozlowski (ed.), *Seed Biology*, Vol. 1. Academic, New York.

Poole, R. W. and B. J. Rathcke. 1979. Regularity, randomness, and aggregation in flowering phenologies. *Science* **203:** 470–471.

Price, M. V. 1978. Seed dispersion preferences of coexisting desert rodent species. *J. Mammal.* **59:** 624–626.

Pudlo, R. J., A. J. Beattie, and D. C. Culver. 1980. Population consequences of changes in an ant-seed mutualism in *Sanguinaria canadensis. Oecologia* **146:** 32–37.

Putnam, L. S. 1949. The life history of the cedar waxwing. *Wilson Bull.* **61:** 141–182.

Queller, D. (in press). Kin selection and conflict in seed maturation. *J. Theor. Biol.*

Queller, D. (unpublished manuscript). The evolution and functions of the endosperm: A kin selection hypothesis.

Rabinowitz, D. 1978. Dispersal properties of mangrove propagules. *Biotropica* **10:** 47–57.

Rabinowitz, D. 1981. Buried viable seeds in a North American tall-grass prairie: The resemblance of their abundance and composition to dispersing seeds. *Oikos* **36:** 191–195.

Rabinowitz, D. and J. K. Rapp. 1979. Dual dispersal modes in hairgrass, *Agrostis hiemalis* (Walt.) B.S.P. (Gramineae). *Bull. Torrey Bot. Club* **106:** 32–36.

Rabinowitz, D. and J. K. Rapp. 1981. Dispersal abilities of seven sparse and common grasses from a Missouri prairie. *Am. J. Bot.* **68:** 616–624.

Rabinowitz, D., J. K. Rapp, V. L. Sork, B. J. Rathcke, G. A. Reese, and J. C. Weaver. 1981. Phenological properties of wind- and insect-pollinated prairie plants. *Ecology* **62:** 49–56.

Racine, C. H. and J. F. Downhower. 1974. Vegetative and reproductive strategies of *Opuntia* (Cactaceae) in the Galápagos Islands. *Biotropica* **6:** 175–186.

Ralph, C. P. 1976. Natural food requirements of the large milkweed bug, *Oncopeltus fasciatus* (Hemiptera: Lygaeidae), and their relation to gregariousness and host plant morphology. *Oecologia* **26:** 157–175.

Rathcke, B. J. and R. W. Poole. 1975. Coevolutionary race continues: Butterfly larvae adaptation to plant trichomes. *Science* **187:** 175–176.

Reichman, O. J. and D. Oberstein. 1977. Selection of seed distribution types of *Dipodomys merriami* and *Perognathus amplus. Ecology* **58:** 636–643.

Rhoades, D. F. and R. G. Cates. 1976. Towards a general theory of plant antiherbivore chemistry. *Recent Adv. Phytochem.* **10:** 168–213.

Rhodes, M. J. C. 1980. The maturation and ripening of fruits. *In* K. V. Thimann (ed.), *Senescence in Plants.* CRC Press, Boca Raton, Fla.

Rice, B. and M. Westoby. 1981. Myrmecochory in sclerophyll vegetation of the West Head, New South Wales. *Aust. J. Ecol.* **6:** 291–298.

Rick, C. M. and R. I. Bowman. 1961. Galápagos tomatoes and tortoises. *Evolution* **15:** 407–417.

Ricklefs, R. E. 1977. A discriminant function analysis of assemblages of fruit-eating birds in Central America. *Condor* **79:** 228–231.

Ridley, H. N. 1930. *The Dispersal of Plants Throughout the World.* Reeve, Ashford, Kent.

Robel, R. J. 1972. Energy content in seeds. *Trans. Kansas Acad. Sci.* **75:** 301–307.

Roberts, R. C. 1979. The evolution of avian food-storing behavior. *Am. Nat.* **114:** 418–438.

Rosenthal, G. A., C. G. Hughes, and D. H. Janzen. 1982. L-Canavinine, a dietary nitrogen source for the seed predator *Caryedes brasiliensis* (Bruchidae). *Science* **217:** 353–355.

Ross, M. A. and J. L. Harper. 1972. Occupation of biological space during seedling establishment. *J. Ecol.* **60:** 77–88.

Russell, M. P. 1962. Effects of sorghum varieties on the lesser rice weevil, *Sitophilus oryzae* (L.), I: Oviposition, immature mortality, and size of adults. *Ann. Entomol. Soc. Amer.* **55:** 678–685.

Rust, R. W. and R. R. Roth. 1981. Seed production and seedling establishment in the mayapple, *Podophyllum peltatum*. *Am. Midl. Nat.* **105:** 51–60.

Salisbury, E. J. 1942. *The Reproductive Capacity of Plants.* Bell, London.

Salisbury, E. J. 1974. Seed size and mass in relation to environment. *Proc. R. Soc. Lond.* **B 186:** 83–88.

Salisbury, E. J. 1975. The survival value of modes of dispersal. *Proc. R. Soc. Lond.* **B 188:** 183–188.

Salomonson, M. G. 1978. Adaptations for animal dispersal of one-seed juniper seeds. *Oecologia* **32:** 333–339.

Sarukhan, J. 1980. Demographic problems in tropical systems. *Botan. Monogr.* **15:** 161–188.

Schaal, B. A. 1980. Reproductive capacity and seed size in *Lupinus texensis*. *Am. J. Bot.* **67:** 703–709.

Schemske, D. W. 1978. Evolution of reproductive characteristics in *Impatiens* (Balsaminaceae): The significance of cleistogamy and chasmogamy. *Ecology* **59:** 596–613.

Schemske, D. W. 1980. The evolutionary significance of extrafloral nectar production by *Costus woodsonii* (Zingiberaceae): An experimental analysis of ant protection. *J. Ecol.* **68:** 959–968.

Schimpf, D. J. 1977. Seed weight of *Amaranthus retroflexus* in relation to moisture and length of growing season. *Ecology* **58:** 450–453.

Schweizer, C. J. and S. K. Ries. 1969. Protein content of seed: Increase improves growth and yield. *Science* **165:** 73–75.

Scott, D. R. 1970. Feeding of *Lygus* bugs (Hemiptera: Miridae) on developing carrot and bean seed: Increased growth and yields of plants grown from that seed. *Ann. Entomol. Soc. Amer.* **63:** 1604–1608.

Seifert, R. P. 1982. Neotropical *Heliconia* insect communities. *Q. Rev. Biol.* **57:** 1–28.

Seigler, D. S. 1979. Toxic seed lipids. *In* G. A. Rosenthal and D. H. Janzen (eds.). *Herbivores: Their Interaction with Secondary Plant Metabolites.* Academic, New York.

Sheldon, J. C. and F. M. Burrows. 1973. The dispersal effectiveness of the achene-pappus units of selected Compositae in steady winds with convection. *New Phytol.* **72:** 665–675.

Sherbrooke, W. C. 1976. Differential acceptance of toxic jojoba seeds (*Simmondsia chinensis*) by four Sonoran Desert heteromyid rodents. *Ecology* **57:** 596–602.

Sherburne, J. A. 1972. Effects of seasonal changes in the abundance and chemistry of the fleshy fruits of northeastern woody shrubs on patterns of exploitation by frugivorous birds. Ph.D. thesis, Cornell University, Ithaca, N.Y.

Sherman, R. J. and W. W. Chilcote. 1972. Spatial and chronological patterns of *Purshia tridentata* as influenced by *Pinus ponderosa*. *Ecology* **53:** 294–298.

Shine, R. 1978. Propagule size and parental care: The "safe harbor" hypothesis. *J. Theor. Biol.* **75:** 417–424.

Silvertown, J. W. 1980. The evolutionary ecology of mast seeding in trees. *Biol. J. Linn. Soc.* **14:** 235–250.

Silvertown, J. W. 1981. Seed size, life span, and germination date as coadapted features of plant life history. *Am. Nat.* **118:** 860–864.

Simms, E. 1978. *British Thrushes.* Collins, London.

Sinclair, T. R. and C. T. de Wit. 1975. Photosynthate and nitrogen requirements for seed production by various crops. *Science* **189:** 565–567.

Skutch, A. F. 1980. Arils as food of tropical American birds. *Condor* **82:** 31–42.

Smith, A. J. 1975. Invasion and ecesis of bird-disseminated woody plants in a temperate forest sere. *Ecology* **56:** 19–34.

Smith, C. C. 1968. The adaptive nature of social organization in the genus of tree squirrels *Tamiasciurus. Ecol. Monogr.* **38:** 31–63.

Smith, C. C. 1970. The coevolution of pine squirrels (*Tamiasciurus*) and conifers. *Ecol. Monogr.* **40:** 349–371.

Smith, C. C. 1975. The coevolution of plants and seed predators. *In* L. E. Gilbert and P. H. Raven (eds.), *Coevolution of Animals and Plants.* University of Texas Press, Austin.

Smith, C. C. and R. P. Balda. 1980. Competition among insects, birds and mammals for conifer seeds. *Am. Zool.* **19:** 1065–1083.

Smith, C. C. and S. D. Fretwell. 1974. The optimal balance between size and number of offspring. *Am. Nat.* **108:** 499–506.

Snell, T. W. 1976. Effects of density on seed size and biochemical composition. *Am. Midl. Nat.* **95:** 499–507.

Snodderly, D. M. 1979. Visual discriminations encountered in food foraging by a neotropical primate: Implications for the evolution of color vision. *In* E. H. Burtt (ed.), *The Behavioral Significance of Color.* Garland STPM Press, New York.

Snow, D. W. 1965. A possible selective factor in the evolution of fruiting seasons in tropical forest. *Oikos* **15:** 274–281.

Snow, D. W. 1971. Evolutionary aspects of fruit-eating by birds. *Ibis* **113:** 194–202.

Snow, D. W. 1976. *The Web of Adaptation.* Collins, London.

Snow, D. W. 1980. Regional differences between tropical floras and the evolution of frugivory. *Proc. Int. Ornithol. Congr.* **17:** 1192–1198.

Snow, D. W. 1981. Tropical frugivorous birds and their food plants: A world survey. *Biotropica* **13:** 1–14.

Sorensen, A. E. 1978. Somatic polymorphism and seed dispersal. *Nature* **276:** 174–176.

Sorensen, A. E. 1981. Interactions between birds and fruit in a temperate woodland. *Oecologia* **50:** 242–249.

Sork, V. L. and D. H. Boucher. 1977. Dispersal of sweet pignut hickory in a year of low fruit production, and the influence of predation by a curculionid beetle. *Oecologia* **28:** 289–299.

Staniforth, R. J. and P. B. Cavers. 1976. An experimental study of water dispersal in *Polygonum* spp. *Can. J. Bot.* **54:** 2587–2596.

Stapanian, M. A. 1982a. Evolution of fruiting strategies among fleshy-fruited plant species of eastern Kansas. *Ecology* **63:** 1422–1431.

Stapanian, M. A. 1982b. A model for fruiting display: Seed dispersal by birds for mulberry trees. *Ecology* **63:** 1432–1443.

Stapanian, M. A. and C. C. Smith. 1978. A model for seed scatter-hoarding: Coevolution of fox squirrels and black walnut. *Ecology* **59:** 884–896.

Stebbins, G. L. 1950. *Variation and Evolution in Plants.* Columbia University Press, New York.

Stebbins, G. L. 1971. Adaptive radiation of reproductive characteristics in angiosperms, II: Seeds and seedlings. *Annu. Rev. Ecol. Syst.* **2:** 237–260.

Stebbins, G. L. 1974. *Flowering Plants.* Belknap, Cambridge, Mass.

Stephenson, A. G. 1982. The role of the extrafloral nectaries of *Catalpa speciosa* in limiting herbivory and increasing fruit production. *Ecology* **63:** 663–669.

Sterk, A. A. 1969. Biosystematic studies of *Spergularia media* and *S. marina* in relation to the environment. *Acta Bot. Neerl.* **18:** 561–577.

Sterk, A. A. and L. Dijkhuizen. 1972. The relationship between the genetic determination and the ecological significance of the seed wing in *Spergularia media* and *S. marina. Acta Bot. Neerl.* **21:** 481–490.

Stiles, E. W. 1980. Patterns of fruit presentation and seed dispersal in bird-disseminated woody plants in the eastern deciduous forest. *Am. Nat.* **116:** 670–688.

Temple, S. A. 1977. Plant-animal mutualism: Coevolution with dodo leads to near extinction of plant. *Science* **197:** 885–886.

Templeton, A. R. and D. A. Levin. 1979. Evolutionary consequences of seed pools. *Am. Nat.* **114:** 232–249.

Thomas, D. K. 1979. Figs as a food source of migrating garden warblers in southern Portugal. *Bird Study* **26:** 187–191.

Thomas, R. L. 1966. The influence of seed weight on seedling vigour in *Lolium perenne. Ann. Bot. (N.S.)* **30:** 111–121.

Thompson, J. N. 1980. Treefalls and colonization patterns of temperate forest herbs. *Am. Midl. Nat.* **104:** 176–184.

Thompson, J. N. 1981a. Elaiosomes and fleshy fruits: Phenology and selection pressures for ant-dispersed seeds. *Am. Nat.* **117:** 104–108.

Thompson, J. N. 1981b. Reversed animal-plant interactions: The evolution of insectivorous and ant-fed plants. *Biol. J. Linn. Soc.* **16:** 147–155.

Thompson, J. N. and M. F. Willson. 1979. Evolution of temperate fruit/bird interactions: Phenological strategies. *Evolution* **33:** 973–982.

Thompson, K. and J. P. Grime. 1979. Seasonal variation in the seed banks of herbaceous species in ten contrasting habitats. *J. Ecol.* **67:** 893–921.

Thompson, P. 1981. Variations in seed size within populations of *Silene dioeca* (L.) Clairv. in relation to habitat. *Ann. Bot.* **47:** 623–634.

Thurston, J. M. 1960. Dormancy in weed seeds. *In* J. L. Harper (ed.), *The Biology of Weeds.* Blackwell's, Oxford.

Tomback, D. F. 1977. Foraging strategies of Clark's nutcracker. *Living Bird* **16:** 123–161.

Tomback, D. F. 1982. Dispersal of whitebark pine seeds by Clark's nutcracker: A mutualism hypothesis. *J. Ecol.* **51:** 451–467.

Turcek, F. J. 1963. Color preferences in fruit- and seed-eating birds. *Proc. Int. Ornithol. Congr.* **13:** 285–292.

Twamley, B. E. 1967. Seed size and seedling vigor in birdsfoot trefoil. *Can. J. Plant Sci.* **47:** 603–609.

Uphof, J. C. T. 1942. Ecological relations of plants with ants and termites. *Bot. Rev.* **8:** 563–598.

USDA. 1974. *Seeds of Woody Plants in the United States. Agriculture Handbook No. 450.* USDA Forest Service, Washington, D.C.

van der Pijl, L. 1957. The dispersal of plants by bats (chiropterochory). *Acta Bot. Neerl.* **6:** 291–315.

van der Pijl, L. 1972. *Principles of Dispersal in Higher Plants* (2nd ed.). Springer-Verlag, Berlin.

Vander Wall, S. B. 1982. An experimental study of cache recovery in Clark's nutcracker. *Anim. Behav.* **30:** 84–94.

Vander Wall, S. B. and R. P. Balda. 1977. Coadaptations of the Clark's nutcracker and the pinon pine for efficient seed harvest and dispersal. *Ecol. Monogr.* **47:** 89–111.

Vander Wall, S. B. and R. P. Balda. 1981. Ecology and evolution of food-storage behavior in conifer-seed caching corvids. *Z. Tierpsychol.* **56:** 217–242.

Venable, D. L. and L. Lawlor. 1980. Delayed germination and dispersal in desert annuals: Escape in space and time. *Oecologia* **46:** 272–282.

Villiers, T. A. 1975. *Dormancy and the Survival of Plants.* Edward Arnold, London.

Vogel, E. F. de. 1980. *Seedlings of Dicotyledons.* Centre for Agricultural Publishing and Documentation, Wageningen, Netherlands.

Wainio, W. W. and E. B. Forbes. 1941. The chemical composition of forest fruits and nuts from Pennsylvania. *J. Agric. Res.* **62:** 627–635.

Waller, D. M. 1979. Models of mast fruiting in trees. *J. Theor. Biol.* **80:** 223–232.

Walsberg, G. E. 1975. Digestive adaptations of *Phainopepla nitens* associated with the eating of mistletoe berries. *Condor* **77:** 169–174.

Waser, N. M., R. K. Vickery, and M. V. Price. 1982. Patterns of seed dispersal and population differentiation in *Mimulus guttatus. Evolution* **36:** 753–761.

Watkinson, A. R. 1978. The demography of a sand dune annual: *Vulpia fasciculata*, II: The dynamics of seed populations. *J. Ecol.* **66:** 35–44.

Webb, D. A. 1966. Dispersal and establishment: What do we really know? *In* J. G. Hawkes (ed.), *Reproductive Biology and Taxonomy of Vascular Plants.* Bot. Sci. Brit. Isles Conf. Rep. 9.

Webber, M. L. 1934. Fruit dispersal. *Malay. For.* **3:** 18–19.

Weiss, P. W. 1980. Germination, reproduction and interference in the amphicarpic annual *Emex spinosa* (L.) Campd. *Oecologia* **45:** 244–251.

Werner, P. A. 1979. Competition and coexistence of similar species. *In* O. T. Solbrig, S. Jain, G. B. Johnson, and P. H. Raven (eds.), *Topics in Plant Population Biology.* Columbia University Press, New York.

Westoby, M. 1981. How diversified seed germination behavior is selected. *Am. Nat.* **118:** 882–885.

Westoby, M. and B. Rice. 1981. A note on combining two methods of dispersal-for-distance. *Aust. J. Ecol.* **6:** 189–192.

Westoby, M. and B. Rice. 1982. Evolution of the seed plants and inclusive fitness of plant tissues. *Evolution* **36:** 713–724.

Wheelwright, N. T. and G. H. Orians. 1982. Seed dispersal by animals: Contrasts with pollen dispersal, problems of terminology, and constraints on coevolution. *Am. Nat.* **119:** 402–413.

Whipple, S. A. 1978. The relationship of buried, germinating seeds to vegetation in an old-growth Colorado subalpine forest. *Can. J. Bot.* **56:** 1505–1509.

Whitham, T. G. (in press). Host manipulation of parasites: Within-plant variation as a defense against rapidly evolving pests. *In* R. F. Denno and S. McClure (eds.), *Impact of Variable Host Quality on Herbivorous Insects.* Academic, New York.

Williams, G. C. 1975. *Sex and Evolution.* Princeton University Press, Princeton, N.J.

Wiens, D. 1978. Mimicry in plants. *Evol. Biol.* **11:** 365–403.

Willson, M. F. 1972a. Evolutionary ecology of plants: A review, II: Ecological life histories. *Biologist* **54:** 148–162.

Willson, M. F. 1972b. Seed size preferences in finches. *Wilson Bull.* **84:** 449–455.

Willson, M. F., P. K. Anderson and P. A. Thomas (in press). Bracteal exudates of *Cirsium discolor* (Muhl.) and *C. flodmani* (Rydb.) as possible deterrents to insect seed-predators. *Am. Midl. Nat.*

Willson, M. F. and N. Burley. (in press). *Mate Choice in Plants: Tactics, Mechanisms, and Consequences.* Princeton University Press, Princeton, N.J.

Willson, M. F. and M. N. Melampy. (in press). The effect of bicolored fruit displays on fruit removal rates by frugivores. *Oikos.*

Willson, M. F. and P. W. Price. 1980. Resource limitation of fruit and seed production in some *Asclepias* species. *Can. J. Bot.* **58:** 2229–2233.

Willson, M. F. and D. W. Schemske. 1980. Pollinator limitation, fruit production, and floral display in pawpaw (*Asimina triloba*). *Bull. Torrey Bot. Club* **107:** 401–408.

Willson, M. F. and J. N. Thompson. 1982. Phenology and ecology of color in bird-dispersed fruits *or* why some fruits are red when they are "green." *Can. J. Bot.* **60:** 701–713.

Wilson, D. E. and D. H. Janzen. 1972. Predation on *Scheelea* palm seeds by bruchid beetles: Seed density and distance from the parent palm. *Ecology* **53:** 954–959.

Wilson, R. C. 1976. *Abronia:* IV: Anthocarp dispersibility and its ecological implications for nine species of *Abronia. Aliso* **8:** 493–506.

Winstead, J. E., B. J. Smith, and G. I. Wardell. 1977. Fruit weight clines in populations of ash, ironwood, cherry, dogwood and maple. *Castanea* **42:** 56–60.

Wulff, R. 1973. Intrapopulational variation in the germination of seeds in *Hyptis suaveolens. Ecology* **54:** 646–649.

Wyatt, R. 1981. Components of reproductive output in five tropical legumes. *Bull. Torrey Bot. Club* **108:** 67–75.

Zimmerman, C. A. 1977. A comparison of breeding systems and seed physiologies in three species of *Portulaca* L. *Ecology* **58:** 860–868.

Zimmerman, M. 1980. Reproduction in *Polemonium*: Pre-dispersal seed predation. *Ecology* **61:** 502–506.

Zohary, D. and D. Imber. 1963. Genetic dimorphism in fruit types in *Aegilops speltoides. Heredity* **18:** 223–231.

EPILOGUE AND SUMMARY

This book is intended as a primer of plant reproductive ecology. The emphasis is placed on plants as individuals and the fitness consequences of divers attributes. Another level of inquiry occurs at the population level—the population effects of changes in the ecology of individuals. Population-level questions about rates of population increase or decrease, rates of changes of gene frequency, control of density, gene flow, interactions among species, and so on, can sometimes be treated independently of the individuals comprising the population but at some point the effect of various pressures on individuals is useful in explaining the processes that lead to population patterns. Interesting and important as the population patterns are, I have not focused on them here, in part because surveys are more easily found elsewhere and in part because the size of this book would become overwhelming (at least from an author's viewpoint). Various aspects of the reproductive ecology of individual plants are subject to intensive contemporary investigation, so it seemed worthwhile to pull together a significant fraction of existing literature to provide a sort of "where are we?" statement. Current and future research will presumably expand our horizons rapidly.

Since this book is itself a summary, to provide a formal summary is to lose many of the fascinating "wrinkles" of observation and interpretation that provide a toehold for future investigation. Nevertheless, a very succinct accounting might be useful at least in indicating the areas covered and a few judgments about the state of our art.

Chapter 1. The life history of an organism is structured by extrinsic probabilities of success, the relative benefits of a particular reproductive tactic, and the intrinsic costs of reproduction. Extrinsic limitations include rate of juvenile and adult mortality, benefits include the rate at which contribution to future generations can increase, and intrinsic costs include both resource drains and mortality risks that are related directly to reproduction. The evolution of life histories is constrained both phylogenetically and developmentally, but the interaction of all these structuring forces is poorly understood. The actual reproductive performance of any surviving individual is often constrained by resources or success in mating.

Both competition and high mortality have been predicted to lead to increased resource allocation to reproduction, but the available data provide no consistent support for such interpretations. High density might favor sexual over vegetative reproduction, but observations are not consistent with this prediction.

Chapter 2. The evolution of sex is still hotly debated; the explanation may be in DNA repair and/or the advantages of recombination in producing varied or highly fit offspring; once male and female roles are differentiated, there is the additional possibility of fitness gain on alternative paths. Outcrossing alone is an inadequate explanation for the evolution of separate sexes; various levels of sex-separation are likely to be linked with the advantages of sexual-role specialization, and sexual selection (both choice by females and competition among males) may be involved. Other possibilities include niche differentiation of males and females (in dioecy) and parental manipulation of sex ratios (of flowers in monoecy, of offspring in dioecy). Sex (in dioecy) is sometimes determined genetically but perhaps more commonly by environmental factors or the condition of the individual. To the extent that environmental sex determination produces the sex that will have the higher fitness in the given situation, this is easily understood; the advantage of genetic sex determination in plants is less clear. Sex ratios vary mightily, sometimes passively as a result of differences in longevity or reproductive expenditures, sometimes actively in response to parental manipulation, and sometimes for unknown reasons. The possible adaptive values of altering offspring or floral sex ratios in plants is virtually unexplored.

Incompatibility systems restrict selfing and inbreeding with relatives; they also limit the number of suitable mates available. Mate choice by plants (particularly by females) is a neglected possibility; several potential mechanisms exist.

Chapter 3. Male and female functions may differ greatly in constraints and in prospects of success. Males (and male functions of hermaphrodites) in partially or completely outcrossing species will be selected to claim as many eggs as possible, whereas females (and female functions) may be selected to be choosy of both the quantity and quality of fathers for their eggs. In many cases, both males and females can increase their RS by improving pollen delivery systems, but females are likely to derive greater benefits from improving the quality of successful pollen and males may be able to package pollen in ways that enhance the capture of ovules.

Animal-pollinated seed plants reward their pollen vectors by several means, but both animals and plants may also cheat the other by stealing rewards or offering false ones. Although the visual signals presented by flowers sometimes favor an exclusive set of pollinators, this is not always true; we are only beginning to understand the complex of factors affecting plant-pollination specializations, either long-term or short.

Pollen shadows tend to be tightly clumped around the pollen donor, with consequences for inbreeding and population structure, the number and variety of accessible mates and opportunities for mate choice (females) and pollen donation (males).

Chapter 4. Parent plants endow their offspring with an initial complement of nutritional reserves, which are generally accepted to be important in seedling establishment. The ecological conditions favoring large endowments include dryness, seedling competition, shade, and no doubt other factors. Counterselection against large size may derive from seed predators or dispersal limitations, but the available evidence is poor. The importance of quality of the seed reserves is little studied.

Parents also protect their offspring, both before and after dispersal, in many ways. Although we have many descriptions of defenses by morphological devices, chemical compounds, animal guards, and elusiveness in time and space, the circumstances favoring one defense over another (or one particular combination of defenses) are generally not understood.

Likewise, although we know of many devices to foster dispersal, the pressures favoring one kind over another and the consequences of each type are virtually unknown. In general, dispersal is adaptive in reaching a site for establishment and sometimes in eluding predators, pathogens, or sibling competition.

After dispersal in space, many seeds wait some time, in some cases a long time, before germinating. During this time, mortality is often very high, but the survivors achieve dispersal in time.

Many biologists have treated plants as rather passive organisms. It has been said that plants just sit there, waiting to die. Others say they just sit there waiting to be counted (or collected). Still others (of which I was once one, alas) view plants chiefly as providing habitat for animals. Whole laboratories deal only with photosynthesis, and plants become intricate machines for processing carbon. I would like to make the case that plants are far more complex than any of these views might admit. It is true that plants are typically sessile and green and that these traits impose some constraints on what plants can do. And it is true that the physiological machinery and the demography of plants provide a wealth of interesting material for study. Nevertheless, because more and more researchers are realizing just how complicated and subtle plant "behavior" may be, I think we will soon see a rapid radiation of fascinating studies in evolutionary ecology of plants.

TAXONOMIC INDEX

SUBJECT INDEX